ECONOMIC GEOLOGY

Economic Mineral Deposits

W0080524

Umeshwar Prasad MSc, PhD, MMGI

Former Director, Geological Survey of India, and
Former Managing Director, Arunachal Pradesh Mineral
Development and Trading Corporation Limited

CBSPD

CBS Publishers & Distributors Pvt Ltd

New Delhi • Bengaluru • Chennai • Kochi • Kolkata • Lucknow • Mumbai
Hyderabad • Jharkhand • Nagpur • Patna • Pune • Uttarakhand

Economic Geology
Economic Mineral Deposits
(Second Edition)

ISBN: 978-81-239-0460-3

Copyright © Author and Publisher

Second Edition: 2000
Reprint: 2003, 2005, 2006, 2008, 2010, 2011, 2012. 2013, 2014, 2015, 2016, 2018, 2019, 2021, 2022, **2024 2025**
First Edition: 1996

Published by **Satish Kumar Jain** and produced by **Varun Jain** for
CBS Publishers & Distributors Pvt Ltd
4819/XI Prahlad Street, 24 Ansari Road, Daryaganj, New Delhi 110 002, India
Ph: 011-23289259, 23266861   Website: www.cbspd.com
    e-mail: delhi@cbspd.com

Corporate Office: 204 FIE, Industrial Area, Patparganj, Delhi 110 092, India
Ph: 011-4934 4934   Fax: 011-4934 4935   e-mail: publishing@cbspd.com;
    publicity@cbspd.com

Branches

- **Bengaluru:** Seema House 2975, 17th Cross, KR Road, Banasankari 2nd Stage, Bengaluru 560 070, Karnataka, India
  Ph: +91-80-26771678/79   Fax: +91-80-26771680   e-mail: bangalore@cbspd.com
- **Chennai:** 7, Subbaraya Street, Shenoy Nagar, Chennai 600 030, Tamil Nadu, India
  Ph: +91-44-26680620, 26681266   Fax: +91-44-42032115   e-mail: chennai@cbspd.com
- **Kochi:** 42/1325, 1326, Power House Road, Opp KSEB, Power House, Ernakulum Kochi 682 018, Kerala, India
  Ph: +91-484-4059061-65,67   Fax: +91-484-4059065   e-mail: kochi@cbspd.com
- **Kolkata:** 147, Hind Ceramics Compound, 1st Floor, Nilgunj Road, Belghoria, Kolkata-700056, West Bengal, India
  Ph: +033-25633055, 033-25633056   e-mail: kolkata@cbspd.com
- **Lucknow:** Basement, Khushnuma Complex, 7 Meerabai Marg (Behind Jawahar Bhawan), Lucknow-226001, UP, India
  Ph: +91-522-4000032   e-mail: tiwari.lucknow@cbspd.com
- **Mumbai:** PWD Shed, Gala no 25/26, Ramchandra Bhatt Marg, Next to JJ Hospital Gate no. 2, Opp. Union Bank of India Noorbaug, Mumbai-400009, Maharashtra, India
  Ph: 022-66661880/89   e-mail: mumbai@cbspd.com

Representatives

- Hyderabad   0-9885175004   • Jharkhand   0-9811541605   • Nagpur   0-8692091830
- Patna   0-9334159340   • Pune   0-9664372571   • Uttarakhand   0-9716462459

Printed at Glorious Printers, Jhilmil Industrial Area, Delhi, India

Preface to the Second Edition

In the second edition, the book has been thoroughly rectified and revised as per needs. The "Introduction" has been fully revised and rewritten, incorporating, therein, the purpose of studies and the scope of the subject, the distribution of elements and outline of mineral deposits giving definitions with a table showing tenors of some important ores and their average grades. Two new chapters, one on "Essential Factors in Mineral Deposition and Ore Genesis" and second on "Mineral Paragenesis and Zoning, Metallogenetic Epochs and Provinces, and Controls of Minerals Localisation" have been added. To illustrate various phenomena all Indian examples have been cited to the extent possible.

For easy and quick identification of economic minerals, physical characters of important minerals have been given in a tabular form and included as Appendix-I. The recent available production figures are given in Appendix-II.

An attempt has been made to design it as a textbook both for elementary and more advanced courses in Economic Geology. The book covers the full courses of the subject for B.Sc. and M.Sc. students in Geology. Efforts have been made to present the facts and figures in tabular forms (a total of 135 tables) as far as practicable to facilitate easy consultation and memorising for examinations. The prospecting guidelines and mining and metallurgical aspects, where deemed necessary, have been given to assist the field workers, technocrats and industrialists. Map showing important mineral deposits of India, which could not be included in the first edition, has been incorporated.

The author expresses his deep sense of gratitude to professors and students of universities and geologists of central and state governments for valuable suggestions and discussions which helped in making the book up-to-date. Special mention may be made of the names of Prof. A.P. Mall, Prof. N.C. Ghose, Prof. S.N. Singh, Prof. N.K. Singh, Prof. Ramesh Shukla, Prof. B.K. Mishra and Prof. Ravindra Kumar of Patna University and Shri Jai Prakash Singh, Mineral Economist, Govt. of Bihar to whom the author is highly thankful for fruitful advices, help in the improvement of the book and encouragements. He is highly indebted to Shri J.K. Shrivastava, Geologist, Geological Survey of India for assistance in correcting the proofs and Mrs. Madhulika Prasad, my wife, for her best cooperation and help without which this write-up would not have been completed.

The author is fully confident that this book will be useful to students of geology and mining, fieldworkers, technocrats, industrialists and others interested in mineral deposits.

Criticism and suggestions are invited so as to enable the author to make the book moré fruitful in its third edition.

Patna   Umeshwar Prasad
Dated : 15th August, 2000

Preface to the First Edition

Minerals are known since time immemorial and their description are recorded in our old literatures. The ancient abandoned mines and slags found at many places in India and elsewhere bear testomony to the mining and metallurgical skill of ancients. Minerals form raw materials of many industries and, thereby, play a vital role in country's economic development. India is fortunate to have a good number of mineral deposits required for its industrial development. After independence of the country special emphasis has been laid to explore different mineral occurrences, and thus a wealth of data has accumulated with discoveries of many new deposits. Further with development in geo-science, new ideas about their use, modes of occurrence, origin and distribution have been put forward. Students of geology and mining, researchers, miners, technocrats, industrialists and others interested in mineral deposits are to hunt a lot to get upto-date data on different minerals. With that in view a concise account of each mineral deposit on its different aspects has been given in the present book of "Economic Geology" written with recent available data and Indian examples.

The processes that lead to the formation of various mineral deposits form basis of understanding the distribution of mineral deposits in different geological milieu, and hence has been dealt earlier to the description of individual mineral deposits. The mineral deposits have been classified broadly into metallic and non-metallic mineral deposits. The metallic mineral deposits have been grouped depending upon their nature and importance into precious, ferrous and ferro-alloy, non-ferrous and allied, light, radioactive and rare metals for the sake of convenience in study and ready reckoner. The non-metallic mineral deposits have been sub-divided into twelve groups, based on their characters and importance, namely mineral fuels, gemstones, abrasive minerals, building materials and dimension stones, industrial minerals, refractory minerals, ceramic minerals, glass manufacturing materials, fertilizer minerals, chemical minerals, mineral pigments, and mineral water and ground water. Efforts have been made to present briefly the world scenario on different mineral deposits. An outline of mineralogy of the important economic minerals has been given in the beginning to have a fundamental knowledge of the individual mineral. This is followed successively by the description of uses to which the concerned mineral is put, modes of occurrence and origin, and its geological and geographical description. The reserves and production figures are incorporated towards fag end of the description. The endeavour has been made, to present the facts and figures in tabular forms as far as practicable to facilitate easy look and consultations. The prospecting guidelines, and mining and metallurgical aspects where deemed necessary, have been included. References have been appended in the last.

The author expresses his deep sense of gratitude to Prof. (Dr.) R. A. Mahto, Patna University, for his encouragements and suggestions given in writing this book. He is highly thankful to Prof. (Dr.) N.C. Ghose, Patna University, and Dr. M.K. Roy, Dy. Director (Geology), Bihar for usual discussions and help in data collections. My sincere thanks are due to Shri J.K. Srivastava, Geologist, Geological Survey of India for his help in correcting the proofs, to Mrs. Madhu, Smriti, Architect, Govt. of Bihar, for tracing the figures, and to Dr. M. Kumar and all others, who helped me in completion of this book and its publication. Finally, I am highly indebted to Mrs. Madhulika Prasad, my wife, who offered her best co-operation and useful assistance without which the book would not have been completed.

I hope, this book will be of great use to the students of geology and mining, research and field workers, technocrats, industrialists and others interested in mineral deposits.

It is my request to readers to write to me about the shortcomings in the book so that the next edition be brought out with all rectifications.

Contents

List of Figures

1. **Fig. 2.1.** Zones of oxidation and supergene enrichment.

2. **Fig. 2.2.** Coarse cellular boxwork with typical triangular and quadrangular patterns, correlated with bornite and chalcopyrite. Loc : Mundatikra, Bastar district, M.P.

3. **Fig. 2.3.** Fine cellular boxwork, Also quadrangular pattern with parallel cell walls, correlated with chalcopyrite. Loc : Mundatikra, Bastar district, M.P.

4. **Fig. 2.4.** Cleavage type boxwork with cellular sponge structure, correlated with galena. Loc : Belbathan, Godda district, Bihar.

5. **Fig. 4.1.** Massive Iron ore. Loc : Deposit No. 5, Bailadila, M.P.

6. **Fig. 4.2.** Laminated Iron ore. Loc : Deposit No. 3, Bailadila, M.P.

7. **Fig. 5.1.** Topaz-vein in kyanite rock. Loc : Lapsaburu, Singhbhum district, Bihar.

8. **Fig. 5.2.** Marble hillock. Loc : Darkala (Kankrauli), Rajasthan.

9. **Fig. 5.3.** Kyanite boulders. Loc : Lapsaburu, Singhbhum district, Bihar.

Introduction

Economic Geology is an important branch of Geology which deals with different aspects of economic minerals being utilised by mankind to fulfill his various needs. The economic minerals are those which can be extracted profitably. The subject provides detailed knowledge of all these minerals, like their origin, process of formation, mutual relations and time sequence, control of localisation, metallogenic epoch and province, and classification, besides their mineralogy, uses, modes of occurrence, geological and geographical distribution, reserves, production and all other related facts.

Studies of all sorts of minerals, economic and non-economic are included in the subject "Mineralogy". Since the subject includes description of a large number of minerals and this number is increasing with research and new finds, this necessitated the development of a separate branch, namely "Economic Geology", independently so that special attention may be possible to be given to individual economic minerals.

PURPOSE OF STUDIES

The purpose of studies of the subject is as follows :

1. The subject "Economic Geology" is aimed to provide detailed description of economic minerals which may be a little more than 200 in number. Besides the detailed description of these minerals, the subject also discusses the proper uses and development of the mineral deposit. The foremost duty of an Economic Geologist is to suggest the suitability of a mineral in the deposit for a particular industry.

2. The reserves of economic minerals in a deposit are limited and their percentages are also variable, which are not replenishable. Once they are extracted fully from a deposit, their reserves are exhausted for ever. This fact requires to be borne in mind while mining a deposit. A wise Economic Geologist prepares a plan of the mineral deposit for the proper utilisation before its extraction.

3. It is observed that certain economic minerals are found in particular rocks, e.g. diamond occurs in kimberlite rock, chromite in peridotite and platinum in norite or peridotite. This knowledge is needed while going for search of a particular mineral. The parent rock of a particular mineral is required to be mapped first, and, thereafter, detailed study of this limited area is taken up making an assessment of the economics of extraction of mineral deposit.

Thus, the objective of the study of Economic Geology is to have full scientific knowledge of the mineral deposits, their utilisation, mode of occurrence and origin, distribution, potentiality and guidelines to explore new deposits. It may be mentioned here that the country which does not produce minerals as per its needs, depends upon another country and remains economically and industrially backward. It is, hence, in the interest of India to stick to proper and full utilisation of its mineral wealth and find out new mineral deposits in which it is difficient. This is possible only when the subject "Economic Geology" is studied in a scientific way and full and proper utilisation of economic minerals are made by taking up researches in the field.

SCOPE

The subject 'Economic Geology' has unlimited scope, since it deals with the basic materials for diverse industries, armaments for protection of country, agricultural implements, transport facilities and all that is needed for human civilisation. All big or small countries want to be self-sufficient in respect of economic minerals and so search for minerals in all possible areas within the country has to be continued. The measure of country's development and civilisation is dependent on large use of minerals in different fields with the applications of the subject of Economic Geology. Thus, there is huge scope of the subject.

DISTRIBUTION OF ELEMENTS

The upper crust of the earth upto 16 km depth is made up mainly of 13 elements, namely oxygen, silicon, aluminium, iron, calcium, sodium, potassium, magnesium, titanium, phosphorus, hydrogen, carbon and manganese.

The attempts to calculate the chemical ∠omposition of the Earth's crust on the basis of the composition of the igneous rocks were made by Clarke (1889), Knopf (1916), Clarke and Washington (1922, 1924), Saderholm (1925), Vogt (1931), Goldsmith (1933, 1937), Grout (1938) etc.

The most prominent among them is the average presented by Clarke and Washington which was based on 5,159 analysis of igneous rocks from all parts of the world and their calculation, even today, is being utilised as a basis for solving various geochemical problems. It is presented in the following table.

Geochemical Affinity

The distribution of elements in earth's geochemical spheres depends on the following :

1. The general chemical behaviour of the elements,

2. The physico-chemical conditions present in the earth's interior, and

3. The origin of geospheres.

The geochemical division of the Earth into a metallic core, a mantle consisting mainly of sulphides and a silicate crust only implies a hypothesis of the distribution of some few major elements that characterise the geosphere, in question. The properties of these elements form the background for the distribution of all other elements.

A study of distribution of the elements in metallurgical products and among the corresponding phases in meteorites along with the information available of the manner of occurrence of elements in minerals and rocks in the upper lithosphere, reveals the general chemical affinity of the elements. The geochemical affinity of an element is understood as the sum of its properties, which are responsible for its observed and supposed distribution. The affinity of an element for oxygen or silica and for sulphur, as compared with the corresponding affinity of iron, will determine the distribution of elements among metallic-iron, iron-sulphide and iron-silicate or oxide.

Chemical composition of Lithosphere based on analysis of igneous rocks (Clarke and Washington)

Chem. constituent	In percent	Remarks
SiO_2	59.14	* Out of the total only 0.48
Al_2O_3	15.34	per cent is left for by far the
Fe_2O_3	3.08	greatest part of the elements
FeO	3.80	
MgO	3.49	
CaO	5.08	
Na_2O	3.84	
K_2O	3.13	
H_2O	1.15	
TiO_2	1.05	
P_2O_5	0.30	
MnO	0.12	
Total	99.52*	

Geochemical Affinity Groups : The division of elements into the three geochemical affinity groups, namely, siderophile, chalcophile and lithophile, characterises their distribution in the molten metal, sulphide and silicate phases respectively. In general, the geochemical behaviour of an element, as indicated by its distribution among with three phases, depends on the conditions under which these phases have been separated. As such the geochemical character is by no means established for all elements, not even under strictly determined conditions. Under given conditions the distribution of the element in the metal, sulphide and silicate phases corresponds to an equilibrium which probably varies with conditions. It may be assumed that the typically siderophile elements are enriched in the nickel-iron core of the earth. The chalcophile elements most likely occur enriched in the sulphide phase of the intermediate basic mantle but might be present to some extent in the nickel-iron core. Among lithophile elements those belonging to the late magmatic stage of crystallisation, are enriched in the Earth's outermost crust.

Applications

The distribution of elements is important in the following ways :

1. It helps in search for minerals as well as for predicting the ore potential of major regions.
2. It provides the average composition of the different rocks and, thus, the local background values are established to be used in mineral prospecting.
3. It helps to evaluate the ore-bearing prospects of intrusive, effusive, sedimentary and metamorphic rocks from their chemical compositions, and that of underground and running water from the distribution of chemical constituents.
4. It works as an index to deposits of associated elements. It is found that some elements when present in high concentrations in the rocks, they form deposits of associated elements or minerals in the same rocks in a given area, e.g. some tin bearing intrusions are enriched with beryllium and rare earths. The distribution of low zirconium and niobium is noted in auriferous intrusions. Platiniferous ultrabasites are distinguished by peculiarities in the distribution of gallium.
5. The distribution of elements points to the relationship between chemical compositions of rocks and ores. For example meta-somatically altered rare-metal granites give rise to deposits of various types, characterised by concentrations of beryllium, tantalum, niobium and yttrium earths.
6. Associate elements work as pathfinders for various mineral deposits. For example, the distribution of Li in granitoids makes it possible to recognise tantalum varieties of granites. Rubidium and caesium often point to the presence of pegmatites, while arsenic points to the presence of auriferous deposits.

7. Important informations are conveyed by some accessory minerals in rocks like biotite, quartz zircon, sphene and rutile which often carry increased concentrations of certain elements pointing to the possible occurrence of corresponding deposits in an area, e.g. the presence of tin in biotite, derived from granites, indicates that the granites are stanniferous. Much of lithium in quartz is evidence of lithium mineralisation in the granite or pegmatite, while the presence of tantalum and niobium in biotile is indicative of a connection between these granites and tantalum-niobium deposits.

8. Scattered elements, like caesium, germanium, thallium, scandium, cadmium, selenium, tellurium, rubidium, gallium, indium, hafnium, rhenium etc., rarely form minerals of their own and are generally recovered as by-products. These elements may be economically important in ores when they average at least three grammes per tonne.

MINERAL DEPOSITS

The mineral deposits of a country are its natural wealth upon which depend its development and prosperity. They include not only metalliferous deposits, but also non-metallic deposits whose value presently is manifold than those of metallic ores. The bulk of lithosphere (99.5 per cent) is made up of only a few minerals, namely quartz, felspar, pyroxene, amphibole, olivine, mica, magnetite and ilmenite. Besides quartz, felspar and mica, kaoline (clay), dolomite, chlorite, calcite and limonite are the other important minerals constituting various sedimentary rocks. The remaining 0.5 per cent is made up of such elements like platinum, gold, silver, nickel, cobalt, lead, zinc, tin etc., which are quite valuable and useful. The local accumulations of these in varying sizes and shapes, such as in various pockets, or impregnations in rocks give rise to different ore and mineral deposits. The geological processes that are responsible for concentration of the diffused elements and widely scattered minerals into economic deposits are quite interesting and are being dealt subsequently. Prior to coming on the process of mineral deposition, certain salient features with regard to mineral deposits, metallic and non-metallic ores, are given in italics below.

Metallic Mineral Deposits

Metallic mineral deposits mean ore deposits from which metals they contain, are extracted. In dealing with metalliferous mineral deposits the concepts of ore, ore minerals, gangue and tenor of ore are required to be understood. *An ore* is the mined out material which is profitably used. A high grade lead deposit in Antarctica may not be considered as ore at the present, since it cannot be commercially utilised. To be an ore it is essential to consider the cost of production and the market value. An *ore-mineral* is the desired metal held in chemical or physical combination with other elements. As for example, chalcopyrite which is mined for copper, occurs in combination with iron and sulphur besides copper. Ore minerals may also be found in free state, of which gold and platinum are examples. Gangue is the discarded material of ore and constitutes mainly the non-metallic materials or the enclosing rock. Thus, an ore is made up of both ore-minerals and gangue. The *tenor of ore* is the metal content of the ore whose lower limit depends upon the economics of production. It varies with ores of different metals, cost of production and price of metals. It does not have any upper limit, higher the content better it is. The tenors of some important ores and their average grades are shown in Table below.

The tenors of some important ores and their average grades are shown in Table :

Table : Tenors and average grades of important ores

Sr.No.	Metals	Tenors of ores	Average grades of ores
1.	Platinum	0.1 ounce/tonne	0.3 ounce/tonne
2.	Gold	0.15 ounce/tonne (4 gm/tonne)	0.2-0.3 ounce/tonne
3.	Silver	10 ounce/tonne (280 gm/tonne)	12-30 ounce/tonne
4.	Copper	0.7%	1.5%
5.	Lead	3%	6-7%
6.	Zinc	3%	10%
7.	Nickel	0.8%	1.5-2%
8.	Tin	0.8%	1.5%
9.	Aluminium	30%	50-55%
10.	Chromium	32%	35-40%
11.	Iron	30%	58-62% (in Hematite)
12.	Manganese	35%	46-50%

Non-Metallic Mineral Deposits

The materials for non-metallic mineral deposits may be solid, liquid or gas. Coal, petroleum, sulphur etc. are some examples. The terms 'ore' and 'ore-minerals' are not employed in the case of non-metallic mineral deposits. The undesired material, if any, in the deposit, is generally referred to as 'waste' and not 'gangue'. This includes enclosing rocks and discarded non-metallic substances which are unfit for use due to certain deficiency or defects. The waste material is removed by handsorting, simple mechanical concentration, flotation or washing. The non-metallic minerals to be usable must meet the definite specifications set for different uses.

1

Essential Factors in Mineral Deposition and Ore-Genesis

The composition of mineralising solutions, temperature and pressure are the three essential factors which play important role in mineral deposition. The formation of mineral deposits takes place by precipitation from the mineralising solutions. The composition of mineralising solutions determines the nature of mineral deposits. The increase in temperature enhances the solubility of materials in solution, while its decrease causes precipitation from aqueous solutions or magmas. The less soluble salts are precipitated first, followed by more soluble ores. The sequence of minerals in a mineral deposit or mineral zoning may be explained by above. The pressure is another factor which plays important role in mineral deposition. In general, the decrease in pressure causes precipitation, while its increase promotes solubility. The gases are comparatively more sensitive to pressure variations, like carbon dioxide held in water under pressure increases solubility of calcium carbonates.

1.1 GEOLOGICAL THERMOMETERS

DEFINITION AND CLASSIFICATION

Proper understanding of origin of mineral deposits and their classification requires the knowledge of formation-temperatures of these deposits. Certain minerals, present over there, give informations with regard to temperatures of their formations and of the enclosing deposits and they are known as geological thermometers. These geological thermometers may be classed chiefly into the following groups based on their preciseness :

1. The thermometers that record fairly accurately the specific temperature condition of formation of deposits;

2. The thermometers that provide an upper or a lower temperature, above or below which the deposits do not form;

3. The thermometers that provide a range of temperature within which the deposits form; and

4. The thermometers that serve as rough indications of temperatures of formation of mineral deposits.

The presence of two or more of less precise geological thermometers in a deposit narrows the range of temperature of formation for the deposits.

METHODS FOR PREPARATION OF GEOLOGICAL THERMOMETRY

The geological thermometry may be prepared by several methods, the details of which are given below :

1. Direct Measurement

This includes direct measurement of temperatures of lavas, fumaroles, hot springs etc. where formations of minerals take place. According to Bowen the earliest minerals of basic rocks, in general, form between 870° and 600°C, decreasing with increase in silica content. High temperature ore mineral (pyrogenic mineral) like chromite forms within the range of magma consolidation.

The surface temperature of Puga hot springs, Ladakh is measured upto 85°C, and as such sulphur, borax and potash which occur there, may have formed at temperatures above 85°C. Presence of native sulphur occurrence in the northern face of the younger cone of the Barren Island Volcano with measured surface temperature of 85°C indicates its temperature of formation to be much above 85°C.

2. Determination of Melting Points and Inversion Points of Minerals

The melting points of minerals indicate the upper limits of temperatures at which they can form. For example, the melting point of galena is known to be 1120°C. This implies that galena of Hesatu-Belbathan belt, Bihar may have formed at temperature below 1120°C. The presence of other ingredients associated with the mineral further lowers the melting point and, thereby, the temperature of its formation. The melting points of some common minerals are given below :

Table 1.1. Melting temperatures of some common minerals

Minerals	Melting temperatures
1. Olivine (Forsterite)	1890°C
2. Anorthite	1550°C
3. Diopside	1391°C
4. Albite	1120°C
5. Antimony	630°C
6. Stibnite	546°C
7. Bismuth	271°C
8. Sulphur	119°C

The inversion points are quite useful temperature indicators. Some of the inversions temperatures are given below :

Table 1.2. Inversion temperatures of some common minerals

Minerals	Inversion temperatures
1. Tridymite inverts to Cristobalite	1470°C
2. Sphalerite to Wurtzite	1020°C
3. Kyanite to Mullite	1000°C
4. High Quartz (β-Quartz) to Tridymite	870°C
5. Low Quartz (α-Quartz) to High Quartz (β-Quartz)	573°C

Thus, the presence of quartz in any deposit indicates its formation below 870°C. The geode quartz, vein-quartz and pegmatite quartz are low quartz and they may have formed at temperature below 573°C.

3. Study of (i) Dissociation, (ii) Exsolution, (iii) Recrystallisation, and (iv) Liquid inclusions in minerals

(i) Dissociation

Minerals that dissociate water content or other volatile constituents or any other mineral at certain temperatures may form excellent geological thermometers. For example, zeolites, when heated, lose water content and thus form geological thermometer. They indicate low temperatures of their formations. The dissociation temperatures of a few minerals are given below :

Table 1.3. Dissociation temperatures of some minerals

Minerals	Dissociation temperatures
1. Tremolite yields diopside	900°C
2. Calcite (at 1 atm) dissociates	900°C
3. Pyrite into Pyrrhotite and sulphur vapour (1 atm)	685°C

(ii) Exsolution

Exsolution is separation of one mineral from another at particular temperature. This helps in preparation of geological thermometry, and the minerals, that separate, constitute geological thermometers. The exsolution or unmixing temperatures of some minerals are shown below:

Table 1.4. Ex-solution temperatures of some minerals

Minerals	Exsolution temperatures
1. Magnetite	800°C
2. Magnetite-Ilmenite	700°C
3. Chalcopyrite-Pyrrhotite	600°C
4. Stannite-Chalcopyrite	500°C
5. Bornite-Tetrahedrite	275°C

(iii) Recrystallisation

This method is adopted principally for native minerals. The recrystallisation temperatures of some of the native minerals are given below :

Table 1.5. Recrystallisation temperatures of a few native minerals

Minerals	Recrystallisation temperatures
1. Copper	450°C
2. Gold	360°C
3. Silver	200°C

This indicates that the above native minerals belong to hypogene class.

(iv) Liquid inclusion

The liquid inclusion in cavities of crystals points to the approximate temperature of formation of the crystals by the amount of contraction of the liquid. By the application of this method the temperature of formation of certain sphalerite was found to be 115° to 135°C. In case of opaque minerals, the temperatures of bursting of inclusions in the mineral powders are considered to indicate the upper limits of temperatures of their formations.

4. Noting Changes in the Physical Properties of some Minerals

Changes in the physical properties of some minerals take place at particular temperatures. Such changes exhibited by some minerals are shown below :

Table 1.6. Physical changes in some minerals at certain temperatures

	Minerals	Physical changes	Temperatures
1.	Limestone	Pigment expelled	605°C
2.	Mica	Pleochroic haloes destroyed	480°C
3.	Smoky quartz	Colour disappears	300°C
4.	Amethyst	Loses colour	240°-260°C
5.	Fluorite	Loses colour	Around 175°C

5. Discerning Association of Diagnostic Minerals

Associations of minerals, signifying one or more geological thermometers, occur repeatedly in some deposits. These minerals may be ranked as high temperature, intermediate temperature and low temperature minerals. The common examples of a few such minerals are given overleaf :

Table 1.7 : Comparative temperatures of formations of some minerals

High temperature	Intermediate temperature	Low temperature
Magnetite	Chalcopyrite	Marcasite
Pyrrhotite	Arsenopyrite	Adularia
Cassiterite	Galena	Chalcedony
Garnet	Sphalerite	Rhodochrosite
Pyroxene	Tetrahedrite	Siderite

The association of two or more such minerals in a deposit may prove to be diagnostic and is taken as a recognised geological thermometer, while the presence of single such mineral itself may give only an approximate temperature of formation.

1.2. ORE-GENESIS

Ore-genesis is faced with the problem to solve the mystery of ore-forming fluids, involving their source, their physical character, their methods of transport and their deposition in ore-deposits. Hutton and Werner laid the igneous and aqueous foundations of all modern theories of ore-deposits. A vast store of geological data that have accumulated, impresses upon the following few observations, throwing light on ore-genesis :

1. Association of ores with igneous rocks,

2. Presence of mineral deposits with gaseous emanations,

3. Origin of ore-minerals from residual liquid of magma.

1. Association of Ores with Igneous Rocks

There seems to be kinship of some sort between ore-concentration and igneous activity. Certain metals and minerals are associated widely with certain kinds of igneous rocks. The examples are of primary platinum and chromite found associated with ultrabasic rocks, tin with granite, diamond with kimberlite, ilmenite and titaniferous magnetite with gabbro and other basic rocks, corundum with quartz-free rocks like nepheline-syenite etc. These associations point to their genetic relationships. The valuable materials of most mineral deposits, whether hypogene or supergene, have directly or indirectly come from magmas. It is said that magmas are the parents, while mineral deposits are offsprings.

According to Graton (1940) it is by progressive enrichment during the consolidation of igneous rocks that the ores were concentrated and were expelled upward through the agency of vast quantities of hot water in the final stages of their cooling. The precipitation of ores took place by replacement or by simple filling of available open spaces in different types of rocks. According to Fenner a magma tends to separate into :

(a) immiscible sulphide liquids that settle and form magmatic sulphide deposits, and

(b) crystals of silicates and oxides that form igneous rocks or ore deposits.

Liquid magma, like any other liquid, under pressure tends to move to the place of least pressure i.e. dominantly upward. In its upward movement, the magma way pry. Some constituents in the magma through crystallisation and differentiation collect as crystal aggregates, and the molten liquids form magmatic oxide and sulphide deposits before final consolidation. Thus, differentiation yields not only different rock types from a common magma but also magmatic mineral deposits of great economic importance and, in part, the mineralising solutions that form the vast majority of metallic mineral deposits.

2. Presence of Mineral Deposits with Gaseous Emanations

The magmas yield gaseous emanations in huge quantities that escape into atmosphere through the following :

(i) Volcanoes,

(ii) Fumaroles,

(iii) Hotsprings etc.

Studies of the constituents of above contribute much towards an understanding of the genesis of related mineral deposits, and they are being described below :

(i) Volcanoes

Volcanoes constitute one of the sources of mineral deposits. Deposits of native sulphur are common in and around volcanic craters. Realgar, glauber salt, cobalt, copper, lead, zinc, tin, tellurium, bismuth and phosphorus, like sulphur, occur as sublimates from volcanoes. Specular hematite, tenorite, sodium, iron, copper, boric acid and other substances have been noted to be emitted by Vesuvius. In India volcano of Barren Island, Andaman is another example.

This establishes a magmatic source of the above materials of mineral deposits.

(ii) Fumaroles

There is better evidence of sublimates and other deposition from fumaroles compared to those of volcanoes where voilence is not congenial for mineral deposition. Fumaroles represent a gaseous phase where substances are carried, and demonstrate the ability of gaseous emanations to collect, transport and deposit the metal. Fumaroles of Etna, namely Vesuvius and Stromboli, have yielded such chlorides as those of sodium, potassium, iron and copper; carbonate and sulphate of sodium; fluorides, sulphides and other compounds of cobalt, manganese, lead, zinc, tin, copper, bismuth tellurium, arsenic, potassium, rubedium, caesium, thallium and others. Most of the elements that make up mineral deposits are found in such emanations. In India sulphur deposits are found at fumaroles of Puga, Ladakh (J & K State) where conditional reserves of native sulphur are estimated at 0.21 million tonnes, all under possible category.

(iii) Hot Springs

Hot springs contain many mineral substances and, thereby, show that such hot waters dissolve, transport and deposit them. When the surface or rain water penetrates great depths, it gets heated and come out in the form of heated water or hot springs along with many mineral substances. The water of springs have possibly been heated by heat liberated from exothermic chemical reactions and radiogenic sources besides being heated by deep circulation.

India is endowed with large number of hot springs. The hot spring of Manikaran (Kulu, Himachal Pradesh) exhibits temperature as much as 98°C, that of Surajkund (Hazaribagh, Bihar) 88°C, that of Sarguja (MP) 98°C, that of Jamnotri (Uttar Kashi, UP) 90°C, that of Agnigundala (Andhra Pradesh) 80°C, and that of Puga (Ladakh, J & K) 85°C, and these hot springs contribute greatly to geothermal energy resource of the country. Some of the spring waters contain negligible amounts of salts and minerals, while others contain appreciable quantities of salts and are highly radioactive. The spring waters of Dwari, Surajkund and Kawa Gandhari (Hazaribagh district) are comparatively high in mineral substances like manganese, lithium, strontium, barium, potash, sodium, fluorine etc. and the total solids estimated in them are 420 ppm, 540 ppm, and 352 ppm respectively. They are also high in radon contents. The hot springs of Puga show conditional submarginal resources of about 74000 tonnes of borax and uneconomic deposits of potash and other salts.

3. Origin of Ore-Minerals from Residual Liquid of Magma

The residual liquids, which result as end phase of crystallisation-differentiation of magma, constitute hydrothermal solution. If the external pressure i.e. rock pressure is less than the vapour pressure of the residual liquid and the confining rock is permeable to the high penetrating power of the vapour under pressure, a vapour phase results, later to condense to a liquid phase. But if the external pressure is greater than the vapour pressure of the residual liquid, no vapour phase will result, and the mother liquors will be alkaline liquids, which may be expelled as such to form rising hydrothermal metallization solutions. The external pressure is high under deep seated conditions and so over there a vapour phase may be absent.

Thus, there are two methods of deposition of ore minerals from hydrothermal solutions :

(i) The hydrothermal solutions leave magma as gaseous emanations that later condense to hydrothermal liquids from which ores are deposited. This is a simple and likely method for transport of valuable constituents from the magma and their deposition.

(ii) The hydrothermal solutions, which leave the magma as attenuated alkaline liquids, carry with them ingredients of mineral deposits and deposit them at suitable sites under favourable conditions.

It seems likely that the ore-genesis took place in both the ways.

Brown, J.S. (1950) gave somewhat different views regarding ore-genesis. According to him ore deposits could not possibly have originated through the agency of water and water played only a minor part in the process. The fundamental reason for this is that the ore-fluid has to travel for many thousands of feet through solid rocks, having a porosity only between one and two per cent and average openings of 0.0005 mm or half microns. Further, hydrothermal theory would require accumulation of enormous volume of water at great depth in the heated crust of the earth to bring the metal to the surface, which seems decidedly difficult. He considers vapour as a major vehicle for these mineral substances to be deposited and entitled it a metallurgical interpretation of ore-genesis.

He advocates for several independent source magmas rather than one. Indications are that these various source magmas accumulate at different depths dependent, in main, on relative density, the lighter ones at the highest level, the heavier ones successively deeper. Outstanding in importance are :

(i) The pegmatitic magmas, which are responsible for the ores of tin and tungsten, various valuable silicates, etc. are the lightest and form at the highest level.

(ii) The iron magmas, which are probably multiple in number and consist mainly either of iron oxides or iron silicates, are heavier than pegmatitic magma but lighter than the sulphides, and form at intermediate depths.

2

Processes of Formation of Mineral Deposits

The geological processes that yield mineral deposits are as follows :
1. Magmatic Concentration
2. Sedimentation inclusive of Evaporation
3. Metamorphism
4. Contact Metasomatism
5. Hydrothermal Processes
6. Oxidation and Supergene Enrichment
7. Sublimation
8. Residual and Mechanical Concentration

The mineral deposits may be classified based on their genesis or the process that operates to form them e.g. Magmatic, Sedimentary etc. These deposits may be referred to as syngenetic if formed at the same time as the rocks that enclose them, and epigenetic in case formed later than the rocks that contain them.

The various processes responsible for formation of minerals deposits are discussed below in brief.

2.1 MAGMATIC CONCENTRATION

The magmatic deposits are formed during different stages of magma crystallisation. Certain metallic oxides, sulphides and native minerals, like chromite, magnetite, millerite, diamond etc., were formed in the early stages of magmatic crystallisation and became segregated by crystal separation to form early magmatic mineral deposits, while others crystallised later than the host rock and accumulated at the original site or injected elsewhere to give rise to late magmatic mineral deposits like pegmatitic deposits containing tin, beryllium, tentalum, niobium, lithium and other ores.

Early Magmatic Deposits

The early formed ore minerals in the deposits may occur as below :
   (i) Disseminated in the enclosing rock e.g. diamond in kimberlite of Majhagawan pipe, Madhya Pradesh and Wajrakarur pipe,

Andhra Pradesh. The diamonds might have crystallised early and were transported with the enclosing magma, and, perhaps, even continued to grow before final consolidation took place in the present pipes.

(ii) Segregated due to gravitative crystallisation differentiation e.g. stratiform and banded graded deposits of chromite in Nausahi-Sukinda area, Orissa and other places. The early magmatic segregation may be due to sinking of heavy early formed crystals to the lower part of the magma chamber or by marginal accumulation.

(iii) Injected into the host rocks or the surrounding rocks e.g. vanadiferous magnetite deposit of Dublabera, Singhbhum district in Bihar. It occurs as veins and lenses within ultrabasics and gabbro. The early formed ore mineral, magnetite, did not remain at the site of original accumulation but have been injected into ultrabasic rocks of Iron-Ore formation.

Late Magmatic Deposits

The late magmatic deposits may be accounted for :

(i) Residual liquid segregation wherein the residual magma with crystallisation becomes progressively richer in silica, alkali and water. It sometimes contains titanium and iron which on crystallisation segregated to form titaniferous magnetite deposit e.g. titaniferous magnetite of Hassan district, Karnataka, which occurs as comformable bands in amphibolites and basic schists. Vanadiferous magnetite deposits of Mayurbhanj, Orissa is another example. Here, it occurs associated with gabbro-anorthosite suite of rocks.

(ii) Residual liquid injection which takes place due to earth's disturbance like igneous intrusion. The residual liquid rich in iron when injected crystallised to form magnetite deposits, e.g. magnetite dyke rock of Kasipatanam, Visakhapatanam district, Andhra Pradesh, where it occurs as cutting across the NE-SW foliation of charnockite-gneiss and also metamorphoses the wall rocks.

(iii) Immiscible liquid segregation in which certain salts in magma under certain conditions separate out an unmixed solutions like oil and water, and segregate to form important mineral deposits. It has been observed that sulphur and silica form two hot immiscible liquids wherein a molten mass consists of various metals. The examples of this category are lead-zinc-copper sulphide deposits of Hesatu-Belbathan belt, Bihar where they occur associated with altered basic-schists in disconnected bodies.

(iv) Immiscible liquid injection when the unmixed sulphide-rich fraction accumulated in the magma chamber, as described above, is squirted out before consolidation towards the places of less pressure, such as shear zones. They intrude the older rocks and enclose brecciated fragments of host and foreign rocks. The nickeliferous chalcopyrite pockets associated with altered basic-schists (chlorite-biotite schist) of Singhbhum Copper belt may be cited as an example. This kind of deposit shows transition to hydrothermal types with enrichment of volatile matters.

2.2 SEDIMENTATION

Sedimentary rocks with valuable mineral deposits like iron, manganese, copper, phosphate, coal, oil shale, limestone, clay, sulphur etc. are formed due to the process of sedimentation. These substances are made up of inorganic and organic materials and their source is the other rocks which have undergone disintegration. Some of the materials like oxygen and carbon-dioxide may have been obtained from atmosphere. Besides source materials, the other factors responsible in resulting sedimentary deposits are gathering of materials by solution, their transportation and deposition at suitable sites.

The minerals in a rock are the important source of sedimentary mineral deposit. For example, the iron-bearing minerals, hornblende, pyroxene, mica etc., in a rock, on oxidation and weathering yield materials for the sedimentary iron-ore deposits. Similarly the source of sedimentary deposits of manganese, phosphate, carbonate etc. are their respective minerals in different kinds of rocks. The earth's crust contains, on an average, 5.05% iron, 0.09% manganese and other elements in different proportions, and these form source materials of the concerned deposits. Occasionally, the materials for mineral deposits are derived from the earlier deposits, such as iron, manganese, copper, phosphate etc.

The weathered materials go into solutions of different kinds. The carbonated water, humic and other organic acids and sulphate solutions form the chief solvents. Humic and other organic acids are derived from decomposition of vegetation, while oxidation of pyrite produces sulphuric acid. The most of the substances that make up sedimentary mineral deposits (except coal) are transported by rivers and subsurface waters. These substances remain in solution until there is appreciable physical or chemical change.

The materials in solution may be deposited mechanically, chemically or biochemically whether in a sea or in a swampy basin or elsewhere. The oxidation, hydrolysis, bacterial or plant action, reaction with other salts, sea-water electrolysis, evaporation etc. are the various agencies causing deposition of different economic minerals when in solution. The mineralogical composition of the resulting deposits, their size, purity and distribution are determined by the conditions of deposition. Sedimentary iron and manganese deposits may result both in fresh and salt water, in bogs, swamps, marshes, lakes, lagoons and in the oceans. Phosphate and sulphur form in marine conditions. Swampy basins are ideal for coal formation like Gondwana and Tertiary coals of India. The shallow water marine condition is quite suitable for majority of sedimentary deposits such as iron-ore, manganese, phosphate rock, sulphur, carbonate rock, clay, etc.

Evaporation

Evaporation is one of the important agencies which brings about deposition of many valuable minerals, once in solution, like gypsum, common salt, potash, nitrates and many other non-metallic minerals. Warm and arid climate is essential to cause evaporation whether in ocean water, lake water or ground water. The concentration of the soluble salts in the bodies of water takes place by evaporation and when supersaturation of any salt is reached, that salt is precipitated. The least soluble salt is precipitated first, and the most soluble last. The sea water with 3.5 per cent salt yields gypsum or anhydrate only when the volume is reduced by evaporation to about one fifth and in the case of common salt it has to reduce to about one tenth. In case of underground water, the solutions are required to be drawn up by capillary actions or other means to air surface for evaporation and deposition of various salts like nitrate salts, iodine and some boron salts, calcium carbonate, sodium sulphate, sodium carbonate etc.

2.3 METAMORPHISM

Metamorphism is an important process to give rise to many new mineral deposits by altering the earlier deposits/rocks. It is by recrystallization and reconstitution of pre-existing rock forming minerals that some valuable mineral deposits are formed. Heat, pressure and water play an important role in bringing about metamorphism. The original texture and structure are obliterated; with the result some of the ore minerals may exhibit streaked, banded and smeared appearances with indistinct boundaries between minerals of different colours. The change, at times, may be so pronounced that the identity of original deposit is entirely lost. Many of the non-metallic mineral deposits of importance are formed in this way e.g. deposits of asbestos, graphite, talc, soapstone, andalusite, sillimanite, kyanite etc. The formation of some of the above deposits is discussed below to understand the nature of recrystallisation or reconstitution or both.

In the case of asbestos formation, chrysotile asbestos is not formed except where there is serpentinization which is an autometamorphic process and may proceed along fractures and fissures. The process involves conversion of olivine to serpentine with addition of water. This alteration may possibly be due to hot residual solutions that emanated from within the intrusives. The serpentine thereafter undergoes change by molecular rearrangement into fibrous form of asbestos. The serpentinised limestone may also give rise to chrysotile asbestos due to metamorphic action. The amphibole varieties of asbestos, namely crocidolite and amosite, may also have originated due to molecular reorganisation in respective minerals themselves.

Graphite is a product of regional or contact metamorphism and carbon of the graphite may have been derived by the sediments from the carbonate rocks. The coal beds which have been altered to graphite are also the result of metamorphism, by which volatile matters of the coal have been removed and the residual carbon is changed to graphite.

Talc is metamorphic product of magnesian-rich minerals in a rock. It is formed at a late stage due to hydrothermal metamorphism, aided to some extent by dynamic metamorphism. Magnesian amphibole or pyroxene when acted upon by CO_2 and H_2O, talc is generated.

The andalusite-sillimanite-kyanite deposits are formed due to regional metamorphism of aluminous silicate rocks like mica-schists and gneisses wherein temperature and pressure play important role. Andalusite is formed under conditions of high temperature and low stress e.g. in thermal aureoles. Sillimanite is produced at high temperature and moderate stress, while kyanite is characteristic of rocks formed under high stress and moderate temperature.

2.4 CONTACT METASOMATISM

Contact metasomatism is a process of formation of new mineral by reaction between the contact rock and the escaping high temperature gaseous emanations with other important materials from the magma chamber. For the deposit of this type, the magma must contain the ingredients of mineral deposits and must be intruded at depth at the contact of reactive rocks. It differs from contact metamorphism in which only effect of heat is involved and role of accession from magma chamber is negligible. Lindgren proposed the term pyrometasomatism which is essentially the same as contact metasomatism.

Process

The process of contact metasomatism starts with recrystallization and recombination of rock minerals in the contact zones, e.g. limestone or dolomite converts to marble, shale to hornfels and sandstone to quartzite. The ore-minerals in such cases result from accession from the magma. Large volume of material may be added and subtracted from the invaded rocks. The reaction apparently commences just after intrusion and continues until well after consolidation of the outer part. A common order of formation of ore-minerals is pyrite and arsenopyrite, followed by pyrrhotite, molybdenite, sphalerite, chalcopyrite, galena and sulpho-salts. In some places, sulphides form contemporaneously with the silicates. The temperature for contact metasomatism may possibly be ranging from 400° to 800°C, or even higher.

Relation to Intrusives

The composition, size, form and depth of formation of the intrusive body play important role in formation of contact metasomatic mineral deposits. In general, intrusives of silicic and intermediate composition, such as quartz-monzonite, monzonite, granodiorite or quartz-diorite, yield mineral deposits. It is because they contain higher water content compared to that in basic rocks, and water in the magma is the chief collector and transporter of metals.

Intrusive bodies of the size of batholiths, stocks and those with gentle dips offer wider zones of reaction and, hence, are more favourable for contact metasomatic deposits. Irregular forms of cupolas and roof pendants also expose greater areas for reaction. Greater the depth of intrusion, lesser is the loss of magmatic emanations and more chances for formation of mineral deposits.

Relation to Invaded Rocks

The composition and structure of invaded rocks determine the nature and the extent of their alteration. Sedimentary rocks particularly carbonate rocks are the most susceptible to changes. They show recrystallization and recombination with foreign materials. The carbonate rock in the contact of the intrusive may be converted to garnet rock, silicate and ore. The shale and slate alter to hornfels with andalusite, sillimanite and staurolite. The sandstone recrystallizes to quartzite with sparse dissemination of ore minerals. The invaded igneous rock is the least affected, especially when the intrusive is also of the same composition. Metamorphic rocks are also not favourable for further alteration and ore localisation.

The structures like bedding, laminations, faults etc. serve as good channel ways for escaping emanations and may produce larger and more widely distributed ore deposits. Bedding planes in contrast to directions across it are more favourable for the magmatic emanations to yield better ore deposits.

Resulting Mineral Deposits

The resulting mineral deposits are found scattered irregularly within the contact aureole or close to the intrusive contacts, having tendency to concentrate towards the gentler dip of the intrusive, e.g. barytes occurs in limestones of Vempalle formation (Cuddapah), Andhra Pradesh, near the contact with basic sills or in the sills themselves. The graphite deposits in Orissa are found along the contact zones of khondalite with granite-gneiss associated with pegmatitic bodies.

The contact metasomatic deposits are mostly small in size, consist of several disconnected bodies and have abrupt terminations. The outlines are irregular with ramifying tongues, projecting outward. The ores are generally coarse in texture. They lack crystal outlines except with a few exceptions like pyrite, arsenopyrite etc. The deposits are characterised by an unusual assemblage of ore and gangue minerals, characteristic of high temperature formation. The ore minerals include oxides such as magnetite, ilmenite, corundum, spinel etc.; native metals/minerals like gold, platinum, graphite etc.; sulphides like those of base metals etc. The gangue minerals consist of garnet, hastingsite, tremolite, wollastonite, diopside, forsterite, anorthite, albite etc.

2.5 HYDROTHERMAL PROCESSES

The term hydrothermal means hot water with possible temperature of 500°C to 50°C. The fluid resulting as an end product of magmatic differentiation, constitutes hydrothermal solution which carries metals originally present in the magma to the site of deposition. The process is responsible for formation of epigenetic mineral deposits i.e. those formed later than the rocks that enclose them. The hydrothermal solution in its journey through the rocks loses heat and metal contents with increased distance. The deposition may have taken place at high temperature (hypothermal deposit), intermediate temperature (mesothermal deposit) or low temperature (epithermal deposit).

Prerequisites

The prerequisites for hydrothermal deposits are :

(i) the availability of enough metal content in the hydrothermal solution,

(ii) presence of solution capable of dissolving and transporting mineral matter,

(iii) available openings in the rocks permitting movement of hydrothermal solution from the source to the site of deposition, and

(iv) chemical reaction causing deposition of ore.

Openings

Pore spaces, crystal lattices, bedding planes, vesicles, cooling cracks, breccias, fissures, shear zones, foldings and warpings, volcanic pipes, solution and rock alterations etc. are the various types of openings in the rocks permitting movement of solution or deposition of ore-minerals. For large deposits vast quantities of solution and fairly large confined channel ways are needed. The flow of solution must be confined to avoid dispersal of mineral matter. Fissures, shears and permeable beds may provide confined channelways. Volcanic breccias, on the other hand, exhibit widespread permeability and the mineralising solution is spread over a large area which result in dispersed ore. Crystal lattices permit diffusion which is a slow process and may not generate large deposits.

Host Rock

Reactive host rocks like carbonate rocks, greenstones etc. are congenial to ore deposition, particularly in the case of replacement deposits.

Deposition

The deposition from hydrothermal solutions is influenced due to chemical changes in solution, reactions between solution and wall rocks or vein matter and changes in temperature and pressure. The reaction between mineralising solution and wall rock gives rise to chemical changes, accompanied by deposition. The solution in its journey loses temperature and pressure which decreases solubility and promotes precipitation. The heat loss is also influenced by nature of opening. Open fissure with straight wall would cause less heat loss than the intricate openings of breccia with large exposed area.

In the formation of ore-minerals, the individual minerals are formed in orderly sequence with quartz coming first, followed by iron sulphides or arsenides, sphalerite, enargite, chalcopyrite, bornite, galena, gold and silver minerals. This sequential arrangement is called paragenesis.

Wall Rock Alteration

The wall-rock alteration is quite common in case of hydrothermal deposits. The nature of mineralising solution like its chemical character, temperature and pressure as well as character and kind of wall rock decide about the nature and intensity of alteration. Sericitization, kaolinization and silicification are the common forms of alterations in the wall rock. High temperature minerals like tourmaline, topaz and amphibole may develop. Basic and ultrabasic igneous rocks are serpentinized accompanied by the production of epidote and chlorite.

Control of Ore Localisation

The ore localization is controlled by the following factors :

(i) *Chemical and physical characters of host rock :* This determines the location, shape and size of opening. For example carbonate rocks permit solution openings and brittle rocks shatter more readily to localize fractures or breccias. Permeability which is necessary in rock for passage of solution and ore localisation, is caused due to pore spaces, fissibility, cleavage planes, brecciation, joints, fractures etc.

(ii) *Structural features :* Structural features like fissures, shears, folds, faults, bedding planes, lamination and unconformity serve important localizers of hydrothermal deposits.

(iii) *Intrusives :* Intrusives being source of ore-bearing fluid, constitute ore loci on a regional scale.

The other factors which have bearing on ore localization are depth of formation, changes in size of rock openings etc.

Cavity filling and replacement are the two types of deposits formed due to hydrothermal processes. Cavity filling is due to deposition of minerals in various types of openings, while in metasomatic replacement or replacement deposit the earlier formed mineral is replaced by the new mineral. In general, replacement deposits are formed at high temperature and pressure, and cavity filling deposits at lower temperature and pressures. However, both types of deposits may also form at the same time and at all temperatures. For example, during the replacement by contact mestasomatism high temperature and pressure prevail, while supergene enrichment may take place at surface temperature and pressure.

1. CAVITY FILLING

The precipitation of minerals from mineralising solution in the cavities or the open spaces in rock forms cavity filling deposit. The walls of the cavity are lined first by the first mineral to be deposited. The minerals usually grow inward with development of crystal faces pointed towards the supplying solution in the form of comb structure. Successive crusts of different minerals may be precipitated upon the first one until filling is complete. This gives rise to crustification and if the cavity is a fissure, a crustified vein is formed. Symmetrical crust may result with similar precipitation on both the walls of the vein, and asymmetrical with unlike crustification on each side. In case of breccia, the crusts surround the breccia and cockade ore is formed. The cavity filling may also give rise to ribbon structue with narrow layers of quartz separated by thin dark seams of altered wall rock. The following types of deposits may result due to cavity filling :

(i) *Fissure vein :* It is a tabular type of deposit, involving formation of fissure itself by stresses operating within earth's crust, and ore forming processes, e.g. fissure filling deposit of magnesite in Salal area, Jammu. These fissure veins may be massive or crustified. They may be simple, composite, linked, sheeted, dilated and chambered. They may be vertical or inclined. Pinches and swells produced by movement along irregular fissures may occur. Several minerals, both ore and gangue, may fill in the fissure. Fissures may occur in groups, and may have formed at the same time or may be of different ages. The depth of fissure veins is quite variable. Some of them continue to depth of several thousand metres like those at Kolar gold mine.

(ii) *Shear zone deposit :* A shear zone with sheet like connected openings, and large exposed surfaces serves as excellent channel ways for mineralising solutions and precipitation takes place as thin plates of minerals or in the form of fine grains, e.g. Singhbhum shear zone deposit.

(iii) *Stockwork :* It signifies a network of small ore bearing veinlets and stringers traversing a mass of rock. The veinlets show crustification, comb-structure and druses, and represent open space fillings. For example, in Zawar area, Rajasthan, veins and stringers of galana and sphalerite traverse dolomite mass and form lenticular bodies. Stockworks of asbestos occur in the Archaean terrain of Barabana area, Singhbhum, Bihar.

(iv) *Saddle reef :* It results when alternating competent and incompetent rocks are closely folded. This gives rise to openings in the crest part of the arch, which is latter filled with ore minerals. The quartz reefs of Hutti gold deposit, Karnataka, and those of Wynad gold deposit, Tamil Nadu are the best illustrations.

(v) *Ladder venis :* This type of deposit forms due to transverse veins or fractures, e.g. magnesite deposit of Mysore, Karnataka, and asbestos deposit of Cuddapah district, Andhra Pradesh.

(vi) *Pitches and flats :* These are formed due to folding of brittle sedimentary beds which gives rise to a series of disconnected deposits. Iron-ore deposits of Bailadila and Chotadongar, Bastar district, Madhya Pradesh, lead-zinc deposits of Baghmari (Katuria), Banka district, Bihar, and talc deposits of Cannore and Kalicut districts, Kerala are the several examples.

(vii) *Breccia filling deposits :* The breccias offer opening spaces in between the angular fragments for deposition by the mineralising solutions traversing through them. These may be volcanic breccia, tectonic breccia or collapse breccia deposits. Wajrakarur kimlerlite pipe, Andhra Pradesh, and fault breccia in Singhbhum Shear Zone, Bihar with copper, lead, uranium and apatite mineralisation may be cited as examples.

(viii) *Solution cavity filling :* Certain solution forming rocks like limestone gives rise to this type of deposit e.g. barytes deposits in Krol limestone of Sirmur district, Himachal Pradesh.

(ix) *Pore space filling :* Many mineral deposits occur as pore space fillings, say in sandstone. Oil, gas and water are the most important among all. Disseminated lead-zinc deposit in gritty conglomeratic dolomite and quartzite of Zawar, Rajasthan is an example of pore space filling.

(x) *Vesicular Fillings :* The vesicular lava flows being permeable, form channel ways for mineralising solution and sites of mineral deposits. Copper occurrences in Dras volcanics, Kargil area, Jammu and Kashmir and in Deccan trap of Maharashtra and Gujarat, and agate, chalcedony, amethyst and opal occurrences in the Deccan trap are the examples of vesicular fillings.

2. METASOMATIC REPLACEMENT

Process

Metasomatic replacement is defined as a process of simultaneous solution and deposition by which earlier formed mineral is replaced by a new one. It takes place when the mineralising solution comes in contact with mineral which is unstable in its presence. The resulting mineral occupies the same volume as that of the replaced mineral and tends to retain the original shape, size and structure. The replacement is by diffusion, a volume for volume interchange. The metasome (replacing mineral) constantly advance against the host which goes into solution. The rate of replacement is dependent upon the rate of supply of new material and removal of dissolved material. If the supply of material for the new mineral stops, other minerals may be deposited at its margins and, thereby, enclosing the earlier formed mineral. Diffusion is an exceedingly slow process which facilitates movement of molecules or ions in solution at the actual front of replacement. This may give rise to only small deposits. For larger deposits, large channel ways are required for transporting large quantities of replacing substances over long distances.

The replacement takes place in stages. The first formed minerals are replaced by the later ones. The sequence formed in common primary metallic minerals is pyrite, enargite, tetrahedrite, sphalerite, chalcopyrite, bornite, galena and silver. Pyrite and arsenopyrite are the first formed metallic minerals, and are replaced by the later formed minerals. In the case of rocks, ferro-magnesian minerals are replaced first, followed by felspar and quartz.

The various requisites for replacement deposits, role of temperature, pressure and host rock, nature of mineralising solution, and control of ore localisation have already been discussed above under hydrothermal processes.

Resulting Mineral Deposits

The replacement deposits may be :

(a) massive,

(b) replacement lodes, and

(c) disseminated.

The massive deposits are of irregular form and of varied size. The replacement may start from a fissure with bold face of massive against the country rock. It may thicken and thin with wavy outline and ramify irregularly in all directions. The host rock is almost completely replaced and the ore shows abrupt ending against the country rock. The ore may retain original texture and structure of the rocks it replaces.

The replacement lodes may be massive or high-grade ore flanked by a fringe of disseminated ore. Here, replacement may start from a fissure with bold front or some prominent centres followed by replacement at many small centres. Like fissure veins, they occur replacing walls of fissures or thin beds. The outlines of the ore body may be wavy, irregular and gradational with the country rock.

The disseminated deposits represent multiple centres of replacement and consist of altered host rock and disseminated ore grains e.g. porphyry copper deposits. The mineralising solution spreads in the host rock and gives rise to scattered specks, grains, blebs and small veinlets of ore scattered within the host rock. The boundaries between the ore and the host rock is vague and gradational. This type of deposit being of low grade and huge, requires large scale mining.

The texture of replacement ores is dependent upon temperature and pressure of formation and degree of replacement. Coarse texture is characteristic of high temperature and pressure, while fine-grained texture is developed at low temperature and pressure. When replacement is complete, holo-crystalline texture is developed. In case of incomplete replacement, the residual host rock forms part of the ore body.

Many precious metals, base metals and rare metals besides ferrous and non-metals deposits are examples of replacement deposits.

Criteria for Identification of Replacement Deposits

The various criterias for recognisation of replacement deposits are as follows :

(a) *Presence of unsupported residuals of host rocks :* The country rock that escaped replacement remained as isolated body with the ore-mass. The residual rock may show bedding or other structural features which are in conformity with those in the wall rock. This constitutes supporting evidence of replacement.

(b) *Preservation of rock structures :* The structure of the replaced body is sometimes faithfully retained in the ore. This constitutes the conclusive evidence of replacement. Such inherited structural elements may be stratification, cross bedding, schistosity, folding, faults, or joints. Fossils and dolomitization rhombs are also sometimes preserved.

(c) *Doubly terminated crystals of ore, transecting rock grains of the enclosing rock :* Such crystals may be microscopic or megascopic in size.

(d) *Mineral pseudomorph :* A mineral after another of different composition is typical of replacement, e.g. chalcocite may replace pyrite cubes and form pseudomorphs.

(e) *Irregular outlines of the ore :* The wavy outlines of the ore with protuberances and embayments into the host rock and extreme irregularity are indicative of replacement.

(f) *Intersection of diversely oriented host crystals :* Small wavy veins of irregular width which transect host rock indicate replacement.

(g) *Absence of crustification :* This is also one of the criterias to recognise replacement deposits.

2.6 OXIDATION AND SUPERGENE ENRICHMENT

The processes of oxidation and supergene enrichment give rise to many large and rich ore-deposits. Both the processes, in general, occur together. Oxidation is operative in the upper part of the ore-deposits above the ground water table, called zone of oxidation. Ore minerals are oxidised by the surface water and produce solvent that dissolves other minerals and carries them down the ground water table. The leaching solution as proceeds downward loses a part of their metallic content within the zone of oxidation as oxidised ore. The secondary or supergene sulphide enrichment takes place when the down trickling solution reaches below the water table and its metallic contents are precipitated as secondary sulphides. Below this zone is the primary or hypogene zone which remains altogether unaffected. The process in ideal situation gives rise to a zone of gossan in the topmost part of the oxidation zone, followed by a supergene enrichment zone and then a primary zone (Fig. 2.1).

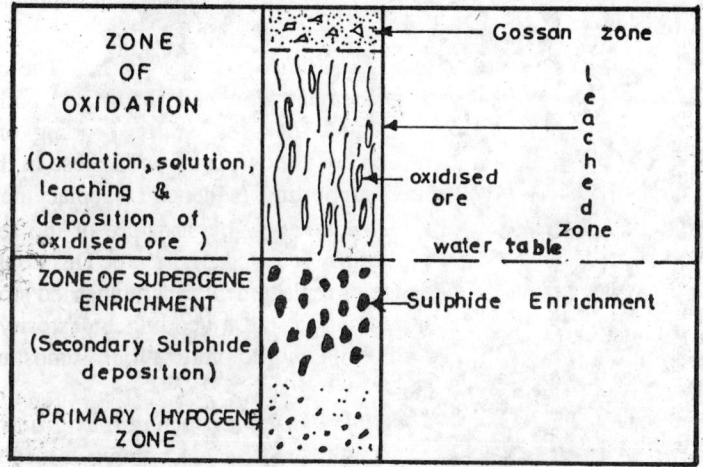

Fig. 2.1. Zones of oxidation and supergene enrichment.

Zones of Oxidation

Gossan : Gossan is the cap-rock of an ore deposit in the form of cellular mass of limonite and gangue formed due to oxidation of an ore and points to the nature of underlying hidden deposits.

Limonite : The limonite which is ubiquitous in the Gossan of the oxidised zone occurs in a variety of colours and structures, each of which is significant. It may have formed from iron-bearing sulphides or iron-bearing rock silicates. The former is distinguished from the latter by its solubility in dilute hydrochloric acid. The limonite so formed may be indigenous or transported. The free acid generated in the oxidation zone, transports iron in ferrous state and deposits it elsewhere, while the indigenous limonite occupies the voids left by former sulphides. In general, seal brown, maroon and orange colours of limonite in the cappings signify copper, while yellow and brick red indicate pyrite. Ocherous orange is suggestive of galena, tan to brown of sphalerite and tan to maroon of molybdenite.

Besides colour, the structure assumed by limonite, called boxwork, helps in diagnosis of earlier minerals. The coarse-cellular boxwork (Fig. 2.2) with ocherous colour is indicative of chalcopyrite, while fine-cellular boxwork (Fig. 2.3) with light brown colour suggests pre-existing sphalerite. Cleavage type boxwork (Fig. 2.4) with ocherous orange colour may possibly be of galena derivative.

Fig. 2.2 Coarse cellular boxwork with typical triangular and quadrangular patterns, correlated with bornite and chalcopyrite.
Loc. : Mundatikra, Bastar district, M.P.

Fig. 2.3 Fine cellular boxwork. Also quadrangular patterns with parallel cell walls, correlated with chalcopyrite.
Loc. : Mundatikra, Bastar district, M.P.

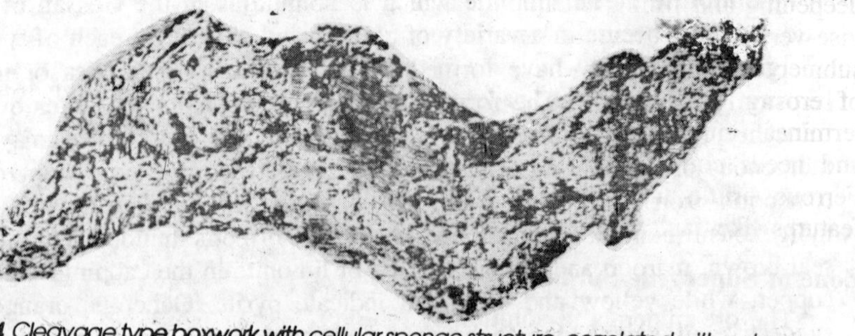

Fig. 2.4 Cleavage type boxwork with cellular sponge structure, correlated with galena.
Loc. : Belbathan, Godda district, Bihar.

Relict sulphides, wall rock alteration and character of voids also help in inferring the underlying hidden deposits. Relict sulphides that remain in the oxidised cropping are the direct clues to the underlying ore deposits. Kaolinization coupled with sericitisation and the presence of indigenous limonite suggest primary sulphide mineralisation which has undergone supergene sulphide enrichment.

Changes in Oxidation Zone

The oxidation causes alteration of minerals, obliteration of structure and leaching of metallic substances. The chemical changes involved in the oxidation zone are formation of sulphuric acid and iron sulphate from pyrite, which is the most common mineral in the metallic deposit, as per reaction given below :

$$FeS_2 + 7O + H_2O = FeSO_4 + H_2SO_4$$

The ferrous sulphate readily oxidises to ferric sulphate and ferric hydroxide as per under-given reaction :

$$6FeSO_4 + 3O + 3H_2O = 2Fe_2 (SO_4)_3 + 2Fe (OH)_3$$

The ferric hydroxide forms limonite, present in the oxidised zone. The ferric sulphate and the sulphuric acid act as solvents for the ore minerals. The chemical reaction involved in the case of chalcopyrite may be expressed as given below :

$$CuFeS_2 + 2Fe_2 (SO_4)_3 = CuSO_4 + 5FeSO_4 + 2S$$

Likewise sulphuric acid also reacts with various sulphides giving rise to sulphates of their metals which trickle downward and deposit their metallic contents as and where conditions are favourable.

Control of Oxidation

Factors controlling oxidation are rise and fall of water table, rate of erosion, climate, time, physical and chemical properties of rocks and structures. Oxidation can take place freely above the water table, while free oxygen is lacking in the ground water and so the oxidation usually ends at the water table. Changes in water table are caused due to valley deepening and filling, erosion, change of climate from arid to humid and vise-versa, and these in turn affect oxidation. A raised water table submerges the oxidised zone. The lowering of water table due to rapid rate of erosion may not keep pace with oxidation and the oxidised zone terminates much above the water table. The process of oxidation is slow and needs considerable time. Nature of rocks also controls oxidation. Ferrous, brittle and fractured rocks favour rapid oxidation. The structural features like fault and shear cause deep and pervasive oxidation.

Zone of Supergene Sulphide Enrichment

The ore minerals in solution which are not precipitated in the zone of oxidation, trickle down below the water table where they are deposited as secondary sulphides. The existing metal in the zone is, thus, enriched by addition from above. This is known as the zone of supergene sulphide enrichment.

The primary requisites for supergene enrichment are :

(i)   oxidation,

(ii) presence of suitable primary minerals to yield necessary solvents,

(iii) permeability of rock to enable the solution to penetrate the oxidation zone,

(iv) absence of precipitants in the oxidised zone,

(v)   zone of no available oxygen where secondary sulphides may be deposited, and

(vi) presence of hypogene minerals below water table to cause precipitation of secondary sulphides.

Method

Certain metals in solution are precipitated as sulphides in presence of other sulphides. Schurmann determined their order of deposition in the presence of any one lower in the series which is as follows : mercury, silver, copper, bismuth, lead, zinc, nickel, cobalt, iron and manganese. The least soluble ore is the first, while the most soluble comes last. To make it more clear, the copper sulphate solution in the presence of any sulphide lower in series, say galena (lead sulphide), is deposited as copper sulphide (chalcocite, covellite) in the zone of supergene sulphide. Manganese sulphide, being most soluble, is not deposited by any sulphide in the series.

Chemical Changes

The chemical changes that take place in the supergene sulphide zone may be expressed by the following reactions :

In the case of copper sulphide :

$$PbS + CuSO_4 = CuS + PbSO_4$$
(Galena)   (Covellite)

$$5FeS_2 + 14CuSO_4 + 12H_2O = 7Cu_2S + 5FeSO_4 + 12H_2SO_4$$
(Pyrite)   (Chalcocite)

In the case of silver sulphide :

$$ZnS + Ag_2SO_4 = Ag_2S + ZnSO_4$$
(Sphalerite)   (Argentite)

The secondary enrichment may be incipient, partial and complete depending upon the degree of replacement of hypogene minerals. It may be selective in case only veinlets of hypogene minerals are replaced or only certain mineral say bornite is replaced and other minerals like chalcopyrite pyrite etc. remain unaffected. In case the supergene sulphides pervade the entire ore body, pervasive enrichment may take place.

Control of Secondary Enrichment

The various factors that control supergene sulphide enrichment are :

   (i) water level,
  (ii) presence of primary ores,
 (iii) nature of enclosing rocks,
 (iv) structural features,
  (v) topography,
 (vi) climate,
(vii) rate of erosion, and
(viii) length of time.

Slow sinking water level coincident with erosion at the rate such that oxidation and enrichment go hand in hand is an ideal condition for sulphide enrichment. Supergene enrichment is dependent upon the presence of primary ore precipitants. Higher is the grade of primary ore, the greater is the degree of enrichment. The nature of enclosing rock also plays an important role. Pliable rocks inhibit enrichment, while cracked brittle rocks favour pervasive and complete enrichment. Carbonate rocks, in general, restrict enrichment by precipitating the ore-minerals in the oxidised zone. Faults and shears permit deeper enrichment, but strike faults at times deflect enriching solutions into barren wall and cause cessation of enrichment. Good enrichment is expected in mature topography, lacking dissected areas. Warm and humid climate, like oxidation, favours secondary enrichment. Rapid erosion is not in favour of enrichment. Considerable length of time is required for complete enrichment of sulphides.

The secondary sulphide enrichment continues so long fresh supplies from above are received and primary ores are present in the hypogene zone. The enrichment ceases to take place in case of :

   (i) complete enrichment,
  (ii) non-availability of primary ores in the supergene enrichment zone,
 (iii) base levelling of the area with stoppage of oxidation,
 (iv) submergence of oxidised and enriched zone beneath water level, and
  (v) burial of secondary enrichment zone under thick cover of sediments or lavas from volcanos.

2.7 SUBLIMATION

Sublimation is a process of mineral deposits associated with volcanism, thermal springs and fumaroles wherefrom volatilised matter is redeposited at lower temperature and pressure. Sulphur and borax of Puga area, Ladakh are examples of such deposits. They are associated with thermal springs and fumaroles. Sodium chloride (common salt) is also formed to some extent by this process. Several other sublimates, such as chlorides of copper, iron, zinc etc., ammonium and various salts of alkali metals formed in this way are quickly washed away.

2.8 RESIDUAL AND MECHANICAL CONCENTRATION

The process of residual and mechanical concentration, yielding mineral deposits, is based on weathering of rock containing economic minerals. The weathering causes mechanical disintegration and chemical decomposition of rock constituents. The unstable minerals undergo chemical decay and are removed in solution, while the insoluble residues remain as such and concentrate. The stable minerals like gold, cassiterite, quartz etc. are freed from the enclosing matrix and mechanically concentrated by moving water or air into placer deposits.

Weathering Process

Weathering is a complex operation, involving both mechanical and chemical action. Mechanical action includes disintegration of rocks due to expansion and contraction under temperature changes, frost actions etc. Chemical action is brought out with the presence of water, oxygen, carbon-dioxide, heat, vegetable and animal life etc. and results in decomposition of rocks. It is highly effective in warm and humid regions. Most rocks, ores and gangue minerals yield to weathering in warm and humid climate and are removed in solution. The kind of rocks also plays an important role. The carbonate rocks are readily dissolved in carbonated water, while quartzites are little affected. Pyrite is readily decomposed to form hydrous oxides of iron and sulphuric acid. The sulphuric acid is the powerful solvent and form solution. The colloidal compounds, resulting from weathering, are aluminium compounds, oxides of iron and manganese etc., and these remain as precipitate, while a few others like colloidal silica are removed in solution. Climate has a great bearing on weathering.

Tropical and sub-tropical climates characterised by alternate wet and dry seasons, hot weather, warm surface waters and abundant supply of organic compounds and bacterial life are favourable for rock decay and removal of material in solution. Lateritic soil results under such conditions. On the other hand in temperate climate, the decay is not complete. The silica is left behind and forms ingredient of clay alongwith hydrous oxides of iron and aluminium. Besides lateritic and clayey soils, some iron and manganese deposits are also the results of weathering.

A. *Residual Concentration*

Residual concentration is the process of accumulation of valuable minerals after removal of undesired material by weathering.

The pre-requisites for the process are :

   (i) presence of rocks containing valuable minerals,
   (ii) climatic conditions which promote chemical decay,
   (iii) country with no too great relief to wash away valuable residues, and

(iv) long continued crustal stability to facilitate residual concentration and prevent destruction of deposits by erosion.

Under the above conditions a limestone deposit with ferruginous impurities may give rise to workable iron-ore deposit by removal of limestone in solution and accumulation of iron-oxides as insoluble residues. The weathering process may also result in new mineral deposit, like felspar of syenite decomposed to form bauxite, while the other constituents go into solution and removed. Likewise residual deposits of manganese, clay, nickel, phosphate, kyanite, barytes, ochre, tin, gold etc. may be formed. The source material, chemical changes and other details of formation may, however, in each case differ considerably. For example, the source material for the kyanite of Lapsaburu, Singhbhum, Bihar is the precambrian kyanite-quartz-granulite rock which by weathering under the warm humid climate caused removal of undesired materials to result a high quality kaynite deposit in the form of massive residual boulders.

B. Mechanical Concentration

Mechanical concentration is a process by which heavy minerals are separated from light ones by moving water or air and concentrated in the form of placer deposits. It, thus, includes two steps :

(i) separation of heavy and stable minerals from mother rock by the process of weathering, and

(ii) their accumulation at suitable site.

The source materials of the placer deposits may be lode deposits including veins and stringers, disseminated ore minerals, rock forming minerals and earlier placer deposits such as buried placers or bench stream gravels. A continuous supply of placer minerals is essential for mechanical concentration. The placer minerals have high specific gravity, and are durable and resistant to weathering.

Process

The placer minerals' after release from the source materials by weathering are washed downslope to the nearest stream or to the seashore. The moving stream water takes away the lighter matrix to the farthest distance, while the heavies sink to the bottom and lag behind in their travel. The short currents and waves also separate heavy minerals from light ones. The heavies, thus, concentrate in the stream or beach gravels in sufficient amounts to form placer deposits. The process of concentration is dependent upon the difference in specific gravity, size and shape of particles. A heavier mineral sinks more rapidly than a lighter one of the same size. The shape and size of particles determine their specific surfaces which decide the rate of settling in water. Lesser specific surface means lesser friction and causes rapid settling, the weights of particles remaining the same. For example, a spherical body having less specific surface compared to a thin platy disc of the same weight, sinks more quickly. Velocity of moving water is the added factor which affects mechanical concentration. With the increase in velocity of water, the transporting power is increased and the material is carried away to a great distance. With its decrease, much of the transported load is dropped. In case the water velocity is too low, the lighter materials will not be removed from the heavier; and if it is too high, the placer minerals will be swept away and lost. Hence, water velocity must be favourable to cause concentration of placer minerals.

Resulting Placers

The placers are known as eluvial in case of their concentration on hill slope, stream or alluvial if concentrated in stream and beach when on beaches. They are known as eolian placers in case of their concentration by wind action. Obviously eolian placers occur in arid regions, e.g. eolian gold placers of Australian deserts. Eluvial placers of tin-ore (cassiterite) are reported in the foot hills of some hillocks of Tongpal-Leda-Kudripal area, Bastar district, M.P. Corundum though in small amounts, occurs as eluvial placers in Bastar and Morena districts of Madhya Pradesh. Stream or alluvial placers are known to have been worked for gold in Sona, Subarnrekha and South Koel basins of Bihar and at numerous places in the country. The beach placers of Quilon district, Kerala; Kanyakumari district, Tamil Nadu and Ganjam district, Orissa are being worked for monazite, zircon, ilmenite and rutile.

Mineral Paragenesis and Zoning, Metallogenetic Epochs and Provinces, and Controls of Mineral Localisation

Mineral paragenesis and zoning, and metallogenetic epochs and provinces are the important factors of controls of mineral localisation, and, thereby, all the three are intimately related. As such all the above have been dealt in the same chapter.

3.1 MINERAL PARAGENESIS AND ZONING

MINERAL PARAGENESIS

The term mineral paragenesis defines the mutual relationships and time sequence of minerals. In other words the individual minerals in mineral deposit of magmatic affiliations are formed in an orderly sequence, and this sequential arrangement is termed paragenesis. The paragenesis is deciphered either by field evidences or by microscopic studies. The mineral parageneses worked out in some of the important mineralised belts are given below :

1. The sequence of mineralisation in the Kolar Gold Field indicates that quartz was introduced first, followed later by recrystallisation of silicates and introduction of sulphides and finally gold was to form.

2. The mineral paragenesis worked out for Khetri copper belt by Das Gupta, S.P. (1964) is magnetite, ilmenite, rutile, hematite, pyrite, arsenopyrite, pyrrhotite, sphalerite, chalcopyrite, cubanite with vallerite, galena, chalcocite, marcasite and late-pyrite. Magnetite was first to crystallise. The presence of late pyrite incrustations and veins, marcasite and chalcocite were noted only locally.

3. The mineral paragenesis in the Singhbhum copper belt shows that the mineralisation in the belt took place over a long period in two stages, separated by an interval. The first to form were apatite (with some tourmaline) and magnetite, the latter continuing for sometime after apatite formation ceased. This was followed by a biotite-chlorite phase which extended beyond the first generation of quartz.

The uranium mineralisation is closely associated with apatite - magnetite rich bands and rocks rich in tourmaline, chlorite and biotite. The period of quiscence which succeeded, came to close with a second phase of silicification towards the end of which a little pyrite developed, followed by pyrrhotite and then by a feeble phase of nickel sulphide. The third and longer phase of silicification was contemporances with the formation of chalcopyrite and was succeeded by a weaker phase of carbonate formation.

Studies of paragenesis in the Singhbhum copper belt indicate that the ore deposition was governed more likely by relief of pressure than by temperature reduction.

The common ores or minerals in all geological periods have been deposited according to some fundamental laws of control over their mineralisation. In a broad way, it is observed that certain silicate mineral assemblage such as felspar, diopside, garnet and quartz were usually the first to form in a particular type of deposit. These were followed in many cases by oxides such as magnetite, hematite and ilmenite, still later sulphides may have entered and formed around or replaced the preceding groups and finally precious metals such as gold and silver may have added to the assemblage in small amounts. This general rule of silicate formation or gangue first, oxide mineralisation next, sulphides following oxides, and precious metals value last, is an established working principle widely used by Economic Geologists.

Efforts have been made to interpret this sequence in the following two different ways :

1. In terms of solubility of minerals in aqueous solutions; and
2. In terms of specific gravity of the minerals.

According to Schurmann and others the most soluble minerals stay in solution longest and the least soluble precipitate first. The Schurmann series of aqueous solutions runs as mercury, silver, copper, bismuth, lead, zinc, nickel, cobalt, iron, manganese, and the rule is that in solution of salts of these metals, a sulphide of any succeeding metal will cause any preceding salt to be decomposed and its metal precipitated as sulphide. Thus, iron-sulphide would be a precipitant for sulphides of most other metals, and, thereby, the sulphide deposition in majority of cases begins with pyrite, the most soluble, and closes with cinnnabar the most insoluble. Later generations of gangue and ore minerals may appear and deposited on older ores.

The above explanation seems exactly the reverse of what the paragenetic sequence should require. In any aqueous solution heavily burdened with sulphides surely the most soluble should travel farthest and be precipitated last, and the least soluble should be dropped first.

Thus, with lead and zinc, we know that zinc compounds are far more soluble than those of lead, and, hence, lead should be precipitated first followed by zinc, but according to paragenesis, this operates otherwise and zinc is precipitated first followed by lead. Likewise copper is much more soluble than lead, yet in ore-deposits it is commonly deposited first.

The above facts necessitated the search for other principle of control of sulphide deposition in terms of specific gravity or density. Bandy, Mark C. recognised this fact and considered density to increase with a decrease in age of formation. In case of sulphides this is true as a broad generalisation. But in case of oxide minerals it would appear that there is a decrease in density with decrease in age, an opposite trend from sulphides. Thus, sphalerite (sp. gr = 4.08) in specific deposits often follows rather than precedes chalcopyrite (sp. gr = 4.48).

The geological succession of some common sulphides based on their sp. gravities and atomic weight of metal components is as follows:

Pyrite, Pyrrhotite, Sphalerite, Chalcopyrite, Enargite, Tennantite, Stibnite, Galena, Boulangerite, Argentite, Cinnabar, Bismuth (native), Silver (native), Gold (native). It may be noted that the paragenetic series is a generalisation, but expresses fairly closely a majority record of observation and opinion.

In case specific gravity is a controlling factor and, in general, the lighter substance escape first and the heavier ones last, then it would seem that this probably implies some sort of stratification in the source reservoir with the lighter substances on top and heavier ores beneath under such conditions that the overlying materials must be eliminated before the deeper and heavier ones can be released. This possibility of such gravity stratification points suggestively towards certain well known facts in the field of metallurgy.

Mineral Zoning

Mineral zoning is the tendency for certain minerals or ores to be deposited at varying distances from a related focus of igneous activity. In general, the higher temperature and least soluble minerals are found nearest the source, while low temperature and most soluble minerals farther out. This points out to a hot centre of mineralisation, and minerals of igneous affiliation constitute a clear-cut indication of kinship with a magma. An ideal sequence of metals presented by W.H. Emmons is, from the parent intrusive outward, 1. Barren zone with quartz, 2. Tin, 3. Tungsten, 4. Bismuth and molybdenum, 5. Gold, 6. Copper, 7. Zinc with some lead, 8. Lead with some zinc and copper, 9. Silver, 10. Barren with quartz, calcite etc., 11. Gold and Silver, 12. Antimony, 13. Mercury, 14. Upper barren zone with chalcedony. No such metal sequence is found in any one deposit. Parts of the sequence may occur in one deposit and other part elsewhere, or some zones may be lacking. The zonal arrangement may be both horizontal and vertical.

A zonal arrangement of minerals is presumed to be a function of their temperature of deposition. But, factors other than temperature are also to be taken into account in deposition of minerals. They are pressure, relative concentrations, reactions with a wall rock, reactions within solution, chemical complexes, relative mobility, volatility etc. The relative concentration of certain substances may cause precipitation of one or other minerals. Garrels has shown that chemical complexes produce zoning in reverse order. Brown, J.S. explains zoning occurring as a result of relative volatility of mineral substances.

Emmons, however, states that zoning is to be expected only with summit cupolas in the roofs of batholiths and not in hood deposits deep in roof pendants. Lindgren points out that in the epithermal veins there is no zoning except a common impoverishment in depth. There are reversals in zoning like zinc may occur below copper and lead below zinc. These reversals are considered by Emmons as due to exceptionally high concentration of certain salts of metals, causing a precipitation of more soluble salts ahead of the less soluble. The process of supergene enrichment also causes reversals in zoning.

Metallurgy establishes the possibility of existence of several independent source magmas which accumulate at different depths dependent on relative density. Thus, the pegmatite magma, responsible for the ores of tin and tungsten, various silicates etc., is the lightest and forms at the highest level, while the iron-magma, consisting mainly either of iron-oxides and iron-silicates, is heavier than the pegmatitic magma but lighter than sulphides, forms at intermediate level.

3.2 METALLOGENETIC EPOCHS AND PROVINCES

Investigations have shown that metallogeny is by no means a series of isolated phenomena in the history of the evolution of the earth's crust. It constitutes characteristic events in the evolution of mobile belts and platforms. Such mineral deposits are related in time and place to periods of crystal evolution and igneous activity that have taken place at definite periods in the earth's history. For this period of mineralisation Lindgren used the term Metallogenetic Epochs, emphasing the timing of certain kinds of mineralisation, such as Precambrian iron-ore deposits of Bihar, Orissa and Madhya Pradesh. The areas which are characterised by preponderance of deposits of some particular type, such as gold, copper or iron, are called Metallogenetic Province. These minerals may have formed by various geological processes besides igneous activities, like sedimentation, metamorphism, weathering etc.

The metallogenetic epochs and provinces, in the Indian context, are described below :

Metallogenetic Epochs

Metallogenetic epochs, as defined above, are specific periods characterised by formation of large number of mineral deposits. It does not mean that all the mineral deposits formed during a definite metallogenetic epochs. In India the chief metallogenetic epochs were :

1. Precambrian
2. Late Palaeozoic
3. Late Mesozoic to Early Tertiary

1. Precambrian Epoch

The Precambrian epoch is the most important the world over because of the great length of time involved and presence of large and varied mineral deposits. The huge sequence of Precambrian metasediments and associated granitoids, gneisses etc. are supposed to have a different metallogenic history in which post consolidated events of magmatism and metallogeny are important. The endogenic deposits have been broadly correlated to ultrabasic, basic, acid-intermediate and post-orogenic acidic phases of magmatism, while exogenic deposits, like iron and manganese, are related to sedimentary, metamorphic and other processes.

This epoch in India is defined by the presence of iron-ore, chromite, manganese ore, gold, copper, lead, sillimanite, gypsum, kyanite, gemstones, etc.

Examples of iron-ore deposits are in southern Singhbhum (Bihar), Keonjhar, Mayurbhanj and Sundargarh (Orissa), Bastar and Durg (Madhya Pradesh), Chanda and Ratnagir (Maharashtra), Dharwar, Bellary, Sendur, Shimoga and Chikmagalur (Karnataka), Goa and other places; that of chromite in Singhbhum (Bihar), Dhenkanal, Cuttack and Keonjhar (Orissa), Mysore and Hassan (Karnataka) and Bhandara and Ratnagiri (Maharashtra); that of gold in Kolar, Hutti and Gadag (Karnataka), Ramgiri and Anantpur (Andhra Pradesh), Wynad (Tamil Nadu) and Kondrakocha (Bihar); that of copper in Singhbhum (Bihar), Khetri and Pur-Banera-Bhindar (Rajasthan), Malanjkhand (Madhya Pradesh), Mailaram, Gani and Agnigundala (Andhra Pradesh), Ingaldalur and Kalvadi (Karnataka); lead and zinc in Zawar, Rajpura-Dariba, Deri, Rampur-Agucha (Rajasthan), Sargipalli, Kesarpur, Karmali (Orissa), Banaskantha and Vadodara (Gujarat) and other places; and Manganese-ore in Balaghat (MP), Bhandara and Nagpur (Maharashtra), Bellary, Chitradurga, Uttar Kannad, Dharwar, Shimoga and Tumkar (Karnataka), Panch Mahal and Vadodara (Gujarat) and other places. There are many more examples of similar deposits.

2. Permo-Carboniferous (Late Palaeozoic) Epoch

Towards the upper Carboniferous (Late Palaeozoic) the Hercynian movement introduced great changes on the suface of the globe and this is marked by mountain building and initiation of sedimentary era. In India the epoch is known by rich coal deposits of Lower Gondwana, like those of Jharia, Bokaro, Karanpura, Giridih, Ramgarh, Auranga, Hutar, Daltonganj, Deoghar and Rajmahal of Bihar; Raniganj, Barjora and Darjeeling of West Bengal, Singrauli, Korba, Chirimiri, Sohagpur, Umaria, Johilla, Bisrampur, Jhilmili etc. of M.P. and Godavari valley of Andhra Pradesh and at many other places. A large number of hypabyssal basic intrusives-dolerites and basalts, and mica-rich ultrabasic rock, mica-peridotite, traverse these coal fields. Other important mineral deposits of this epoch are fireclay, iron stone and ochre, which occur within the Gondwana formation.

3. Late Mesozoic to Early Tertiary Epoch

This epoch is dominated by fissure eruptions of basaltic lava flow (Deccan Trap) which now occupy over 500,000 sq.km. area in western and central India with semi-precious stones like rock crystal, amethyst, agate, carnelian, onyx and other varieties of chalcedony. Rare copper mineralisation is noted in the trap rocks.

Considerable igneous activities of this epoch marked by granites, granodiorite, basic and ultrabasic rocks occur in extra peninsular regions, comprising main Himalayan ranges, Manipur-Meghalaya and Andaman-Nicobar islands. These are associated with occurrences of fluorite, copper, lead, zinc, chromite, magnesite, clay asbestos etc. in Ladakh area of Indus ophiolite belt; magnetite, nickel chromite, asbestos, magnesite and talc occurrences in Manipur-Nagaland and base metals, nickel and chromite in Andaman-Nicobar areas.

Metallogenetic Provinces

The metallogenetic province is known by the name of dominant and specific mineral, such as Gold Province, Copper Province, Iron-Ore Province or Manganese Ore Province. It may comprise mineralisation of more than one epoch, each superimposed upon the other, but essentially of the same type.

India affords many examples of metallogenetic provinces. A few impotant ones are enumerated below :

1. Gold Province of Karnataka - Andhra Pradesh - Tamil Nadu (Hutti-Kolar-Anantpur-Godag-Wynad Gold Province).
2. Copper Province of Singhbhum.
3. Copper Province of Khetri-Pur Banera-Bhinder.
4. Lead-Zinc Province of Hesatu-Belbathan.

5. Iron-Ore Province of Southern Singhbhum-Keonjhar-Sundergarh-Mayurbhanj.

6. Iron Ore Province of Drug-Bastar-Chanda-Ratnagiri.

7. Iron Ore Province of Karnataka-Goa.

8. Manganese Province of Balaghat-Bhandara-Nagpur.

The description of the above povinces are given below.

1. Gold Province of Karnataka-Andhra Pradesh-Tamil Nadu

It covers gold fields of Kolar, Hutti and Gadag of Karnataka, Ramgiri, Gooty, Bisanatham and Gavanikonda of Andhra Pradesh and Wynad, Cherambadi and Bensibetta of Tamil Nadu. By far the greatest part of the Province is occupied by Dharwars, represented by metamorphosed mafic volcanic rocks altered to schistose, massive, granular and fibrous amphibolites.

The other rock types are muscovite-biotite-schist, quartzites and granulites, with Peninsular gneisses and granites. All these rocks and the auiferous reefs are intruded by dolerite dykes, and the final phase of igneous activity is marked by a few pegmatites.

The structural metallogenic zones of the Province are characterised by mineralisations of the intermediate-late-final stages of development of the mobile belt [Au, (As, Fe, Cu,Pb, W), Pb, (Ag, Zn, Cu, Fe) and may be defined principally as :

(i) Gadag-Chitaldrug-Mysore zone.

(ii) Ramgiri-East Bangalore zone.

(iii) Hutti-Kolar-Mamandur-Vellore-Terupati zone.

The gold is associated mostly with quartz-lodes or quartz-veins and sulphide bearing reefs within schist and belongs to high temperature (hypothermal) hydrothermal class.

2. Copper Province of Singhbhum

The Copper Province of Singhbhum is localised in the shear zone, which extends from Duarpuram in the west to Barahagora in southeast for over 128 km with a width ranging upto 5 km and is moulded along the northern and the noth-eastern margin of Singhbhum granite massif. The copper-ores of Singhbhum seems to be related to tongues of granite which intrude the schists. The ores occur as veins in the granite and the neighbouring mica-schists, quartz-schists, and hornblende-schist or epidiorite. These veins are well developed along a zone of overthrust, where they form well defined lodes. This province [Cu, Au, (Fe, Ni, U)]; Pb, (Cu, Au, Fe, Zn); U, Th, Ce, (La, Pr, Nd, Sm, Be)] indicates intermediate-late-final stages of development of the mobile belt.

The Province represents hydrothermal and pegmatitic mineralisation in refolded synforms and dislocated antiforms of predominantly metamorphosed sedimentaries of the greywacke suite and associated metavolcanics.

3. Copper Province of Khetri-Pur-Banera-Binder

There are two main zones of copper mineralisation in the Province, namely Khetri zone and Pur-Banera-Bhinder zone. They run for about 80 km in the northeast and 135 km in the southeast part respectively of the state. The potential areas for copper are in Madhan Kudan, Kalihan and Chanmari in the Khetri zone and Dariba in Pur-Banera-Bhinder zone. Pre-Aravalli, Aravalli and Delhi groups of rocks are found in the area and the copper mineralisation are along favourable structural zones related to different orogenic movements. The Province shows basic volcanics of the earlier stage of development of the mobile belt and an acidic megmatism of the intermediate-late stage. The mineralisations associated with earlier stage are not very predominant, while the mineralisations of the late stage have given rise to the following important structural metallogenic zones; namely :

   (i) Singhana-Khetri-Babai zone [Cu, (Fe, As, Co)].

  (ii) Anjari-Dariba-Bairat-Jadawas zone [Cu, (Fe, Ni, Co), Pb, (Zn, Sb)].

 (iii) Kishangarh-Ajitgarh-Ambamara zone [(Pb, Cu, Ti), Cu].

   All these zones occupy highly contorted and refolded synforms, flanked by granites and gneisses.

4. Lead and Zinc Province of Hesatu-Belbathan, Bihar

The Province extends in WNW-ESE for about 250 km from Hesatu (Hararibagh district) in the west to Belbathan (Godda district) in the east and has a width of about 50 km. There are over four dozen occurrences showing evidences of lead-zinc, copper mineralisation, covering the whole Province, some of the important occurrences are at Baraganda, Chandio, Bhalakdiha, Ganganpur, Damgi, Toolsitanr, Baghmari and other places. The mineralisation is associated mostly with Precambrian tremolite-actinolite schist, amphibolite and calc-granulite which occur as pockets within Chotanagpur granite-gneisses.

5. Iron Ore Province of South Singhbhum-Keonjhar-Sundargarh-Mayurbhanj

About 15 to 20 million tonnnes of annual Iron-Ore production is from this Province. Deformed and folded belts of Precambrian metasediments (flysch and volcanics) along with gneisses and granites of at least two generations (Chotanagpur gneiss and Singhbhum granite) cover a major part of the area and indicate the development of interior areas of a mobile belt. The Province embraces iron-ore deposits of Noamundi, Notoburu, Jamda, Gua, Jhilingburu and Pansirabuda and others of south Singhbhum

(Bihar) and Thakurani, Bolaria, Joda, Malantoli etc. of Keonjhar, Baliapahar, Barsua, Bonai etc. of Sundargarh, Gorumahisani, Salaipat, Badampahar and Simlipal hill of Mayurbhanj; Daitari, Tamka and Kansa of Cuttack, and Hirapur and Umarkot of Koraput (Orissa). The banded Iron-Ore formation with very good hematite iron-deposits are found. The important structural metallic zones delineated in the area are given below :

(i) Cuttack-Keonjhar-Singhbhum-Mayurbhanj Zone [Cr, (Fe, Ni, Co); Fe, (Va, Ti); Pb, (Cu)].

(ii) Singhbhum-Keonjhar zone (Fe, Mn).

(iii) Mayurbhanj-Sambalpur-Koraput-Sundargarh zone (Fe, Mn).

The above are structural metallogenic zones of the sedimentary-metamorphic deposits of iron and manganese. Titano-vanadiferous iron ore type, recognised in south-east Singhbhum and Mayurbhanj is classed as liquid magmatic deposit and have formed during early crystallization of basic plutonic rocks.

6. Iron-Ore Province of Chanda-Drug-Bastar

This iron-ore Province embraces important iron-ore deposits of Lohara, Pipalgaon, Asola and Dewalgaon (Chanda district); Dhalli and Rajhara (Drug); and Rawghat, Parrekaro and Bailadila (Bastar district) and produces about 12 to 15 million tonnes ore annually. This represents metamorphosed flysch type of association with exogenic mineralisation in Banded-Hematite-Quartzite of the early stage of development of mobile belt. The following structural metallogenic zones have been recognised in the Province :

(i) Pipalgaon-Lohara-Asola-Dewalgaon zone.

(ii) Dhalli-Rajhara-Parrerkao zone.

(iii) Bailadila-Kandpai-Parotwada-Taki zone.

7. Iron-Ore Province of Karnataka-Goa

This is another iron-ore province which produces altogether 46% of Indian production i.e. annually about 25 million tonnes of iron-ore. This represents structural metallogenic zones of sedimentary metamorphic deposits, considered to be corresponding to the early stage of development of mobile belts - iron-ores in greywacke association. Banded iron-ore formation with good hematite iron-ore deposits is prevalent. Magnetite iron-ore (Titano-venadiferous) is found in Tumkur district and may be classed as magmatic deposit, formed during early crystallisation of basic plutonic rocks. Bellary-Hospet, Shimoga, Chitradurga, Chikmaglur, Kudramukh and Bababudan hills are the important iron-ore deposits found in the Province.

8. Manganese Province of Balaghat-Bhandara-Nagpur

This Province falls in Madhya Pradesh and Maharashtra states and together produces about 40 per cent of manganese-ore annually out of total Indian output. They represent gondite type of deposits associated with metamorphosed Dharwar rocks which are characterised by predominantly quartzite-carbonate associations with subordinate gneisses and granites of at least two generations. The exogenic deposits of Mn in the Province are associated with an early phase of sedimentation and as such may also be largely pre-orogenic.

3.3 CONTROLS OF MINERAL LOCALISATION

The localisation of economic minerals is controlled by the following factors :

1. Structure
2. Stratigraphy
3. Physical and chemical properties of host rocks
4. Relation between igneous rocks and associated ores
5. Mineral paragenesis and zonal distribution of mineral deposits
6. Metallogenic epochs and provinces.

1. Structure

Structures play an important role in ore-localisation. These may be of two types : (i) Regional, and (ii) Local or Detailed.

(i) Regional Structural Control

Regional structures demonstrate the broader localisation of ore-beds or mineral provinces. They are influenced by orogenic movements, igneous intrusions and major faults. The regions of orogenic or mountain building movements are the places of crustal movements, dislocations and igneous intrusions that yield ore-forming fluids as well as several kinds of channel ways for mineralising solutions. It is, thus, most of the epigenetic mineral deposits occur in the regions of orogenic or mountain building movements. Folding, igneous intrusion, faulting and metallisation are related and occur in the area in the sequence.

Large regional faults or shear zones provide master channel for the mineralising solution moving up from depths. The example is that of the Singhbhum thrust which provided channel ways for the mineralising solution of copper for over 128 km of the Singhbhum copper belt.

(ii) Detailed Structural Controls

Smaller detailed structural controls are of prime importance in mineral localisation. They provide various types of rock openings, like foliation, bedding planes, lineation, shears, joints etc. for passage of mineralising solutions and determine the immediate localisation of ore. These structural details receive utmost attention during mineral explorations and mining operations.

2. Stratigraphy

Stratigraphic controls are by far the most important in localisation of oil, water and many sedimentary mineral deposits, like coal, iron-ore, mangnanese, phosphate, limestone, etc. They may also be divided into two types : (i) Regional and (ii) Detailed.

(i) Regional Stratigraphic Controls

These include geosynclines, basins, plateau margins, unconformities etc. Geosynclines are the sites of thick accumulations of sediments and deposition of sedimentary ores, rocks and fuels of economic value. The sedimentation and deposition is followed by uplift, folding, faulting and igneous intrusions which provide channnel ways for mineralising solutions.

Plateau margins are also known for extensive sedimentation that later were strongly folded, faulted, intruded and mineralised. Unconformities indicate surface of erosion which are favourable loci for accumulation of residual concentration deposits like residual iron and manganese and for localisation of oil, gas and water pools.

(ii) Detailed Stratigraphic Controls

Bedding, lenses, impervious covers, impervious bases etc. signify detailed stratigraphic controls. Bedding planes offer ways for movement of underground fluids, and, thereby, they control localisation of oil, gas, water and many metallic and non-metallic mineral deposits. They are also favourable sites for replacement and contact metasomatic deposits. Cross bedded strata are more favourable.

Lenses of sandstone localise oil, water and ores, while carbonate lenses are suitable for replacement deposits. Impervious covers, like shale, help to localise many mineral deposits beneath it, since they form effective barriers to ascending mineralising solutions. They also constitute cap rocks for oil-pools. Impervious rocks likewise serve as barriers for descending fluids. In this way they help to form oil pools in synclines. Similarly the descending supergene mineralising solution gives rise to supergene sulphide and oxidised ore deposits on such impervious strata.

3. Physical and Chemical Properties of Host Rocks

Congeneal host rocks, judged from their physical and chemical characters, are essential for mineral localisation. The permeability, brittleness, grain-size etc. are the physical properties which control mineral localisation. Permeability offers connected openings in the rock for the mineralising solution to reach the site of deposition. In some epigenetic mineral deposits where channel ways are absent, permeability alone is the main factor of mineral localisation. Many of the disseminated ore bodies owe their origin to the permeable character of the host rock. The crackled enclosing rock and porosity permit permeation of the ore-fluids.

Thus, the localisation of oil pools, ground water supplies and many oxidations and sulphide enrichments take place in permeable rocks.

Brittleness of the rocks is another physical character which causes the rocks to crackle readily under slight stress, resulting in high permeability. Thus, the rocks become favourable to ore deposition. Rhyolite, quartzite, limestone and silicified rocks come under this category.

Chemical character of host rock exhibits a dominant role in the localisation of epigenetic mineral deposits. Reactive rocks, like carbonate rocks, namely limestone, dolomite etc., greenstones, namely chlorite-schist, tremolite-actinolite rocks etc. are congenial to ore deposition. For example, lead and zinc deposit at Zawar (Rajasthan) occurs in dolomite, while in the Hesatu-Belbathan belt (Bihar) tremolite-actinolite rock is the main host rock for lead-copper mineralisation; chlorite-schist shows affinity for copper mineralisation in Singhbhum copper belt and for gold in Kolar-Hutti gold fields. This may also be designated as lithological control of mineralisation.

4. Relation Between Igneous Rocks and Associated Ores

There is certain kind of genetic relationship between specific types of igneous rocks and particular kind of mineralisation. So, there is consistent associations of some ores with certain type of igneous intrusives. This is another form of control of mineral localisation. Some of the igneous rocks and their associated ores are given below.

Table 3.1. Igneous rocks and associated ores

Rock type	Associated ores and minerals	Examples of occurrences
1. Kimberlite	Diamond	Wajrakarur (Andhra Pradesh) and Panna (Madhya Pradesh). Kimberlite pipes.
2. Ultrabasic Intrusives, Peridotite, Pyroxenite and Dunite rocks	Chromite	Jojohatu, W. Singhbhum (Bihar); Nausahi, Keonjhar (Orissa); Sukinda, Cuttack (Orissa); Drass, Ladakh (J & K State) etc.
3. Gabbro-Anorthosite	Titaniferous magnetite	Bisoi-Rairangpur, Mayurbhanj (Orissa)
4. Nepheline syenite	Corundium	Khammam district, Andhra Pradesh
5. Granite and Granitic Pegmatite	Tin (Cassiterite)	Koraput, Orissa and Tosham, Haryana.
	Tungsten	Quartz and pegmatite veins at Degana and Sirohi, Rajasthan
	Uranium	Granitic intrusions in Singhbhum Copper belt, Bihar

Batholithic type of igneous intrusion shows association of ore deposits in their specific portions. According to Emmons, W.H. the ore deposits tend to be concentrated in the roof pendants of the batholiths, in the upper parts of stocks and in the surrounded invaded rocks, pointing to a specific control of ore-localisation.

5. Zonal Distribution of Mineral Deposits

In a zonal distribution high temperature minerals lie close to the centres of igneous activity i.e. the magmatic source, while the low temperature minerals occur in more distant zones. An ideal sequence of metals given by Emmons, is, from the parent intrusive outward. (1) barren zone with quartz, (2) tin, (3) tungsten, (4) bismuth and molybdenum, (5) gold, (6) copper, (7) zinc, (8) lead, (9) silver, (10) barren zone, (11) gold and silver, (12) antimony, (13) mercury and (14) upper barren zone. It is not necessary that full sequence be present in a deposit. Some of the zones may be absent. It is, thus, that the zonal arrangement controls ore localisation.

6. Metallogenic Epochs and Provinces

Most minerals are, generally, associated with the periods of crustal disturbances and orogenic movements. As such, these mineral deposits are related in time and place to the events of crustal and igneous activity which constitute metallogenic epochs (time) and provinces (place). Thus, the metallogenic epochs and provinces function as controllers of mineral localisation.

3A

Classification of Mineral Deposits

Classification of mineral deposits has been attempted by many authors, but none of them have been unanimously accepted. The genetic classification is considered to be valuable for the working geologist, since it can be applied directly to the field, points to the cause of ore-formation and other features of individual deposits.

Lindgren (1911) classified mineral deposits in two main sub-divisions :

    (i) those formed by mechanical concentration, and
    (ii) those formed by chemical reactions in solutions.

The basis of this classification is the temperature and pressure of formation. The principal features of the classification are given below :

    (i) Deposits by mechanical process
    (ii) Deposits by chemical process

A. In Surface Water

1. By reactions
2. Evaporation

B. In Rocks

1. Concentration of mineralisation contained within rocks :
    (a) by weathering
    (b) by ground water
    (c) by metamorphism
2. By Introduced Substances :
    (a) without igneous activity
    (b) related to igneous activity
        A. By ascending waters
          1. Epithermal deposits — 50° to 200°C temp. and medium pressure.
          2. Mesothermal deposits — 200° to 300°C temp. and high pressure.
          3. Hypothermal deposits — 300° to 500°C temp. and very high pressure.

    B. By direct igneous emanations
      1. Pyrometasomatic deposits — 500° to 800°C temp. and very high pressure.
      2. Sublimates — 100° to 600°C, low to medium pressure.
    C. In Magmas by Differentiation
      1. Magmatic deposits — 700° to 1500° temp., very high pressure.
      2. Pegmatites — 575°C ± temp., very high pressure.

The above is the most accepted classification. Bateman (1942) gave another genetic classification based on the various processes of formation of mineral deposits, which is shown below :

Processes	*Deposits*
1. Magmatic concentration	(a) Early magmatic — Disseminated crystallization, segregation, injection.
	(b) Late magmatic — Residual liquid segregation, residual liquid injection, immiscible liquid segregation and immiscible liquid injection.
2. Sublimation	Sublimate
3. Contact metasomatism	Contact metasomatic
4. Hydrothermal processes	(a) Cavity filling
	(b) Replacement
5. Sedimentation	Sedimentary
6. Evaporation	Evaporites
7. Residual and Mechanical concentration	(a) Residual deposits
	(b) Placers
8. Surfacial oxidation and supergene enrichment	Oxidised, supergene sulphide
9. Metamorphism	Metamorphosed and metamorphic deposits

The classification based on genesis is meant only for the technocrats and is not of any commercial importance. Most common in use may be the classification based on the nature and most prevalent use of minerals which is proposed as below :

A. Metallic Mineral Deposits

    I. Precious metals, e.g. gold, silver, platinum.
    II. Ferrous and Ferro-alloy metals, e.g. iron, manganese etc.
    III. Non-ferrous and allied metals, e.g. copper, lead etc.
    IV. Light metals, e.g. lithium, magnesium etc.
    V. Radio-active metals, e.g. uranium, thorium.
    VI. Rare metals, e.g. palladium, selenium etc.

B. Non-Metallic Mineral Deposits

  I. Mineral fuel, e.g. coal, petroleum.

  II. Gemstones, e.g. diamond, ruby etc.

  III. Abrasive minerals, e.g. corundum, garnet etc.

  IV. Building materials and the dimension stones e.g. marble, granite etc.

  V. Industrial Minerals, e.g. mica, asbestos, etc.

  VI. Refractory Minerals, e.g. fireclay, graphite etc.

  VII. Glass manufacturing materials, e.g. quartz, silica etc.

  VIII. Ceramic minerals, e.g. clay, felspar etc.

  IX. Fertilizer minerals, e.g. phosphorite, sulphur etc.

  X. Chemical minerals, e.g. rock salt, borax etc.

  XI. Mineral pigments e.g. ochre, umber etc.

  XII. Mineral water and ground water.

The above classification has been followed in this book while describing the mineral deposits.

4

Metallic Mineral Deposits

4.1 PRECIOUS METALS

1. GOLD

Gold was the earliest metal to be mined by the mankind. The references of gold mining are seen in holy scriptures like the Rig Veda, the Puranas, the Shastras, the Hebrew scriptures, and the Greek and the Roman literatures. The present production of gold in India (2039 kg, 1991-92) is about 0.1 per cent of the total world production (2156.6 tonne, 1991). The Republic of South Africa continued to be the leading producer with an output of about 40 per cent of the total world production followed by Russia (about 17 per cent), USA (7 per cent), Canada (6.6 per cent), Australia (5 per cent) and China (4 per cent).

Mineralogy

Gold is bright yellow in colour with resplendent metallic lustre, quite soft (readily scratched with a knife), resistant to tranish, highly malleable and ductile, and can be beaten to a leaf so that it transmits greenish blue rays. Its hardness is 3, sp. gr. 19.32 and melting point 1063°C. It dissolves in aquaregia and fuming HCl. Gold is found in nature in native state as irregular masses, stringers, scales, nuggets, and other forms. It crystallises in isometric system, the common crystal forms being octahedron, dodecahedron and cube. Moss gold, wire gold and dendritic forms are ascribed to incipient crystallization. It is also found in nature as tellurides. Nearly all gold contains some silver. Natural amalgams of gold-silver, gold-palladium, gold-rhodium, gold-antimony, gold-bismuth etc. are found in some deposits.

Uses

Most of the gold produced goes into the monetary reserve and forms a monetary base for currency. The other popular utilisation of gold is in jewelry. In India some gold is used in the textile industry for 'Zari' work.

It is also used in dentistry, chemical industry, medicine, scientific laboratory equipments, electrical components and inlay for crockery. Besides the above, it is employed in the preparation of glass bangles, lettering, interior decorations, and plating.

Mode of Occurrence and Origin

Gold occurs mainly in two forms :

  (i) lodes, and
  (ii) placers.

(i) Lodes Deposits

The lode deposits are primary in nature. The primary gold deposits occur in intrusive rocks (dyke rocks) having composition of diorites, quartz-diorites and granites and their metamorphic equivalents. Gold is found commonly associated with sulphides of non-ferrous and related metals like chalcopyrite, sphalerite, galena, arsenopyrite, pyrite and antimonite. Quartz and limonite are the main gangue associates. The iron hats (gossan) or limonites, at times, contain appreciable quantities of gold. Most frequently gold occurs in quartz-veins, where yellow brown or blue quartz has been found to be favourable carrier of gold (Kundrakocha, Singhbhum district, Bihar).

The lode deposits are mostly formed through igneous emanations during the last stages of chilling of magma which came up along some openings like fissures, faults, fractures, shear zones and folds to upper layers of the earth's crust. During cooling of magma gold crystallises in native state or in combination with other elements like Ag, Cu, Hg, Sb, Bi, Se, Te, As and S depending upon the physico-chemical conditions then prevailing. Since the bulk of deposits are formed at the end stages of differentiation, they are hydrothermal in origin with accummulation of gases and water which function as the carrier, e.g. Kolar gold field. A few lode deposits, however, are formed at various stages of the process, and are referred to as magmatic segregation deposits (Utah Gold Hill, Montana Gold Curry etc.) and contact metasomatic deposits (Montana Cable Mine, British Columbia Nickel Plate Mine etc.).

The origin of auriferous veins of the Kolar Gold Field was assigned to lateral secretion by Smith (1889). Hatch (1901) and Pryor (1923), however, regarded their deposition from gold bearing solutions. Narainswamy *et al.* (1960) attributed the origin to be high temperature (hypothermal) hydro-thermal mineralisation.

(ii) Placer Deposits

Gold occurs as placers, deposited together with alluvium, sands, gravels and conglomerates. Auriferous placers, also called 'alluvials', are the earliest of the gold deposits to have been worked. The gravels of

Nilambar Valley, Kerala contain gold along with magnetite, zircon and garnet. Gold in beach placers, generally, occurs associated with minerals like magnetite, ilmenite, cassiterite, garnet, zircon and monazite. In Assam placers, the gravels of Brahmaputra contain garnet, magnetite, zircon, topaz, platinum, osmium and iridium with very fine gold.

In placers deposits, natural agencies play important role in breaking the rock containing gold, its transport and concentration. Considerable quantities of gold are reported to have been won from placers in California, Victoria and Newzeland of Tertiary and Recent ages. While most of the detrital gold is deposited at a distance from the source rock, a portion is, however, accumulated on the hillslope to form eluvial or residual deposit (Victoria, nuggets of gold found). The home for most productive placers is in valleys or gullies of small streams. Workable deposits of placers indicate the gradient of auriferous water courses to be between 10-15 m per km. Gold is concentrated in the bottom of the stream bed above the bed rock and below the gravel, as it is 5 to 6 times heavier than the ordinary rock. The concave side of the stream beds and the downstream sides of the rocky bars are also ideal sites of gold deposition.

Distribution

Almost all gold productions of India have come from the vein-deposits (lode deposit) and very little from the placers. The vein-deposits may be further classed into :

(i) Principal deposits (active mines) which are at the moment producing gold, e.g. the Kolar and the Hutti gold fields;

(ii) Potential deposits which at one time or other have produced gold and may hold promise of turning out to be commercially important, e.g. Anantpur gold field (Andhra Pradesh), Gadag gold field (Karnataka), and Wynad gold field (Tamil Nadu);

(iii) Minor occurrences which have not been fully assessed or have been explored to some extent and found to be unimportant, e.g. numerous gold occurrences and small old prospects.

The geological-cum-geographical distribution of gold is given below in the Table 4.1.

Table 4.1—Distribution of Gold in India

Age	Types of deposits	Category	Locality and geological details
Dharwars	A. Lode Deposit	1. Principal Deposits (Active Mines)	(a) Kolar (Karnataka) : Gold is associated with Champion quartz-lodes and sulphide-bearing reefs, especially Oriental lode within schist-belt, belonging to hypothermal class.
			(b) Hutti (Karnataka) : Gold is associated with quartz-reefs within metabasalts, represented by greenstones passing into chlorite-schist.

(Contd.)

Age	Types of deposits	Category	Locality and geological details
		2. Potential Deposits (Abandoned Mines)	(a) Ramgiri, Anantpur distt., Andhra Pradesh: The auriferous belt, 150m-200m wide and comprising quartz-vein zone, is spread over a strike length of about 15 km within schistose rock (sericite-chlorite phyllite).
			(b) Gadag, Karnataka : The gold is associated with vein-quartz, occurring in en-echelon fashion mainly within greenstone for about 50 km.
			(c) Wynad, Tamil Nadu : The gold is associated with quartz reefs within biotite-gneiss and interbanded hornblende-granulite, and, at places, within magnetite-quartzite.
			(d) Kundrakocha, Bihar : The gold mineralisation is associated with quartz-veins within cherty phyllite of Iron-ore formation.
		3. Minor occurrences	These include numerous occurrences and small old prospects. A few may be named as : Gooty (Anantpur dist.), Bisanatham (Chittoor) and Gavanikonda (Kurnool) of Andhra Pradesh; Sithaura (Nalanda), Sonapet (Ranchi) and Pahardiha, Lowa and Mysara (Singhbhum) of Bihar; Alech hills (Jamnagar) of Gujarat; and Kojhikoda and Cannore districts of Kerala.
Pleistocene & Recent	B. Placer Deposits		Gold washing has been carried out in the alluvial and gravel beds of many rivers in parts of Assam, Bihar and H.P. The rivers, Subansiri, Lohit, Dihang, Buri Dihing and Janglu, Poni of Assam-Arunachal; Sona, Subernarekha and South Koel (Singhbhum), Sonapet, Karkari, Nakari and Jumar (Ranchi) of Bihar; and several streams in different parts of the country have yielded gold and are worth mentioning.

Resources

India's recoverable reserves of gold metal are estimated at 101 tonnes, concentrated almost entirely in Karnataka and Andhra Pradesh. The break up of the ore/metal reserves is given in the **Table 4.2.**

Table 4.2—Reserves of Gold in India (Recoverable Reserves as on 1.4.1990)

State	Reserves of Ore in million tonnes and metal in tonnes				Remarks
	Proved	Probable	Possible	Total	
Andhra Pradesh					
(i) Ore	1.396	1.583	3.861	6.840	Primary gold
(ii) Metal	6.0	7.4	18.3	31.7	reserves
Bihar (Kundrakocha)					
(i) Ore	—	—	0.008	0.008	Primary gold
(ii) Metal	—	—	0.1	0.1	reserves
Karnataka					
(i) Ore	6.855	6.063	0.695	13.613	Primary gold reserves
(ii) Metal	32.6	29.4	2.9	64.9	
Kerala					
(i) Ore	—	2.552	22.198	24.750	Both primary and
(ii) Metal	—	2.3	2.1	4.4	placer gold reserves
Madhya Pradesh					
Ore	—	—	1.7	1.7	Primary gold reserves
India Total - Reserves					
(i) Ore	8.251	10.198	26.162	45.211	
(ii) Metal	38.6	39.2	23.3	101.1	

In Andhra Pradesh, gold resources of Chigargunta (Kuppam Taluk, Chittoor district) have been explored by Mineral Exploration Corporation (MEC) in 1990-91 and reserves of 0.516 million tonnes of gold-ore with an average grade of 4.34 g/t gold have been estimated. The gold samples obtained incidental to prospecting and exploratory operations carried out in the State have yielded 193 kg, 292 kg and 286 kg of gold during 1989-90, 1990-91 and 1991-92, respectively.

In Karnataka the assessment of gold prospects in Hassan and Chitradurga districts is being carried out in Hutti-Muski, Nuggubali and Chitradurga - schist belts, and a total of 2.18 million tonnes of probable reserves of ore with average grades varying from 2.7 g/t to 4.09 g/t have been estimated during 1990-91. The detailed exploration is also in progress in Dharwar, Kolar and Raichur districts.

In Kerala primary gold is reported from Kozhikode and Cannore districts with possible reserves of 0.29 million tonnes, while placer gold from Kozhikode district has probable reserves of 2.55 million tonnes (2 tonnes gold metal) and possible reserves of 22.20 million tonnes, totalling 24.75 million tonnes of ore (4.4 tonnes metal).

In Madhya Pradesh, in the Pandripani area of Raipur district, gold values in the range of 0.1 to 4.1 g/t over 150 m strike length with width varying from 0.90 to 3.5 m have been noted. In the Suda area of Sidhi district, geochemical sampling has shown gold values ranging between 2.56 and 17.5 g/t over 1500 m strike length.

Production

The productions of metal in 1989-90 and 1990-91 were 2163 kg and 2207 kg, respectively which decreased to 2,039 kg in 1991-92 and to 1865 kg in 1992-93. The entire production was from the public sector. The average grade of ore mined was 4.30 gm/tonne.

The recovery of gold from copper slime at Ghatsila Copper Smelter of HCL is significant. It was at 111 kg in 1986, declined by 8% compared to that in 1985. During 1989-90, 1990-91, 1991-92 and 1992-93 the recovery was 127 kg, 192 kg, 283 kg and 316 kg, respectively.

Two public undertakings, namely :

(i) Bharat Gold Mine Ltd. (BGML) - a Central Govt. undertaking, and

(ii) Hutti Gold Mines Company Ltd. (HGML) - a Karnataka State Govt. Undertaking,

are producing gold in the country. The BGML obtains ore from Nandydroog Champion Reef and Mysore Mines of Kolar Gold field in Karnataka and Yeppaman mine of Ramgiri Gold field in Andhra Pradesh. The HGML operates two mines; Hutti in Raichur district, and Manglur in Gulmarga district of Karnataka.

Mining

All the gold mines in the country are underground. The mining method adopted in Nandydroog mine was cut and fill over hand stoping with sand fill. In Champion mine underhand stoping in upper levels and slope driving in deeper levels were carried out. In Mc Taggart West Lode New Trial Shaft underhand stoping with granite masonry was adopted. In Hutti mines, mainly shrinkage and partially cut and fill methods of stoping were followed.

PROSPECTING GUIDES

1. Host Rock and Associated Minerals

Intrusive rocks of composition diorites, quartz-diorites and granites are found to be suitable host rocks of primary gold. The associated minerals

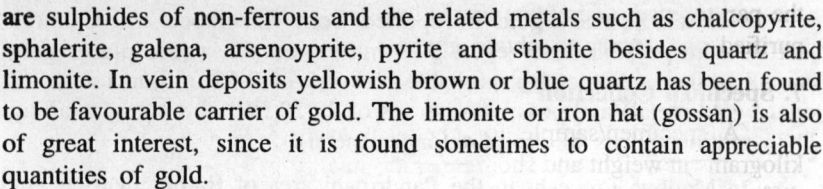

are sulphides of non-ferrous and the related metals such as chalcopyrite, sphalerite, galena, arsenoyprite, pyrite and stibnite besides quartz and limonite. In vein deposits yellowish brown or blue quartz has been found to be favourable carrier of gold. The limonite or iron hat (gossan) is also of great interest, since it is found sometimes to contain appreciable quantities of gold.

2. Rock Examination

Besides the above, all the outcrops of quartz-veins, silicified rocks and quartz boulders should be examined carefully for visible gold specks or sulphide minerals. Free gold is often found confined to fine cracks, and while examining the rocks, this should be borne in mind.

3. Soil-Cover Areas

In soil-cover areas, the soil under uprooted trees, the loose material thrown up from animal burrows and rain-rill may be examined for mineralised fragments of quartz or other rocks. The gravels of the river channels are required to be looked for mineralised fragments.

4. Pebble and Boulder Tracing

This is a useful technique to locate gold occurrences. It is possible to find the primary gold lode by tracing the mineralised quartz fragments/pebble or any other rock in the river gravel upstream. When they cease to reappear, the search must be continued up the nearest side stream. As one proceeds further up, the fragments/pebbles or boulders of mineralised rock should occur more frequently and become more angular. In this way one can reach the source rock.

5. Panning

Panning is the principal prospecting method in case of placer gold. Samples for panning should be taken from the lowest layers of gravels (near the bed rock, if possible) in the stream channels at intervals of 200 to 300 metres. Shallows, bars (head parts), down stream side of rocky barriers, cancave sides of the stream beds, erosible banks, crevices in bed rocks and spots between ridges are the ideal sites for gold accumulation. Samples should be panned on spot. The presence of associates of gold or heavy minerals in the concentrate is a good sign for placer gold prospecting. The concentrate should be dried up and stored in a sample bag for chemical analysis. In case of presence of gold, the sampling intervals should be reduced accordingly.

6. Use of Suitable Pan

Panning should be carried out in a clean wooden pan, free of grease, and the pan should be shaken until the gold and the black minerals get separated. In case of very fine placer gold, a little of mercury is put on the pan to ensure complete extraction, and, later on, the gold amalgam is purified of mercury to obtain gold.

7. Specimen Collection

A specimen/sample for chemical analyses should be one to two kilograms in weight and should include small fragments of rock taken from different parts of the ore-body.

8. Gold Value

The gold values of 2 g/t and above are suggestive of the area for detailed exploration, while the values above 4 g/t are indicative of economic deposit worth-mining.

2. SILVER

Silver is a precious metal next to gold, and is known to man since time immemorial. The world mine reserves of silver is estimated at around 420 million kg. (420 thousand tonnes) out of which over 85% reserves are from eight principal countries - U.S.A., Canada, Mexico, Peru, Russia, Africa, Japan and Australia. The silver mining operations are through both open pit and underground operations. Silver is being extracted through a flotation process which results in the recovery of silver during lead, copper, zinc and gold smelting. The world production of silver is roughly about 14 thousand tonnes per year. In 1989 a total of 14,452 tonnes of silver was reported to be produced. Mexico continued to be the leading producer of silver contributing 16%, followed by USA (14%), Peru (13%), Russia (10%), and Canada (9%).

Mineralogy

Silver is white, lustrous metal; highly malleable and ductile. It is unaltered by dry or moist air. Its sp. gr. is 10.5 and melting point about 1000°C, and is well known metallic conductor of heat and electricity. The main minerals of silver with their characters are given below in the Table 4.4.

Table 4.4—Silver Minerals and their Characters

Minerals	Chem. Comp.	Chief Characters
Native Silver	Ag (Usually associated with small amounts of other metals - Cu, Au, Pt, Hg, Pb, Zn, Bi etc.)	Usually filiform, arborescent or massive, colour and streak-silver white, metallic, malleable and ductile, H = 2.5-3, sp. gr. = 10.1-11.1.
Argentite (Silver glance)	Ag_2S	Blackish grey in colour and streak, metallic lustre, H = 2-2.5, sp. gr. = 7.19-7.36.
Stephanite-(Brittle · Silver Ore)	Ag_5SbS_4	Iron black in colour and streak, brittle, H = 2-2.5, sp. gr. = 6.26

(Contd.

Minerals	Chem. Comp.	Chief Characters
Prousite (Light red Silver Ore)	Ag_3AsS_3	Commonly granular and massive, cochineal red in colour and streak, H = 2-2.5, sp. gr. = 5.55-5.64
Pyrargyrite (Dark red Silver Ore)	Ag_3SbS_3	Commonly massive, black to cochineal red colour, cochineal red streak, H = 2-3, sp. gr. = 5.7-5.9.
Cerargyrite (Horn Silver)	AgCl	Usually massive and wax like and also in encrustations, pale shades of grey, sometimes greenish or bluish, streak shinning, H = 2-3, sp. gr. = 5.8.
Hessite	Ag_2Te	Lead grey, metallic, sectile, H = 2.5, sp. gr. = 8.4.

There are many more minerals of silver and are beyond the scope of the book to describe. Silver also occurs in solid solution with gold and base-metal sulphides.

Silver is obtained from its ores mainly by amalgamation with mercury or by cyanidation if present in finely divided state. If not finely divided, the ores are concentrated and smelted. But largely silver is derived from smelting of lead, zinc and copper ores which are argentiferous.

Uses

The larger uses of silver are in coinage, plate and jewelry. Some amounts are employed in electroplating, soldering and bearing. Silver salts are used in medicine, photography, for colouring glass and for various subordinate purposes. The addition of a small amount of copper produces an alloy having lower melting point, a greater hardness and affording a sharper casting than pure silver.

Mode of Occurrence and Origin

Silver ores occur in a variety of ways, such as in veins, stringers and disseminations, as replacement deposits, contact metamorphic deposits or as alluvials. The upper parts of silver deposits or lodes are weathered, and in several cases presence of gossan is prominent wherein minerals like cerargyrite (AgCl), bromyrite (AgBr) and iodyrite (AgI) are present. Below this haloid zone, a zone of secondary enrichment with native silver and rich secondary sulphides may be developed. The primary deposits with usually much poorer ore exists below the secondary sulphide enrichment zone.

Hydrothermal solutions might be responsible in bringing about the replacement or cavity filling deposits. Massive and lode replacement of silver-lead ores are numerous,·but most of the world's silver is won from fissure veins of mesothermal and epithermal types.

Distribution

In India there is no silver deposit as such. It is found associated mainly with lead and zinc and to some extent with copper and gold. Some of the galena of Rajasthan, Andhra Pradesh, Uttar Pradesh and Bihar are argentiferous. Lead-zinc sample from Zawarmala mine, Udaipur district, Rajasthan assayed 18 ppm silver. The lead concentrate gave 610 ppm silver, while the zinc concentrate analysed 81 ppm silver. The lead occurrences in Hesatu-Belhathan belt, Bihar are found to be argentiferous showing silver up to 200 ppm. An occurrence of argentiferous galena with silver up to 0.1% has been noted at Birgana in Pauri-Garhwal district, Uttar Pradesh.

Production

In India silver is recovered as a by-product from the smelting of lead, zinc, gold and copper. The production of silver made this way during 1990-91, 1991-92 and 1992-93 were about 35, 38 and 47 thousand kg, respectively. About 90% of the output accrued from lead and zinc smelters, while the remaining 10% were from copper and gold smelting. The entire production of silver was in the public sector. The details of the producers with locations of plants are given in the Table 4.5.

Table 4.5—Principal Producers of Silver in India

Producers	Location of plants	Source
1. Hindustan Zinc Ltd.	Visakhapatnam, Andhra Pradesh	Lead concentrate produced in Andhra Pradesh
2. Hindustan Zinc Ltd.	Debari Zinc Smelter, Rajasthan	Zinc concentrate produced in Rajasthan
3. Hindustan Zinc Ltd.	Tundoo Lead Smelter Dhanbad, Bihar	Lead concentrate produced in Rajasthan
4. Hindustan Copper Ltd.	Maubhandar Copper Smelter, Singhbhum, Bihar	Copper slime from copper smelting
5. Bharat Gold Mines Ltd.	Kolar, Karnataka	Produced during Gold refining
6. Hutti Gold Mine Ltd.	Raichur, Karnataka	Produced during Gold refining

3. PLATINUM

Introduction

Platinum was first found in the alluvial deposits of river Pinto in Colombia, S. America from whence it was taken to Europe in 1735 by the Spanish mathematician Antonio de Ulloa. Here, it first received its name as Platinum from 'Platina', silver like (Platina del Pinto). No workable platinum deposit exists in India. The country depends on imports from other countries. The Ural mountains of Russia where platinum was discovered in 1819, is the world's most important mining district, while Canada, S. Africa, Australia, S. America and Alaska are the other main producing countries.

Mineralogy

Platinum belongs to the group of metals, consisting of palladium, osmium, iridium, rhodium and ruthenium. It occurs in nature in fine scales and minute grains in the following forms :

1. Native state — Element Platinum (Pt)
2. Arsenide — Sperrylite ($PtAs_2$)
3. Arseno-sulphide — Cooperite (Pt $(As, S)_2$)
4. A new group of platinum-bismutho tellurides and sulphursides of iridium and rhodium have been identified.
5. Alloyed with other metals of its group.

Platinum is a greyish white lustrous metal, with white steel-grey streak, and is having hardness of 4-4.5, sp. gr. 21.46 (pure), melting point 1760°C and boiling point 4300°C. It is malleable and ductile with high tensile strength and is resistant to acids and other chemicals. It displays hackly fracture like gold.

For testing platinum, Messrs. Johnson Matthey & Co. Ltd. has recommended a solution of conc. HNO_3 = 3/4 oztroy, conc. HCl = 1¼ oztroy and potassium nitrate = 1/20 oztroy, on application of which platinum (95% and above purity) does not indicate any colour, while palladium gives medium brown colour. To distinguish it from stainless steel a saturated solution of ferric chloride ($FeCl_3$) is used. This solution has no effect on platinum, while stainless steel is attacked and leaves a grey stain.

Uses

Platinum is not only precious, but also a strategic metal as is used as a catalyst in the production of higher grade petrol. The demand for this metal is increasing with the re-emergence of the oil industry. It is an industrial catalyst of great importance because of its use in the contact process for the manufacture of sulphuric acid and oxidation of ammonia to nitric acid.

It is used in laboratory apparatus and in electro-chemistry. In alloys it is used in dentistry, and jewelry as for diamond settings in U.S.A. It has the same co-efficient of expansion as glass into which it can be scaled. It is used in resistance thermometer which gives accurate results upto 1000°C.

The chemical industry uses it in pressure vessels because of its excellent anti-corrosive property, while its resistance to acid renders it ideal for electrodes in electrolytic oxidation fields. Fibre glass is produced by forcing molten glass through the platinum nozzles. Another valuable outlet of platinum-gold and platinum-rhodium alloys is as spinnerets with several hundred fine diameter holes for making rayon fibre viscose. Another of the best known applications of platinum and its alloys is in thermocouples, providing an accurate measurement of liquid steel temperatures upto 1700°C. In electrical engineering their major uses are for contacts in voltage regulations, thermostats, relays and contacts in high tension magnets. Also, spark plugs equipped with platinum alloys have a long life and resist fouling.

Mode of Occurrence and Origin

The platinum metals occur genetically in the following five modes :

1. Early Magmatic Concentrations :
   (i) *Disseminations* : Sparse disseminations with chromite in dunite e.g. Urals, Alaska, Colombia.
   (ii) *Segregations by fractional crystallisation* - e.g. Rustenburg, S. Africa.
2. Late Magmatic Concentrations
   (i) *Immiscible liquid segregations* - e.g. Vlackfontein, S. Africa.
   (ii) *Immiscible liquid injections* - e.g. Sudbury, Canada.
3. Contact Metasomatic deposits - e.g. Potgietersmast, S. Africa.
4. Hydrothermal - e.g. Waterburgs, S. Africa & Sudbury, Canada.
5. Placers - e.g. Urals, Colombia, Alaska.

Platinum metals are primarily found as orthomagmatic ore deposits in ultrabasic rocks, usually associated with chromite and nickel-ores. The erosion of the ultrabasic parent rocks has yielded placer deposits in Urals, Alaska and Colombia. In Ural mountains it occurs as sparse disseminations with chromite in dunites. In South Africa, the ore is concentrated in the :

(a) hortonite - dunite pipes of the differentiated Bushveld Igneous Complex,

(b) Chromitite, and

(c) Nickel - copper sulphides of the Complex.

In the Merensky Reef, Rustenburg, South Africa, the platinum ores are associated with the chromite segregations, by fractional crystallisation, in the norite zone.

The platinum-ores associated with the nickel - copper sulphide deposits of Vlackfontian, Rostenburg, South Africa, are regarded as late magmatic immiscible liquid segregations, and the ores associated with the massive copper-nickel sulphide body of Frood Mines at Sudbury, Canada are possibly late magmatic immiscible liquid injections.

In the area north-west of Potgieterscrust, South Africa contact metasomatic deposit of platinum metals has been reported. Here, the Bushveld igneous rocks have intruded dolomite and ironstones with development of diopside-grossularite skarn at the contact, which is impregnated by platinum bearing yellow sulphides with unusually large crystals of sperrylite. The Waterburg deposit is a very unusual type of high temperature hydrothermal or pneumatolytic origin where platinum occurs in a vuggy quartz-vein with chalcedony, specularite, pyrite and chrome-mica.

Metallogenetic Epochs

Platinum occurs in the following three metallogenetic epochs :

1. **Late mesozoic to Early Tertiary :** Occurs in Zambales, Phillipines as the 'Alpine' type. The minor occurrence in Dhangawan bauxite, M.P., India may correspond to this.

2. **Late Palaeozoic :** Occurs in Alpine type of ultramafic intrusions related to Hercynical folding movement in Ural Mountains, Russia; and dunitic rocks of New Zealand.

3. **Precambrian :** Occurs in stratiform type of Bushveld Igneous Complex, Transvaal, South Africa and as late magmatic injections in Frood Breecia, Sudbury, Ontario, Canada, cutting earlier noritic intrusion. The placer occurrence reported in Assam, West Bengal and Bihar may correspond to this.

Distribution

No workable platinum deposit has so far been located in India. However, a few stray occurrences have been reported from different parts of the country. They are being described delow :

1. Assam

Platinum associated with gold was reported by Dalton & Hannay in the sands of the Noa Dihing river (27°33' : 96°00') and examined by Mallet (1882). The platinum in this river is probably derived from the ultrabasic suite of rocks of the Patkoi Range.

2. Bihar

Dunn (1937) recorded the occurrence of platinum associated with gold in the sands along the Gurma river near Dhadka (22°48' : 86°30'). The origin of such platinum may possibly be connected with the basic intrusives north of Dalma lava.

3. Karnataka

The gold washings of the Kolar mines have shown traces of platinum.

4. Madhya Pradesh

H.L. Chibber found traces of platinum in a sample collected from the bauxite deposit of Dhangawan ($22°35'$: $80°11'$) on Jabalpur-Katni road. Three samples of this bauxite, analysed in USA, were reported to contain 22, 15 and 11 gm of platinum per tonne of ore.

5. Tamil Nadu

The stratiform magnesio-chromites laminated with ultramafics of Sittampundi complex have indicated the presence of platinum and palladium to the extent of 300 ppb and 1000 ppb respectively. The presence of mineral braggite (Pt, Pd, Ni) S and cooperite Pt (As, S)$_2$ has been confirmed by optical properties. No economic deposit of Pt has been confirmed.

6. West Bengal

While examining a sample of alluvial gold from the Guram river, Mallet (in P.K. Chatterji, 1937) detected minute grains of platinum, the largest scale weighing 0.005 gr. The gold from other rivers also yields minute traces of platinum.

Prospecting

In view of the precious nature and strategic importance of platinum and to meet the indigenous demands, there is an urgent necessity to take up concentrated exploration programme for locating workable platinum-ore deposit in India. While launching the prospecting/exploration programme the following basic points may be taken into consideration :

(a) Platinum is geochemically relatively immobile compared to other elements.

(b) It has a very high sp. gravity.

(c) Antimony, arsenic and bismuth are common associates and as such may be used as path finders.

(d) The stratiform chromite laminated with ultramafics, could possibly be best suitable locale for concentration of platinum metals with the exception of Alpine type chromite deposit.

(e) The crystallising environment for platinum group of elements in any particular mafic - ultramafic igneous complex could be inferred largely on the presence of magnesio-chromite, crystallising at high temperature (about 1300°C) and oxygen fugacity of 10^{-7} atoms.

4.2 FERROUS AND ALLIED METALS

1. IRON

Introduction

Iron is the most vital metal in human use. It constitutes 5.05% of the crust material and holds third position in abundance after silicon and aluminium. It is rarely found in native condition except in meteorites and some eruptive rocks. It enters into large number of rocks forming silicates and is widely available as oxides. The mineral containing iron must be mineable at profit in order to be called an 'iron-ore.' The total world production of iron-ore in 1990 was about 1,008 million tonnes to which India's contribution was about 55.5 million tonnes (contributing 5.5%) with 6th position in the world production.

Mineralogy

The chief ores of iron are its oxides and carbonates, the important of them are named below in the Table 4.6.

Table 4.6—Chief Ores of Iron

Minerals	Chem. Comp.	Iron (Fe%)	Remarks
Oxides			
Hematite	Fe_2O_3	69.9%	Steel grey to red colour, streak cherry-red, sp. gr. 5-6
Magnetite	Fe_3O_4	72.4%	Blue black or brown black colour and streak, sp. gr. 5.17
Turgite	$2Fe_2O_3.H_2O$	66.1%	Brown to red colour, sp. gr. 4.2-4.7
Goethite (Limonite)	$Fe_2O_3.H_2O$	62%	Brown to yellow, sp. gr. 3.3-4.3
Carbonate			
Siderite	$FeCO_3$	48.2%	Ash grey to brown streak, sp. gr. 3.96

Uses

Iron found its industrial use first time in 800 B.C. which marks the commencement of "Iron-age". The Iron-age culminated in "Steel age" in the nineteenth century with steel gaining the importance. The iron-ore is basically used for the extraction of iron-metal in the form of cast-iron, wrought iron, steel and iron-alloys which have their own particular uses. The iron metal finds its use both in home and outside. It has a major role in building construction, farm tools, machines, rails, railway wagons, automobiles, ships etc. besides domestic utensils. The other use of iron-ore is in coal washery as a medium for flotation. It is also used in certain industries like cement to make up the proportion of iron in the raw material, under water cable sheathings, powder metallurgy etc. The quantity so used in these ways is insignificant compared to its use in iron smelting. Iron-

)re is mainly utilised for making pig iron, sponge iron and steel. Lumpy iron-ore is utilised after preliminary washing and screening for sizing. The fines cannot be utilised as such and hence these are subjected to agglomeration and utilised in the form of pellets or sinters. The use of pellets as feed in blast-furnace has several advantages because of its uniform size, known composition and strength. In recent years, there is a trend for more utilisation of pellets or sinters as compared to natural ores. A major thrust has been given to sponge iron industry in the recent years, since manufacture of sponge iron requires non-coking coals as reducing agents.

Classification of Iron-Ore Deposits

While preparing iron-ore maps of the world, the International Committee under International Geological Congress standardised the classification of iron-ore. A broad break-up is given below :

1. **Metamorphic banded deposits :** The major iron-ore deposits of India fall within this group. They are typically sedimentary or volcano-sedimentary and metamorphosed rocks, consisting of rich iron-ore and siliceous (chert) bands. They have been differently named in different countries, like Itabirites. Taconites, Jaspilites, Banded hematite quartzites, Metamorphosed iron-ores etc. Indian iron-ore deposits of Bihar-Orissa belt, Bailadila (M.P.) and those of Karnataka are typical examples. The metamorphosed equivalent of this rock is banded magnetite-quartzite.

2. **Continental sedimentary deposits :** These are assumed to have formed in fresh water (fluvial or lagoonal) or under brakish swamp (lacustrine) conditions. Ironstones of Raniganj and Auranga Coalfields are typical examples of this type of deposit. The ore occurs as layers and lenses of siderite in Gondwana rocks, generally altered to limonite near the surface.

3. **Marine sedimentary deposits such as oolites, detrital, placer and mixed type :** The type area is in Lorrain, France. The Indian example is magnetite deposits of coastal regions such as at Travancore associated with ilmenite and heavy mineral sand.

4. **Volcano-sedimentary deposits :** These deposits are related to the volcanic rocks of diabase group of initial geosynclinal magmatism. The ore-bearing horizons are sub-marine and are bedded. Insignificant minor pockets of iron-ore in Dras-Thasgam area, Ladakh are the Indian example.

5. **Liquid magmatic deposits :** They are formed during early crystallization of basic plutonic rocks, mainly by-gravitational differentiation. The typical Indian occurrence is that of titano-vanadium bearing iron-ore deposits of Mayurbhanj district, Orissa.

6. **Intrusive magmatic deposits :** They are related to alkaline rocks of the Precambrian shields. Apatite - magnetite rocks of Singhbhum represent this type.

7. **Contact metasomatic deposits :** These are found at the contact aureoles of granitoid intrusions within the limestone and are widely distributed. Main ore mineral is magnetite. Magnetite deposits of Kotwalsa, Vishakhapatnam district, Andhra Pradesh, and of Gore and Biwabathan, Palamau district, Bihar are examples.

8. **Polymetallic skarn ore deposits :** These occur associated with sedimentary deposits which are later affected by regional metamorphism. The iron-ores lie in the basal sequence of acid volcanics with intercalations of limestone, dolomite and clastic sediments. The iron-ore deposits at Iron Springs, Utah and. Hanover, N. Mexico are the examples. No Indian example is recorded.

The Indian iron-ore deposits can be conveniently divided into the following six basic types based on their origin and mineralisation :

Fig. 4.1 Massive iron-ore
Loc. : Deposit No. 5, Bailadila, M.P.

(i) *Banded ferruginous formation of Precambrian age :* The ores can be classified into massive ore, laminated ore and blue dust (Fig. 4.1 and 4.2). The major iron-ore deposits of the country are associated with this formation.

(ii) *Sedimentary iron-ores* of sideritic or limonitic composition e.g. found in coalfields of Bihar and West Bengal, and Tertiary formations of Assam.

(iii) *Lateritic iron-ore :* e.g. associated with Deccan trap, also widespread in the country.

(iv) *Apatite - magnetite rocks* of Singhbhum Copper belt.

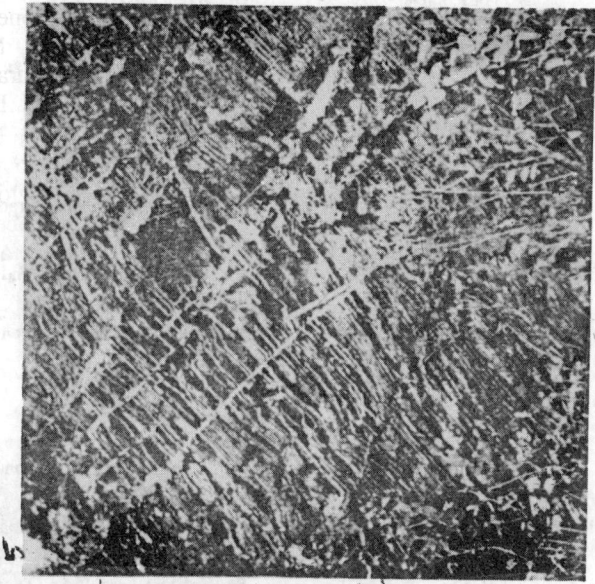

Fig. 4.2 *Laminated iron-ore*
Loc. : deposit No. 3, Bailadila, M.P.

(v) *Titaniferous and vanadiferous magnetites :* e.g. S.E. Singhbhum
of Bihar, Mayurbhanj district of Orissa and Tumkur district of
Karnataka.

(vi) *Fault and fissure fillings of hematite :* e.g. hematite of Valdurthi
and Ramalkota in Kurnool district and also in Cuddapah district.

Distribution

The geological and geographical distributions of iron-ore in India are
given in Table 4.7 below.

Table 4.7—Geological Distribution of Iron-Ores in India

Formation	*Types of Deposits*	*Occurrences/Deposits*
Precambrian		
Basic & Ultrabasic	Titaniferous and Vanadi-ferous magnetite	SE Singhbhum (Bihar), Mayurbhanj (Orissa) and Tumkur (Karnataka) deposits
Granodiorite Granites	Apatite-magnetite rock Magnetites (Residual)	Singhbhum copper belt and Mayurbhanj deposits, Assam Jantia Hills
Banded Iron-ore Formation	Hematites (massive, shaly, powdery etc.)	Singhbhum (Bihar); Bonai, Keonjhar & Mayurbhanj (Orissa); Poonch (J & K); Bastar, Durg and Jabalpur (MP); Chanda and Ratnagiri (Maharashtra); Dharwar, Bellary, Sundur, Shimoga and Chikmagalur (Karnataka); Goa etc.

(Contd.)

Formation	Types of Deposits	Occurrences/Deposits
Banded Iron-ore Formation (Metamorphosed)	Magnetite-Quartzite	Guntur (A.P.), Salem and Trichinapalli (Tamil Nadu), Shimoga (Karnataka), Mandi (H.P.)
Bijawar/Gwalior Cuddapah	Hematite & Ferruginous Quartzite	Bijawar, Gwalior, Indore and Rewa (M.P.); Mohindargarh (Haryana); Jhunjhunu, Sikar and Jaipur (Rajasthan); and Cuddapah (Andhra Pradesh)
Gondwana		
Barakar-Mahadeva	Ironstone	Birbhum (W. Bengal), and Auranga coal field (Bihar)
Ironstone-shale	Ironstone and siderite	Raniganj coal field (W. Bengal)
Triassic		
Sirban Limestone	Hematite & limonite	Udhampur (J & K)
Jurassic		
Rajmahal trap (Inter-trappean beds)	Ironstone	Rajmahal (Bihar) and Birbhum (West Bengal)
Eocene and Miocene	Ironstone	NE districts of Assam, Kumaon (U.P.), Travancore and Malabar (Kerala)
Quaternary	Laterite	Several states of India. Derived from different formations including Deccan trap

Table 4.8—Geographical Distribution of Iron-ores in India

State	District	Locality	Remarks
Andhra Pradesh	Guntur		Magnetite ore is found.
Bihar	Singhbhum	Noamundi, Notoburu, Jamda, Gua, Jhilingburu, Pansirabuda etc.	Titaniferous and vanadiferous magnetite is found.
		Pathargora, Kumaria, Khariatola, Singri etc.	Apatite-magnetite ore is found.
	Palamau	Gore Pahar and Biwabathan near Daltonganj	These are magnetite deposits.
		Rajabar, Balunagar and Morwai, Auranga coal field	Ironstone with siderite is found.
Goa			Banded iron-ore formation with good hematite ore.
H.P.	Mandi	—	Magnetite-ore is found.
Karnataka	Bellary-Hospet — (Sandur) Shimoga Chitradurga Chikmaglur		Banded iron-ore formation with good hamatite iron-ore deposits.

(Contd.)

State	District	Locality	Remarks
	Tumkur	Kudramukh Bababudan Hills	Magnetite iron-ore deposits are found.
M.P.	Bastar	Bailadila, Rowghat, Parrekaro near Kanker	Banded iron-ore formation with good hematite-ore deposits.
	Durg	Dhalli, Rajhara	-do-
	Chanda	Lohara, Pipalgaon, Asola, Dewalgaon etc.	Of these only Lohara deposit is large.
	Jabalpur	Majgaon, Agaria, Saroli, Lora, Jauli, Kanhara hills etc.	The ores are poorer and contain iron varying from 45 to 60 per cent.
Orissa	Keonjhar	Thakurani, Bolaria, Joda, Banspani, Gandhmardan, Bhadrasahi, Koira, Malangtoli etc.	
	Sundargarh	Buliapahar, Badamgarh Pahar, Mankar nacha, Lusi-Rontha, Khandahar, Barsua, Bonai-Keonjhar belt	Banded iron-ore formation with very good hematite iron deposits.
	Mayurbhanj	Gorumahisani Pahar, Sulaipat-Badampahar, and Simlipal hills	
	Cuttack	Daitari, Tamaka and Kansa	
	Koraput	Hirapur, Umarkot	
	Mayurbhanj	Bisoi-Rairangpur, Bisoi-Joshipur and Baripada-Udasa	Titaniferous magnetite is found.
Rajasthan	Jaipur	Udaipur, Alwar & Bundi	Banded iron-ore formation with hematite ore.
Tamil Nadu	Salem, Tiru-chirapalli		Magnetite-ore is found.

Reserves

India is endowed with large resources of iron-ore. As per assessment by the United Nations, India possesses 29 billion tonnes of crude ore. However, recoverable ore reserves are limited to less than 50% of the crude. The recoverable reserves of iron-ores in the country as on 1.4.90 are placed at 9,602 million tonnes of haematite and 3,143 million tonnes of magnetite.

Production and Consumption

The total production of iron-ore during 1990-91 was about 56 million tonnes, of which iron-ore lumps contributed 42%, fines 46% and iron-ore concentrate 12 per cent. Iron-ore industry is basically an export oriented one. Of the total despatches of 54 million tonnes, 31 million tonnes (57%) were reported for export. Japan, Korea Republic, Romania and China are the main countries where iron-ore is being exported. The average metal content of the ore produced was about 64% Fe. Goa remained to be the leading producer accounting for 24%, followed by Madhya Pradesh and Karnataka 22% each, and Bihar and Orissa 15% each. The remaining quantity was contributed by Maharashtra, Andhra Pradesh, Rajasthan and Haryana. The production figures of iron-ore and its products during 1989-90 to 1992-93 are given below in the Table 4.9.

Table 4.9—Indian Production of Iron-Ore and its products, 1989-90 to 1992-93 (in million tonnes)

	1989-90	1990-91	1991-92	1992-93
Iron-ore	55.4	55.5	57.5	58.1
Pig iron	12.1	13.0	14.6	-
	14.3	15.3	15.4	-
Semi-finished steel	2.0	3.8	4.1	-
Finished steel	9.2	9.6	10.2	-
Sponge iron	0.298	0.632	0.475	-

During 1989-90, about 20 million tonnes of iron-ore was consumed in different industries. The producers of pig-iron and steel are the main consumers of iron-ore (about 95% of total consumption). The sponge iron units, cement, alloy steels, ferro alloys, foundries, electrode industries and coal washeries are the other consumers of iron-ore.

2. MANGANESE

Introduction

Manganese (Mn) is widely distributed in earth's crust in the form of oxides, carbonates and silicates. Although known to man in one form or another for a long time, it was not until 1774 that the metal was first isolated by Scheeli. The manganese ore attracted the attention of industry for the first time as raw materials for medicines, paints etc. at the end of 18th century. With the growth of steel industry and the introduction of new techniques, the demand of manganese-ore increased and current world production exceeds 26.4 million tonnes (1990). India is self-sufficient in this mineral and produces about 1.5 million tonnes (1990-91) per year of Mn-ore and ranks 7th in the world production. Republic of S. Africa is one of the major holders of Mn-ore reserves and dominated the western world market. Russia, the other major holder of Mn-ore reserves, no longer exports it to West.

Mineralogy

In nature, manganese does not occur in the native state, but it may be produced by electrolysis, or by reduction of its oxides (Thermit Process). It is a pinkish, grey metal, very hard and brittle, melting at about 1260°C, boiling point 1980°C and having sp. gr. of 7.2.

Manganese-ore is usually a black lumpy mass, which may be compact or loose, often sooty, and generally contains inclusions of gangue which is usually white, yellowish grey or yellowish brown clay. It is homogeneous and frequently consists of several minerals. Though there are numerous manganese minerals (more than 150) occurring in nature, the most common and important ones are described below in the Table 4.10.

Table 4.10—Common Manganese Minerals and their Characters

1. Oxides

Pyrolusite (MnO_2)	Dull, fine-grained, brittle or sooty black masses, black or bluish black streak, metallic or opaque lustre, hardness 2 to 2.5, often soils finger, sp. gr. 4.8.
Psilomelane (Hydrated Oxide of Mn)	It is black with brown streak, having hardness of 5 to 6 and sp. gr. of 3.7 to 4.7. Loose, soft, dark brown variety is known as wad.
Manganite ($MnO.OH$)	It is distinguished from Psilomelane by lower hardness (4).
Brownite (Mn_2O_3)	It is hard, and brownish black in colour and streak. It oxidises to pyrolusite and psilomelane.
Hausmanite (Mn_3O_4)	It occurs together with brownite, but is somewhat higher in Mn-content (72%). It is brownish black in colour with red-brown streak. Its hardness is 5-5.5 and has sp. gr. of 4.86.

2. Carbonate

Rhodocrosite ($MnCO_3$)	The crystalline - granular variety which has a distinct pink colour, is comparatively rare. Fine-grained, porous mass is white, grey white or green grey. Its streak is white, hardness 3.5 to 4.5 and sp. gr. 3.45 to 3.6. It quickly oxides to black hydrous oxide of manganese.

3. Silicate

Rhodonite ($MnSiO_3$)	Occurs in solid compact masses of intense pink colour which becomes black on exposure. Its streak is white, hardness 5.5 to 6.5 and sp. gr. 3.4-3.6. It is sometimes used as an ornamental stone, and also for imparting a violet colour to glass.

Pyrolusite and Psilomelane are the two most common manganese ores.

Uses

Mn-ore is an important raw material in iron and steel metallurgy where it is used in the form of ore as such, and as ferro-manganese. It improves the strength, toughness, hardness and workability of the steel, acts as deoxidiser and desulphuriser and also helps in getting ingots free from impurities. About 90 to 95% of world production of Mn-ore is used in metallurgy of iron and steel.

Mn-ore is also used for manufacturing dry cell batteries and chemicals. Manganese-dioxide, used for manufacturing dry cell batteries, functions as a depolarizer of hydrogen which collects as a non-conducting film on the carbon electrode, thus facilitating the continuous flow of current within the battery. The sulphide of manganese is used in the manufacture of salts and in calico printing. Manganese-chloride is used in cotton textile as a bronze dye. Manganese salts are also used in photography, and in leather and match industries. Pyrolusite is generally used in glazes for pottery and making coloured bricks. It is also used as driers for oils, varnish and paints in glass industry for decolourising green glass, medical preparations, carbon-monoxide gas masks etc.

Specifications

The specifications of Mn-ore by some important industries are given below :

(i) *Iron and Steel industry :*

Mn-content	28 to 35% (in exceptional cases, 25% min.)
Fe	15 to 25%
SiO_2	6 to 13%
Al_2O_3	5 to 8%
P	0.30% (max.)

(ii) *Ferro-manganese industry :*

The important consideration is the high Mn : Fe ratio (7 : 1 min.) and very low content of Phosphorus. The Indian standard specification, for different grades of ore for the industry, in question, are as follows :

	(a) Grade 1	(b) Grade 2	(c) Grade 3	(d) Grade 4
Mn	48% (min.)	46-48%	44-46%	40-44%
Fe	7% (max.)	7.5%	9%	12%
SiO_2	8% (max.)	9%	10%	12%
P	0.12% (max.)	0.15%	0.15%	0.15%

(iii) *Dry cell battery :*

MnO_2 = 80% (min., preferably 84% min.)

Fe = 3% (max., preferably 2% max.)

Acid insolubility = 2% (max.)

Moisture = 3% (max.)

Cu = 0.1% (max.), Ni = 0.1% (max.), CO = 0.2% (max.)

The size requirement is 200 mesh. Ore with predominant gama structure is needed.

(iv) *Chemical industry :*

High grade material is used.

For Potassium Permanganate :

MnO_2 = 80% (min.), SiO_2 = 5% (max.)

Fe_2O_3 = 10% (max.), Ore size = 200 to 250 mesh.

(v) *Glass industry :*

MnO_2 = 80% (min.), preferably 88% (min.)

Fe_2O_3 = 5% (max.), preferably 0.75% (max.)

SiO_2 = 2.8 (max.), Al_2O_3 = 1.1% (max.)

BaO = 1.3% (max.), CaO = 0.4% (max.)

MgO = 0.4% (max.).

Mode of Occurrence

Mn-ores in India occur mainly in two types of deposits :

(a) Bedded deposit, and

(b) Lateritoid deposits.

Bedded Deposits

They are sheet like bodies of oxide-ores overlying Precambrian rocks, rich in Mn-content like gondites and kodurites. They are conformable, tabular bodies, sinuous bands and lenses of variable thickness and lateral extent, following the trend of enclosing rocks e.g. Tirodi area, Balaghat district, M.P. The bedded deposits may further be classified as follows :

1. *Syngenetic deposits :*

   (i) *Primary oxide :* Mostly braunite with or without sub-ordinate manganese silicate e.g. near Chaibasa (Bihar) occurs as thin lenticles parallel to bedding, associated with basal rocks of Kolhan formation.

   (ii) *Gondite deposits :* Mostly quartzite with spessartite and rhodonite and other manganese bearing silicates, minor rhodochrosite and subordinate bauxite e.g. Bhandara and Nagpur (Maharashtra), Balaghat (M.P.) and Gangpur (Orissa). The ore-bodes were originally laid down in oxidising conditions as syngenetic sediments as a part and parcel of Sausar Formation and were later regionally metamorphosed to form compact recrystallised ore bodies.

(iii) *Kondalite suite* : Mostly hybrid, consisting of garnet-granulite and garnetiferous quartzite, and usually phosphoric. The composition of rock varies widely from acid to ultrabasic. It is developed in the eastern coast of India.

2. *Supergene deposits* : They may represent oxidised braunite, oxidised gondite and khondalite deposits. The rocks consist of lithomarge, ochres and wad. The ore bodies in them are irregular and occasionally attain large dimensions as at Garbham and Kodur (Andhra Pradesh). The ores consist mainly of psilomelane with some pyrolusite, braunite and mangan-magnetite and are high in iron and phosphorus and low in silica.

Lateritoid Deposits (or Detrital Deposit)

The lateritoid deposits contain appreciable quantities of iron and vary from manganese to iron-ores through every gradation, the minerals present being pyrolusite, psilomelane, wad, and also limonite and earthy hematite. The deposits in Goa, Sandur (Karnataka), Keonjhar (Orissa) represent this type.

Genesis

The bedded deposits are believed to have been formed by the metamorphism of manganiferous sediments, the impure portions forming silicates, and the pure ones such primary ores as braunite, psilomelane, hollandite, sitaparite, vredenburgite (a mixture of hausmanite and jacobsite) etc. In the intermediate zone of metamorphism (staurolite-kyanite zone) gondite and manganese silicates are predominantly developed, while the coarse-grained braunite with subordinate hollandite and jacobsite and manganese pyroxenes are characteristic of sillimanite zone (high grade) of granulite facies. The garnetiferous quartz and garnet granulite are developed in the granulite facies. Various agencies like pressure, heat and water act upon rocks and minerals, causing recombination and recrystallisation of the ingredients into new minerals that are stable under the imposed conditions.

The lateritoid deposits have resulted from the surface alteration of manganiferous Dharwarian rocks such as phyllites, schists and ferruginous quartzites. The undesired constituents of rocks are removed during weathering resulting in residual concentration of Mn-ores. The concentration is due largely to a decrease in volume, effected almost entirely by surfacial chemical weathering. The Mn-oxides resist solution and, thereby, these accumulate to form economic deposits. Further the radiating and colloform structures and complex admixtures so characteristic of manganese oxides suggest that most of the residual manganese oxide compounds have been deposited as colloids or as colloidal mixtures. The manganese hydroxide sol is negative and is precipitated by positive ions. Mixing of sols brings about precipitation of manganese compounds in bewildering complexity.

Distribution

The Mn-ore deposits are associated with the Archaean rocks with the exception of occurrences of true laterites in Goa. The major Mn-ore deposits are in Orissa, Karnataka, Madhya Pradesh and Maharashtra. Some deposits also exist in Andhra Pradesh, Goa, Gujarat and Bihar. The geographical distribution of Mn-ore deposits with chief geological characteristics are given in the Table 4.11 below.

Table 4.11—Mn-ore Deposits in India

State	District	Chief Geological Characteristics
Andhra Pradesh	Adilabad	The Mn-ores occur as thin lenses with chert and jasper within limestone. Deposits are of minor nature with low phosphorus content.
	Srikakulam	Associated with kodurite-rock (Garnet granulite and garnetiferous quartzite) forming a part of Khondalite formation, and are formed mainly due to supergene enrichment. Four distinct belts have been demarcated with a total reserves of 1.6 million tonnes. The ore is of low grade with high phosphorus.
	Visakhapatnam	Associated with kodurite of Khondalite formation.
Bihar	Singhbhum	Associated with rocks of Iron-ore and Kolhan formation as lenticles parallel to bedding or as lateritic material at the surface.
Goa	To the south of Nissan Val. 363m, and Central and Northern part of Goa	The deposits are of lateritoid type found at or near the surface. The iron-content is inversely proportional to Mn in the ore.
Gujarat	Panch Mahal Vadodara	The ores are partly lateritoid and partly primary, associated with less metamorphosed Dharwars.
Karnataka	Bellary (Sandur basin) Chitradurga, Uttarkannad, Dharwad, Shimoga, Tumkur	The deposits are of varying dimensions, associated with limestones, schistose grits, and ochery-schists of the Shimoga-Chitradurga schist belt of Dharwar group. The ores are of lateritoid type and the individual deposits are lenticular and impersistent.
Madhya Pradesh	Balaghat	They represent gondite type of deposits, associated with metamorphosed Dharwar rocks
Maharashtra	Bhandara, Nagpur	The ore bodies are banded braunite-quartzite and grade on to quartz spessartite - rhodonite bearing gondite. Weathering has given rise to residual enrichment deposits.

(Contd.)

State	District	Chief Geological Characteristics
	Ratnagiri	They represent secondary enrichment deposits associated with lateritised Dharwarian metasediments, composed of quartzite, banded hematite-quartzite and phyllite. The ores are generally more ferruginous.
Orissa	Sundargarh & Keonjhar (Bonai-Keonjhar area)	The ore bodies occur as lenses or in irregular shape in shales, brecciated cherts, and laterites capping them, belonging to Iron-ore group of rocks. The ore deposits are for the most part epigenetic enriched under syngenetic conditions. About 90% of Orissa production comes from this area.
	Koraput, Kalahandi & Bolangir (Koraput-Kalahandi-Patua area)	The ore bodies are associated with khondalite suite of the Eastern Ghats group.
	Sundargarh (Gangpur area)	The deposits are associated with gonditic rocks.
	Sambalpur	The deposits are associated with laterites on the metasediments.
Rajasthan	Banswara	The deposits are associated with laterite on the metasediments.

Reserves

The reserve estimate of Mn-ore in India is placed at about 176.5 million tonnes (as on 1.4.1990) against the world reserves of about 3,900 million tonnes. Out of the total Indian reserves, about 18% is in the grades of more than 46% Mn, about 26% in the grades of 35% to 46% Mn, and the rest belongs to inferior grades or unclassified.

Karnataka has the main share (36.6%), followed by Orissa (23.1%), Madhya Pradesh - Maharashtra belt (20.2%) and Goa (13.4%). The remaining reserves are distributed in Andhra Pradesh, Bihar, Gujarat, Rajasthan and West Bengal.

Production

The production of Mn-ores in India was about 1.4 million tonnes in 1988 which is gradually increasing with passage of time.

Orissa continued to be the leading producer contributing about 36%, followed by Madhya Pradesh (21%), Maharashtra (19%) and Karnataka (16%). The balance 8% was produced by other states together (Andhra Pradesh, Bihar, Goa and Gujarat). The mining is mainly carried out by open cast and to a limited extent by underground method. The underground mines are located in Madhya Pradesh and Maharashtra operated by M/s. Manganese Ore (India) Ltd. in the public sector. All the underground mines are mechanised.

There are ten major ferro-manganese plants in operation in the country with a total installed capacity of 342.6 thousand tonnes per year. These produced about 143 thousand tonnes ferro-manganese in 1988.

The production figures for 1989-90 to 1992-93 are given below in the Table 4.12.

Table 4.12—Indian Production of Manganese Ore, and Ferro-manganese 1989-90 to 1992-93

	1989-90	1990-91	1991-92	1992-93
Manganese ore (in million tonnes)	1.45	1.49	1.55	1.87
Ferro-manganese (in thousand tonnes)	176	201	206	—

3. CHROMIUM

Introduction

Chromite is the only source of chromium metal, required by the alloy steel industry. India has been one of the important producers and exporters of this mineral for the past several decades. Recently, some restrictions have been imposed on the export of chrome-ore. The chrome-ore lumps with Cr_2O_3 less than 42 per cent, the chromite fines, and the friable ores are allowed to be exported on merits. With the industrial advancement, the demand for chromite has increased and so its production. The world chromite production increased to 13 million tonnes in 1990 from about 11.6 million tonnes in 1987. To the above India contributed about 940 thousand tonnes (7.2%) and ranked 3rd in the world production. The total world reserves of chromite are placed at 6,800 million tonnes.

Mineralogy and Classification

Chromite is iron-black or brownish black in colour with brown streak, faint submetallic lustre, and uneven brittle fracture and is having hardness of 5.5 and sp. gr. of 4.5 to 4.8.

It has the chemical composition of $FeO.Cr_2O_3$ and theoretically contains 68% Cr_2O_3 and 32% FeO; but Al_2O_3, Fe_2O_3, MgO, CaO and SiO_2 replace some Cr_2O_3, reducing Cr_2O_3 content to as little as 40 per cent. Based on chemical composition the chromites are classified as :

   (i) High grade ore with very low iron-oxide and other impurities e.g. Sukinda-Nausahi (Orissa) with high Cr_2O_3 and very low Al_2O_3 and SiO_2.

   (ii) Ferriferous chromite with Fe_2O_3 more than 30% e.g. Jojohatu (Bihar) ore, Pauni (Maharashtra) ore, and some ores from Sukinda-Nausahi (Orissa).

   (iii) Aluminous chromite with Fe_2O_3 more than 20% and less than 30% e.g. Kondapalli (Andhra Pradesh) and Sittampundi (Tamil Nadu).

(iv) Siliceous Chromite with high SiO_2 content e.g. some of the chromites of Sukinda (Orissa), Pauni (Maharashtra) and other places.

While the Cr_2O_3 and Fe_2O_3 have reciprocal relationships, Al_2O_3 varies from zero in Sukinda - Nausahi ore to as high as 18% in Kondapalli, Byrapur, Pauni and to 30% in Sittampundi ores. The Cr : Fe ratio is 3 : 1 for Sukinda ore; 2 : 1 for Chaibasa and Kondapalli ores, 1.5 : 1 for Sittampundi ore; 1.6 : 1 for Pauni, Byrapur and Sindhuvalli ores and for Tertiary deposits.

Based on above observations, the Indian deposits may be classified into :

1. Those with Cr : Fe = 3 : 1 and above, Al_2O_3 nil.
2. Those with Cr : Fe = 2 : 1 average and Al_2O_3 considerable.
3. Those with Cr : Fe = 1.5 : 1 and less, with practically low Al_2O_3, SiO_2 etc.

Uses

Chromite is mainly used in metallurgical, refractory and chemical purposes. The metallurgical uses consist of a variety of alloys, mainly with iron, nickel, and cobalt. Ferro-chrome of different types like high carbon ferrochrome, medium carbon ferrochrome and low carbon ferrochrome; and silico-chrome and charge-chrome are the principal alloys. The ferro-chrome and charge chrome units consume about 80% of chromite produced. The stainless steel contains 18 per cent chromium and 8 per cent nickel. The stellite is another special chrome-alloy with tungsten, molybdenum and cobalt. This is used in high speed tool steel. Chromium plating is quite popular. The presence of chromium makes the alloys strong, tough, hard and resistant to oxidation, corrosion, abrasion, chemicals, electricity and high temperature break down. The great strength of chrome steels allows reduction in weight of metal in automobiles, aeroplanes and trains. For metallurgical purposes chromite with Cr : Fe ratio of more than 2.5 : 1 is required.

Chromite is an important raw material for refractories that withstand temperatures about 2000°C. Generally refractory bricks are made by using chromite and magnesite. These bricks are used mainly as lining in basic open hearth furnaces. For refractory purposes the chrome-ore should be compact, massive and hard with a minimum length size of 2 cm. Cr_2O_3 should be 40% (min.), SiO_2 6% (max.), FeO 18% (max.) and CaO 1% (max.). Cr_2O_3 and Al_2O_3 contents combined should be atleast 58-60 per cent. Lower Fe_2O_3 and SiO_2 contents and high Al_2O_3 and MgO in chromite is preferred. Lime is detrimental.

Chromite is also used for manufacturing chemicals which in turn are used in pigment manufacture, leather and textile treatment, dyeing, bleaching and oxidising agents.

Mode of Occurrence

The Indian chromites occur in :

(a) banded, massive,

(b) banded, crystalline,

(c) massive,

(d) crystalline, and

(e) disseminated forms.

The Kondapalli (Andhra Pradesh) and Chaibasa (Bihar) ores are banded, massive; while the Sukinda and Nausahi (Orissa) ores are banded, crystalline. The disseminated ores are found in Kondapalli (A.P.) and Nausahi (Orissa) areas. Structurally, they are mainly of two types as follows :

(i) Banded or stratiform type, as in Sukinda and Nausahi areas.

(ii) Lensoid, as in Kondapalli, Chaibasa, Pauni and some lodes of Sukinda and Nausahi.

Thayere (1969) classified the chromite deposits, based on their mode of occurrences, into the following two types :

1. Stratiform Type Deposits

The bands and layers indicate gravitative settling. Graded bedding and variation of grain size are more frequent. Pyroxenite and peridotite are more common rock types, though dunite is also, at times, found. They tend to be ferriferous and less aluminous. The Cr : Fe ratio is about 2 : 1 or less. They occur in the Precambrian or Archaean geosynclines, e.g. the deposits of S. Africa, Southern Rhodesia and U.S.A. belong to this type. In India, the deposits of Sukinda-Nausahi, Chaibasa, Sittampundi, Kondapalli, Pauni, Byrapur and Sindhuvalli areas are associated with pyroxenite, peridotite and dunite of Precambrian/Archaean age and represent stratiform type of deposits. All the deposits occur in long bands or lensoid bands simulating stratification and some of these are associated with anorthosite, gabbro, vanadiferous magnetite bodies and nickeliferous ultramafic layers. They have Cr. Fe ratio around 2 : 1 except in case of Sukinda-Nausahi ore, where it is above 3 : 1.

2. Podiform Type Deposits

They occur mostly in nodular forms and are associated with dunite and serpentinite. They are more aluminous with Cr : Fe ratio exceeding 2 : 1 and occur along the Tertiary geo-synclines. The deposits of Greece, Cuba, Urals and Pakistan are of this type. In India the deposits of this type are in Ladakh (J & K) and Manipur. They are associated with dunite and serpentinite and are emplaced into the Mesozoic-Tertiary geo-synclines along the cores of folds.

Origin

The chromite deposits are characterised by their close relationship with ultrabasic rocks (deep seated intrusive igneous rocks), and have originated from concentration of early magmatic crystallisation. They have long been considered unimpeachable illustration of early magmatic segregation. This is supported by field and microscopic evidences at many places. However, some workers cite evidences of late magmatic or even hydrothermal chromite. In some places disseminated grains have undergone residual concentration.

Distribution

A. Geologic

The Indian chromite deposits are distributed in the following three types of formations :

- (i) The Eastern Ghats Group of rocks,
- (ii) The Iron-ore Group - Dharwar Group of rocks,
- (iii) The Tertiary formations.

(i) The Eastern Ghats Group of Rocks

The chromite is associated with hypersthenite, bronzitite and diopsidite (pyroxenite) as lensoidal bodies with charnockite. The lenses have their long axes parallel to the regional trend of the country rock, indicative of a broad fold with N-S axis. The chromite is emplaced during the Eastern Ghats orogeny. Andhra Pradesh and Tamil Nadu deposits belong to this group.

(ii) The Iron-Ore Group - Dharwar Group of Rocks

The chromite is associated with pyroxenite and peridotite, and occurs as lenses and bands within Iron-ore group of rocks in Singhbhum district, Bihar, and in Dhenkanal, Cuttack and Keonjhar districts, Orissa. They are emplaced and folded along with the Iron-Ore Group rocks during the iron-ore orogeny. The chromite deposits of Karnataka and Maharashtra are associated with pyroxenite, peridotite and dunite, enclosed by Dharwar schists. They are emplaced along the axial planes of anti-formal folds during the Dharwar orogeny.

(iii) The Tertiary Formations

The chromite deposits of Ladakh (J & K) and Manipur come under this category. In Ladakh chromite occurs in association with serpentinite and dunite within the Dras volcanics of Cretaceous age. In Manipur the chromite is associated with ultrabasic rocks intrusive into the Tertiary Formation.

B. Geographic

The geographic distribution of chromite deposits with chief characteristics are given overleaf in the Table 4.13.

Table 4.13—Geographic Distribution of Chromite Deposits in India

State	Districts	Locality	Chief Characteristics
Andhra Pradesh	Krishna	Kondapalli	The deposits are in N-S trending en-echelon lenses.
Tamil Nadu	Salem	Sittampundi Complex	The deposits occur as lensoid bodies in anorthosite within the country rock, charnockite.
Bihar	Singhbhum	Jojohatu, near Chaibasa	Chromite occurs in small NE-SW trending lenses and bands.
Orissa	Cuttack Dhenkanal	Sukinda and Katpal	Chromite occurs with nickel-ore in bands and lenses within the folded limonitised ultramafic rocks along deep marginal fractures.
	Keonjhar	Nausahi	Chromite occurs in bands and lenses within the serpentinite, peridotite, pyroxenite, gabbro, vanadiferous magnetite and anorthosite, resembling the well known stratiform complexes of the world.
Karnataka	Mysore Hassan	Sindhuvalli Byarapur	Chromite occurs as N-S trending bands and lenses associated with ultrabasics within Dharwar schists.
Maharashtra	Bhandara Ratnagiri	Pauni Kankauli	Chromite is found in altered ultramafic rocks intruded along synclinal folds in Sakoli metasediments.
Jammu & Kashmir	Kargil	Dras	Chromite is associated with serpentinite and dunite within Dras Volcanics of Cretaceous age.
Manipur	Ukhrul, Manipur East	Sirohi peak, Nepali Basti, Tengopal - Moreh Road	Sporadic occurrences in peridotite - and serpentinite.

Reserves

The recoverable reserves of chromite in the country as on 1.4.1990 are assessed at 88.35 million tonnes. Orissa is by far the richest as it contains 97% of the reserves. Of the total reserves, about 35% are of metallurgical grade. Reserves of refractory grade chromite may be about 5% of the total and the supplies of this grade at present come from Orissa, Manipur, Maharashtra and Karnataka.

The total geological resources as on 1.4.1990 are placed at 182 million tonnes, with about 104 million tonnes of in situ reserves and about 78 million tonnes of conditional resources. The largest share of the total geological resources is accounted for by Cuttack district of Orissa.

Production

The productions of ferro-chrome and charge-chrome which are the two main mineral based products, were about 43 and 97 thousand tonnes, 1992-93. Practically the entire production was from Orissa (96.4%) and Karnataka (3.5%). Insignificant amounts of production come from Andhra Pradesh, Maharashtra and Manipur. High grade chromite (above 47% Cr_2O_3) constitutes 48%, medium grade (40 to 47% Cr_2O_3) 34% and low grade (below 40% Cr_2O_3) 18%. The principal producers of chromite are :

   (i) Orissa Mining Corpn. Ltd.,

  (ii) Tata Iron and Steel Co. Ltd.,

 (iii) Ferro Alloys Corpn. Ltd. (Bhubaneswar, Orissa),

 (iv) Mysore Minerals Ltd., and

  (v) Misrilall Mines Pvt. Ltd. (Chaibasa, Bihar).

The production of ferro-chrome and charge-chrome which are the two main mineral based products, were about 43 and 97 thousand tonnes, respectively in 1988. There were three major plants each for ferro-chrome and charge-chrome in operation. The ferro-chrome plants are located at Garividi (Andhra Pradesh), Jaipur road (Orissa), and Bhadravati (Karnataka) with a total capacity of 52,000 tpy; while the charge-chrome plants are at Randia, Rayagada and Bamnipal (all in Orissa) with a total installed capacity of 145,000 tpy. One new charge-chrome plant with a capacity of 62,500 tpy is under installation at Choudhar, Cuttack district, Orissa. The complex will also have 2,50,000 tpy briquetting plant.

The productions of these alloys for 1989-90 to 1991-92 are given below in the Table **4.14.**

Table 4.14—Production of Ferro-Chrome and Charge-Chrome for 1989-90 to 1991-92 (in thousand tonnes)

	1989-90	*1990-91*	*1991-92*
Ferro-chrome	52	58	83
Charge-chrome	85	92	110

Mining

The Chromite mining is mostly done by open-cast method except at the Byarapur mines (Karnataka) and Boula and Kathpal mines (Orissa). At Byrapur in Hassan district of Karnataka, chromite ore-body extending to 300 m. depth is being mined by underground method since 1967. Here cut-and-fill-method of stoping is practised. In Sukinda valley chromite had been mined upto the depth of 63 m. by open cast method. Operations at most of the mines are semi-mechanised. Equipments, like dozers for removing the overburden and shovels and loaders alongwith dumpers for moving ore and overburden are employed at some mines.

4. VANADIUM

India's resources of vanadium ore are in the form of vanadium-bearing titaniferous magnetite ore. The red mud generated during the production of alumina from bauxite usually contains 0.3 to 0.4% V_2O_5. Vanadium is obtained mainly as a by-product or co-product. The world production of this metal is about 0.036 million tonnes per year to which India's contribution is negligible. S. Africa, U.S.A., Russia, China and Finland were the important producers. In Canada efforts are being made to extract vanadium from flyash, generated by commercial oil sand plants. The total world reserves of vanadium metal is about 16.56 million tonnes, where India's reserves stand at 0.021 million tonnes.

Mineralogy

Vanadium is soft, malleable and silver white metal, melting at about 1720°C. It has a great affinity for oxygen which demands its use in metallurgy. It is extracted from its ores by smelting in electric furnace and then reducing (thermit process) to produce ferro-vanadium with some 30% vanadium. Vanadium pentoxide is produced by treatment with calcium. The chief vanadium minerals with their characters are given in the Table 4.15 below.

Table 4.15—Chief Vanadium Minerals

Minerals	Chem. Composition	Properties
Patronite	VS_4 (possibly) complex composition	Black, Amorphous
Roscoelite (Vanadium mica)	Silicate in which vanadium, to a certain extent, replaces Al of muscovite.	In minute scale, clove-brown to greenish brown, H = 2.5, sp. gr. 2.97
Carnotite	Vanadate of uranium and potassium	Powdery or minute crystal plates, canary yellow
Vanadinite	Chlorovanadate of lead	In crusts and prismatic crystals; colour ruby red, orange-brown or yellowish; streak white or yellowish; hardness 3, sp. gr. 6.7-7.1

Uses

Vanadium imparts toughness to steel and so is used in the manufacture of special steels. A little of vanadium added to any steel helps to remove oxygen and nitrogen, and gives uniform grain size. For high speed steels, 4 to 5 per cent vanadium is used. It has gained its importance in production of materials for defence purposes. For engineering and structural works, it is added to chromium, molybdenum, tungsten etc.

It forms special alloys with gold used in dentistry, with copper and zinc for bushings in aeroplane, and with aluminium in aeroplane manufacture.

Vanadium salts are used for various chemicals, printings of fabrics and in electrical, ceramic, paint and dye industries. It works as a catalyst in the manufacture of sulphuric acid.

Mode of Occurrence and Origin

Vanadium does not occur in nature in free state and is found with other elements in the form of sulphides, silicates, vanadates etc. It occurs mainly with other ores, e.g. vanadium-bearing titaniferous magnetite. The ore is found associated with altered basic and ultrabasic rocks in the form of veins, stringers and lenses.

Economic deposits of vanadium are mostly in weathered form, with or without residual concentration. Some of the deposits may possibly have formed by concentration through organism. In India the ore occurs in the form of vanadium-bearing titaniferous magnetite, and is also obtained from red-mud, a residue from the alumina industry.

Distribution and Reserves

Important deposits of vanadium bearing titaniferous magnetite are located in Karnataka, Orissa and Maharashtra. It has also been reported from Singhbhum district of Bihar. Important deposits of vanadium-bearing magnetite are located in Shimoga and Hassan districts of Karnataka, Bhandara district of Maharashtra and Mayurbhanj district of Orissa. The total recoverable reserves of these ores are placed at 13.34 million tonnes of ore, with an estimated vanadium metal content of 20,387 tonnes (Table 4.17). In addition, the conditional resources are placed at 10.91 million tonnes of ore with 22,068 tonnes of metal. The distribution of vanadiferous magnetite and reserve position are given in the Tables 4.16 and 4.17, respectively.

Table 4.16—Distribution of Vanadium-Bearing Titaniferous Magnetite

State	Locality	Geological details
Bihar	Dublabera, Lango, Sindurpur and Kumardubi of Singhbhum distt.	Associated with basic and ultrabasic rocks of Iron-Ore group (Precambrian)
Karnataka	Shimoga distt.; Nuggehali, Hassan district; Santepet, Mulemane, Suryakalyanigudda, Masanikere, North Kanara distt.	Associated with basic schists and occurs as bands in ultrabasic rocks of Precambrian age
Maharashtra	Mahalgaon, Khursipur, Bhandara distt.	Associated with amphibolites or metamorphosed pyroxenites
Orissa	Bisoi-Rairangpur, Joshipur and Baripeda-Udala in Mayurbhanj distt., also in Keonjhar and Balasore distt.	Occurs in association with gabbro-anorthosite rocks in Precambrian metamorphites

Table 4.17—Recoverable Reserves of Vanadium as on 1.4.1990
(in tonnes)

States	Recoverable Reserves							
	Proved		Probable		Possible		Total	
	Ore	Metal	Ore	Metal	Ore	Metal	Ore	Metal
Karnataka	500000	700	4000000	5600	4342000	6087	8842000	12387
Maharashtra	-	-	1292000	1420	668000	735	1960000	2155
Orissa	2370875	5472	-	-	163920	373	2534795	5845
All India Total	**2870875**	**6172**	**5292,000**	**7020**	**5173920**	**7195**	**13336795**	**20387**

Production

India has to depend on the imported vanadium pentoxide for its ferro-vanadium industry. The productions of ferro-vanadium in 1987, 1988 and 1989 stood at about 71 tonnes, 54 tonnes and 69 tonnes, respectively. There is no regular production of vanadium ore, except through occasional mining of vanadiferous magnetite by VISL from Masanikere deposit, about 40 km from Bhadravati in Karnataka. The ore containing 1.04% V_2O_5 and 53.1% Fe was successfully treated for producing ferro-vanadium at VISL plant. VISL had earlier completed the construction of ferro-vanadium project involving 100 tonnes per year capacity. Catalyst (India) Private Ltd. recovers vanadium from red-mud, a residue from alumina industry. The main ferro-vanadium plants are :

1. Electro-Ferro Alloys (P) Ltd., Gujarat,
2. Visvesvarya Iron and Steel and Danceli Ferro Alloys Ltd., Karnataka,
3. India Thermit Corpn. Ltd., U.P.,
4. Bharat Pulverising Mills Pvt. Ltd., Maharashtra,
5. Utkal Ferro Alloys (P) Ltd., Orissa,
6. Mehra Ferro Alloys, Punjab, and
7. R. Sen and Co., West Bengal.

Vanadium catalyst required in sulphuric acid manufacture is produced by Catalyst (India) Private Ltd., Thane, Maharashtra, which partly used vanadium obtained from vanadium sludge. VISL and India Thermit Corpn. Ltd. consume indigenous ore, like vanadium magnetite and vanadium obtained from red mud.

5. MOLYBDENUM

The production of molybdenum gained momentum only in the 20th century with its large scale use in iron and steel industry. The world production of molybdenum metal went up to 117,000 tonnes in 1989, the chief producing countries were U.S.A., Chile, Canada and Russia. Its demand is increasing with growing steel market. The world reserves of molybdenum metal have been estimated at about 11.3 million tonnes, the major reserve holding countries are U.S.A., Chile and Canada.

Mineralogy

Molybdenum does not occur in free state in nature. The main ores of molybdenum is molybdenite (MoS_2). To a certain extent it is also extracted from wulfenite ($PbMoO_4$). The chief characters of the above ores are given below in Table 4.18.

Table 4.18—Chief Characters of Molybdenum Ores

Ores	Chem. Composition	Chief Characters
Molybdenite	MoS_2	Usually in scales, also massive and foliaceous; lead grey in colour with greenish lead-grey streak; H = 1-1.5, sp. gr. 4.7-4.8.
Wulfenite	$PbMoO_4$	Wax yellow, orange yellow, yellowish grey or brown; streak-white; waxy or admantine lusture; H = 3, sp. gr. 6.3-7.

Molybdenum metal is produced from molybdenite directly in electric furnace or by reduction of its oxides by carbon or aluminium. The metal has a white or greyish colour with sp. gr. of 8.6 and melting point 2500°C.

Uses

The principal use of molybdenum is in the production of special steel and for this ferro-molybdenum alloy is frequently used in place of metal. Molybdenum trioxide or calcium molybdenate are also, at times, employed in the steel and alloy industries.

Besides alloy steel and iron-steel industries, ferro-molybdenum is also consumed in electrodes and foundry. In steel it works like tungsten, but needing only one half to one-third as much to produce similar result. It enhances the strength of steel and cast iron and also increases the ductility. The molybdenum metal is also needed in certain electrical apparatus. Ammonium molybdenate is used as a reagent in the detection of phosphates.

Mode of Occurrence and Origin

Mobybdenum is rare in occurrence, but is widely distributed in the crustal rocks, averaging 2.3 gm/ton in granitic rocks (Goldsmidt, 1958), 2.4 gm/ton in basaltic rocks (Rankama and Sahama, 1950) and 0.001 to 0.0059 gm/ton in sea water (Fairbridge, 1972). Molybdenite is a widely distributed ore and frequently occurs in small veins or scattered in tiny flakes through the rocks so that a concentration by table dressing, oil flotation etc. is necessary. Wulfenite is found in the oxidised portion of lead and molybdenum-bearing deposits.

The incidence of molybdenum in a variety of Precambrian crystalline rocks of the southern part of Indian peninsula, like granite, pegmatite, meta-pyroxenite, metanorite, charnockite and khondalitic rocks, has been brought out. But, the promising prospect is genetically related to acid magmatism e.g. Karadikutam, Madurai district, Tamil Nadu. The molybdenum ore of Climax deposit, Colorado and Querta deposit, New Mexico which are the world's leading producers of molybdenite, is genetically related to quartz monzonite and granites of Tertiary age (acid magmatism).

The contribution of molybdenite from copper deposits has shown significant increase in the recent years, but these are also found related to pegmatites and quartz-veins. The molybdenite mineralisation in the copper belts of Singhbhum, Bihar, and of Malanjkhand, M.P., is intimately associated with the fine veins of quartz and pink felspars which have filled up the shears and minor fractures, and the hydro-thermal solutions which brought about alkali metasomatism in area, were responsible for collection, transport and deposition of molybdenum at favourable locii.

Distribution and Reserves

There are a number of occurrences of molybdenite mineralisation in India where explorations have been carried out in the past. A few of them are mentioned below in Table 4.19.

Table 4.19—Molybdenite Occurrences in India

State	District	Areas where explorations were carried out	Remarks
Arunachal Pradesh	Dibang Valley	Estaline	
Bihar	Singhbhum	Rakha Copper Deposit	Contains an average of 0.11% molybdenum
Karnataka	Kolar	Yogavakota	-
Kerala	Alleppeye	Kunnathukara	-
	Waynad	Ambalavayal and Kalpetta	-
Madhya Pradesh	Balaghat	Malanjkhand Copper deposit	Contains 0.04% recoverable molybdenum besides bismuth, arsenic and cadmium
	Raipur	Naumera and Bundeli	-
	Dharampuri	Harur - Alangayum	Tentative reserves of 1.6 m. tonnes of ore averaging 0.175% molybdenum have been estimated
Meghalaya	Khasi and Jaintia hills	Umpyrtha	Contains Mo along with Cu, Pb, Zn and W
Nagaland	Phek	Washello, Lachambalee and Ziphu	-
Rajasthan	Dariba-Rajpura	Lead-zinc deposit	Contains Mo along with Bi, As and Cd.
Tamil Nadu	North Arcot	Narasinghapuram	-
	Madurai	Karadikutam	Contains 0.02 to 0.14% recoverable molybdenum

The recoverable reserves of molybdenum as on 1.4.1990 in the country are given below in the Table 4.20.

Table 4.20—Recoverable Reserves of Molybdenum as on 1.4.1990 (in tonnes)

State	Recoverable Reserves							
	Proved		Probable		Possible		Total	
	Ore	Contained MoS_2	Ore	Contained MoS_2	Ore	Contained MoS_2	Ore	Contained MoS_2
Karnataka	-	-	-	-	900	2	900	2
Madhya Pradesh	-	-	-	-	8000000	2762	8000000	2762
Tamil Nadu	-	-	36000	62	-	-	36000	62
All India reserves	-	-	**36000**	**62**	**8000900**	**2764**	**8036900**	**2826**

The all India conditional resources are estimated as 2,523,212 tonnes of ore and 1,740 tonnes of contained MoS_2.

Production

There is no regular production of molybdenum ore in the country. The molybdenum concentrates are, however, produced as by-product intermittently from Jaduguda mine, where it is associated with uranium ore. There are, however, certain units in the country, namely Utkal Ferro Alloys Pvt. Ltd., Orissa; Mehra Ferro-Alloys, Punjab; Electro Ferro Alloys Pvt. Ltd., Gujarat and India Thermit Corpn. Ltd., U.P. which are producing ferro-molybdenum. The productions of ferro-molybdenum during 1988-89, 1989-90 and 1990-91 were about 184.8, 184.3 and 179.5 tonnes, respectively.

6. TUNGSTEN

Tungsten (Wolfram) was first recognised and named by Scheele in 1781, and isolated in 1783. It is a strategic metal, and the acceleration and deceleration in the search for tungsten in India coincided with the peak and ebb of the two world wars. The first discovery of tungsten deposit in India was in 1907 at Aragaon in Nagpur district, Maharashtra. The world reserves of tungsten-ore is estimated at about 3.55 million tonnes (metal content) out of which only China is having 1.56 million tonnes. China continued to be the world's largest producer followed by USSR. In 1989, the world production of tungsten concentrate was about 43 thousand tonnes (metal content) out which China produced about 21 thousand tonnes and USSR 9,300 tonnes.

Mineralogy

Tungsten does not occur in native state. In nature it occurs only in chemical compounds with other elements. The mined out ores of tungsten are readily concentrated by means of Jigs and Tables because of the high specific gravity of tungsten minerals. This metal is obtained by reduction of its ores with hydrogen, carbon or aluminium. The metal is produced in the form of a greyish black powder with a sp. gr. of about 17. Chemically, pure tungsten (W) is a refractory silver-white metal. It is highly resistant chemically and practically insoluble in hydrochloric and sulphuric acids, but will dissolve in nitric and hydro-fluoric acids to form wolframic acid. It becomes malleable at temperature of about 1600°C and can be drawn into fine filaments.

The principal tungsten bearing minerals with their chemical compositions and properties are given below in the Table 4.21.

Table 4.21—Principal Tungsten Minerals

Minerals	Chemical Composition	Properties
Tungstate		
1. Wolframite	(Fe, Mn) WO_4 Isomorphic mixture of huebnerite ($MnWO_4$) and Ferberite ($FeWO_4$). Minerals with 5.9 to 17.6% Mn (by weight) are regarded as Wolframite	Usually brown to black in colour, chocolate brown streak; metallic lustre, brilliant shining on cleavage surface and dull on other surfaces; perfect cleavage parallel to clinopinacoid; H = 5-5.5; sp. gr. = 7.1-7.9
2. Scheelite	$CaWO_4$	Yellow-white or brownish, sometimes orange yellow; white streak; brittle; H = 4.5-5; sp. gr. = 5.9-6.1.
Oxide		
3. Tungstite (Tungstic ochre)	WO_3	Bright yellow or yellowish green; result from alteration of ores of Tungsten.

Uses

The steel industry is the biggest consumer of tungsten taking about 80 per cent of the world's output of this metal. It is used chiefly for the manufacture of high grade tool steel and heat resistant alloys for high speed cutting tools, springs, gun barrels, pneumatic tool component, internal combustion engine valves etc. The addition of small amount of tungsten to steel (1 to 10%) greatly increases the latter's toughness, strength and resistant to heat and acids.

It forms numerous alloys with iron, nickel, cobalt and molybdenum. Compounds of tungsten with carbon (carbides) and boron (borides) are extremely hard. A large quantity of tungsten (about 18%) goes into carbide and boride alloys, which are used in many industries, as for example for the manufacture of drill rods and bits, cores for anti-tank shells, cutters etc. In 1985 nine units with installed capacity of 432 tonnes were engaged in the manufacture of tungsten carbide sintered products.

A small amount of tungsten output (2 to 5%) is consumed by the electrical engineering industry for filaments for incandescent bulbs, X-ray tubes, radiovalve components and electrodes for contact and atomic welding. It is also used by the chemical industry for paints and varnishes, and also for fire and water-proofing textiles.

Mode of Occurrence

The tungsten deposits occur in the form of veins, stockworks, skarns and placers. The vein-deposits are numerous and occur among intrusive granitic masses, generally in their convex parts, and in rocks overlying the intrusive body.

Stockwork deposits occur in outcrops of granitoids, and are usually large, measuring hundreds of metres in length and tens to hundreds of metres across. Mineralisation is confined to a network of small cracks, filled with quartz streaks containing wolframite, pyrite, arsenopyrite, scheelite, molybdenite, beryl and sometimes cassiterite. The metal content of the ore is usually low (a few fractions of one per cent) but the reserves, as a rule, are large aggregating tens of thousands of tonnes. Despite their low wolframite content, deposits of this type are attractive economically because they can be worked by the cheap strip mining technique.

Skarn deposits are usually found in the contact zones of intrusive bodies. The ore bodies are of irregular shapes. In some places mineralisation is disseminated through the entire thickness of skarns, while in others it occurs in patches. In skarns, mineralisation is usually represented by scheelite, molybdenite, and partly by tinstone, gold and chalcopyrite. The ores carry varying amounts of tungsten. The deposits are often large and are of great economic importance.

Placer deposits are loose or cemented accumulation of fragmental materials, containing grains or crystals of heavy resistant minerals such as scheelite, wolframite, tinstone and gold. Tungsten placers derive through the destruction of vein, stockwork and skarn deposits, as well as of rocks in which these minerals occur as fine impregnations.

Genesis

Tungsten mineralisation is universally associated with granitic rocks and occurs in or near them. All the granitic rocks in the younger orogenic belts have higher tungsten contents in comparison to the similar rocks of older orogenies. The mineralisation associated with older orogenies has dominantly "Scheelite Province" with gold, wherein the former is apathetic to latter, in general. Where gold values decrease, the scheelite mineralisation increases, for example in Canada, South Rhodesia and Kolar gold field (India). The mineralisation associated with younger orogenies has dominantly "Wolframite (Ferberite) Province" for example in Egypt, Uganda and Transval, northern Australia and India. The separation of wolframite and scheelite in distinctly different metallogenic provinces, has been recognised and depends on H_2S partial pressure during mineralisation, higher pressure facilitating scheelite formation.

The study of tungsten mineralisation in terms of Plate Tectonics indicates that practically all productive tungsten occurrences are located either on the boundaries of two orogenic cycles (convergent and divergent plate boundaries) or on the points which could be called "tripple junction" wherein one or two rifts may prove to be abortive in the present stage of geological setting. Such junctions are the locations of hotspot activity and magma generation. Tungsten ore may have formed under conditions of medium to high temperature by gaseous emanations or hydro-thermal solution. The deposits so formed are fissure veins, replacement, contact metasomatic and pegmatitic. The tungsten deposit of Agargaon, Nagpur district, Maharashtra is hypothermal, formed in the temperature ranges of about 400-270ºC and has polyascendant character.

It is inferred from the study of fluid inclusions that the ore-bearing solutions were dominantly potassic-chloride at high temperature during greisenization and formation of quartz-tungsten vein. At lower temperature when sulphides and carbonate minerals deposited, the solutions were dominantly sodi-bicarbonate.

Distribution

The distribution of tungsten-ores in India is given below in the Table 4.22.

Table 4.22—Distribution of Tungsten-Ores in India

State	Locality	Details of Mineralisation	Genetic classification of mineralisation
Andhra Pradesh	Burugubanda (17º18' : 81º47'), East Godavari district	Tungsten mineralisation is present in quart-rich pegmatite traversing graphitic gneisses. The wolframite content varies from 0.3 to 1 per cent by weight. Scheelite as fine disseminations is intimately associated with wolframite.	Pegmatite deposit
	8 km ENE of Ettipali (17º28' : 81º15'), Chittoor district	Tungsten mineralisation is in pegmatite intruding Khondalite suite of rocks. In the Bisanatham mine scheelite tailings contain 0.04 to 0.06 per cent WO_3 and the reserves are estimated to be about 7,000 tonnes.	Pegmatite deposit

(Contd.)

State	Locality	Details of Mineralisation	Genetic classification of mineralisation
Bihar	Tata Nagar (22°47′ : 86°12′), Singhbhum district	A small vein of wolfram was mined between 1916-1918 on the northern side of a small hill down to a depth of 30 m from an inclined shaft. The occurrence is not of economic importance. The vein occurred between quartzite and mica-schists at the contact.	It is high temperature metamorphic differentiation type formed through mobilisation of tungsten from epidiorite, and deposited along the bedding contact
Karnataka	Kolar and Hutti Gold mine	Scheelite occurs in gold-quartz lodes intergrown with quartz as thin lenses, veins, streaks and blebs. Tailing dumps form important source of scheelite recovery.	Hydrothermal quartz-scheelite veins. Pneumatolytic replacement type
Madhya Pradesh	In Raigarh distt. Lawakra area	Tungsten mineralisation analysing 0.91 to 6.65% WO_3 for about 9 km has been traced.	
Maharashtra	Agargaon (21°06′ : 79°29′) Nagpur district	Mineralisation occurs for 1400 m. long, 10 to 115 m wide and over 100 m deep zone with an ore reserves of 2.23 million tonnes upto 50 m depth. It contains 2015 tonnes of wolframite and sheelite concentrate of 65% grade.	Hypo to Meso-thermal quartz-veins, attributed to post Sakoli granites
	Kuhi (21°01′ : 79°22′) - Chapegarhi (21°05′ : 79°26′) area, Nagpur district; 10-16 km south west of Agargaon deposit	Two parallel zones of tungsten mineralisation; one extending for a strike length of 1.65 km with a width between 10 m and 85 m and depth beyond 170 m and average grade of 0.11% W; and other occurring for 1 km strike length with a width of 300 m, have been delineated. It is restricted to the zones of intense pneumatolytic activity and greisens within Sakoli meta-pelitic and migmatic gneisses.	Pneumotolytic type. Boron metasomatic serves as a useful indicator
	Paladi (21°08′ : 79°53′), Chandori (21°11′ : 79°54′) and Dahegaon-Pipalgaon (20°46′ : 79°55′), Bhandara district	The mineralisation is associated with vein-quartz and pegmatite within chlorite-muscovite-schist. Incidence of scheelite as specks has been recorded.	Vein-quartz and pegmatite type
	Kosamtondi-Baga-wan area, Bhandara district	Scheelite mineralisation in two bodies of calc-silicate rock, 180 m x 30 m & 200 m x 50 m, have been delineated.	At the contact of pegmatite and gneiss.

(Contd.)

State	Locality	Details of Mineralisation	Genetic classification of mineralisation
Rajasthan	6 km NE of Sirohi, Sirohi district (Balda, Dera-ka-Bera)	The mineralisation is associated with pneumatolytic quartz-veins intruding Erinpura granite and quartz-mica-schist of Aravalli Super Group. Individual occurrences of wolframite and scheelite are low grade - low tonnage type with grade between 0.1 to 0.2%. WO_3 and tonnage 0.05 to 0.15 million tonnes. Exploration in Balda yielded 0.15 million tonnes of tungsten ore analysing 0.2% WO_3 in two blocks with 50,000 tonnes of high grade fluorite upto 25 m depth. Scheelite ore reserves of 57,000 tonnes of 0.22% WO_3 are estimated at Dera-ka-Bera upto 10 m depth.	Pneumatolytic quartz-veins
	Degana, Nagaur district	Mineralisation is in quartz and pegmatite veins and stockworks in Malani granite. Quartz-veins are greisenised and contain more wolframite near xenoliths of host Aravalli rocks. Pellets, grains, needles and lump crystals of wolframite have been noted. The alluvial placer surrounding Rawat hill, also contains wolframite.	Hydrothermal endogenic deposit
	Motiya, Pali district	Wolframite mineralisation has been traced intermittently for over 300 m.	
Uttar Pradesh	Chamoli and Pauri districts	Sporadic tin-tungsten mineralisation, known by the presence of wolframite, scheelite and cassiterite, with traces of columbite and tantalite in quartz and pegmatite-veins traversing metasediments around Dudhatoli gneissic complex has been recorded.	Hydrothermal deposit. (The hydrothermal solutions emanated from the granitic body and migrated to the low pressure zones)
West Bengal	Chandapathar, Bankura district	Tungsten ore (wolframite) occurs sporadically as lenses, pods and stringers in the quartz-veins within mica-schist unit flanking a granitic body. The tungsten mineralisation does not show any continuity beyond the zones of oxidation and is limited in three small lenses within a cummulative strike length of 500 m	Mesothermal quartz-veins

(Contd.)

Grades and Reserves

At Degana, Rajasthan, the WO_3 value in vein-deposits varies from 0.25 to 0.54%, while in gravel deposit it averages 0.04%. In Sirohi deposit WO_3 value ranges between 0.02 and 2.2%. In West Bengal, the Bankura deposit contains an average of 0.1% WO_3. In Maharashtra, the deposits contain 0.01 to 0.19% WO_3 at Kuhi, 0.13 to 0.38% WO_3 at Kobana, and 0.48% WO_3 at Paladi-Dahegaon-Pipalgaon, the average being 0.17% WO_3. Gold-ore at Mysore mine contains 0.43% WO_3. The tailing dumps at the Kolar Gold Field analyse 0.035 to 0.18% WO_3.

The all-India recoverable reserves of tungsten-ore have been estimated at 23.894 million tonnes or 55.026 thousand tonnes in terms of WO_3 content.

Production

The production of tungsten concentrate in 1988 was only 38,317 kg compared to 51,304 kg in 1985, 45,258 kg in 1986 and 51,116 kg in 1987. Apart from the produce from Degana mine in Rajasthan, recovery of scheelite of 65% WO_3 was obtained from Karnataka while treating tailing sands by BGML. The entire production of tungsten concentrate was from the public sectors.

Rajasthan State Tungsten Development Corpn. Ltd. was the only producer of Tungsten concentrates in the country during 1989-90 to 1992-93. The productive mines are located in Nagaur and Sirohi districts of Rajasthan. The total production figures of tungsten concentrates for the years 1989-90, 1990-91, 1991-92 and 1992-93 stood at 21,333 kg, 20,881 kg, 7755 kg and 3,696 kg, respectively.

Prospecting Guides

1. The host rocks of tungsten mineralisation are relatively siliceous and potassic with appreciable quantities of modal muscovite, biotite and normative corundum and closely resemble the "S-type granite (derived from partial melting of pelitic/psammo-pelitic sediments).

2. Quartz-topaz-tourmaline-fluorite assemblage can indicate a wolfram mineralisation. Wall rock alteration, greisenisation, silicification, sericitisation, tourmalinisation, kaolinisation, fluoritisation and propylitization, when carefully studied, yield clues to mineralisation.

3. Plumesitic (high alumina) leucogranites are usually tungsten bearing. A reliable threshold value for a mineralised granite is a minimum Pb/Sr ratio of 4. A relatively high abundance of Sn, F, Li, Rb, Zn, Cu, Pb with increased Pb/Sr; Li x 10^3/F and Ba/Sr ratios and depletion of Sr, Ba, Zr and lower values of K/Pb, Ba/Pb and Zn/Sn are indicative of mineralisation.

4. TiO_2/Ta ratio varies systematically with differentiation trends in a whole rock. Increased concentration of F in highly differentiated granite rocks are noticed with tin-tungsten mineralisation.

5. Mineralised granite may be identified from a barren one by alumina/alkali ratio in the whole rock with threshold value of 1.4 and also by plotting mol. ($Al_2O + K_2O + \frac{1}{2}CaO$) ratio greater than 1.1 and by plotting normative corundum percentage against differentiation index, normative quartz percentage and Si-Niggli values, respectively.

7. NICKEL

The name nickel was derived from abbreviation of Swedish "Kopparnickel" means "false copper". The nickel has been known since the ancient times. The most ancient coins with an admixture of nickel are those of the Bactrian Kingdom, which existed approximately 200 B.C. The most of the world resources of nickel are centered in Canada, Cuba, New Caledonia and Russia. Indian requirements of nickel as well as its alloys were met through imports. A small quantity of nickel sulphate used for plating is produced during copper refining by HCl at Ghatsila smelter in Bihar. The world resources of nickel metal have been estimated as 121 million tonnes. The world mine production of nickel metal was 910 thousand tonnes (1989), the principal producers were USSR, Canada, New Caledonia, Australia, Indonesia and Cuba.

Mineralogy

Pure nickel is silver white and has a metallic luster. It is malleable and ductile, and does not oxidise in air even when heated upto 500°C. Nickel does not occur in native state. The important nickel ores with their chief characters are given in the Table 4.22 below.

Table 4.22—Principal Nickel Ores

Ores	Chem. Com.	Chief Characters
1. Pentlandite	(Fe, Ni)S	Bronze yellow with black streak, metallic, H = 3.5-4, sp. gr. = 5.0
2. Millerite	NiS	Brass yellow to bronze yellow with greenish black streak, metallic, usually in capillary crystal, H = 3-3.5, sp. gr. = 5.3-5.6
3. Niccolite	Ni As	Pale copper red with pale brownish black streak, metallic, H = 5-5.5, sp. gr. = 7.3-7.6
4. Garnierite	Hydrated nickeliferous magnesium silicate	Apple green to nearly white; amorphous, soft and friable; adheres to tongue; H = 3-4, sp. gr. = 2.2-2.8

Pure nickel is obtained by reduction of its oxides or by the Mond process which consists of the formation of volatile nickel carbonyl produced by passing carbon monoxide over heated nickel oxide, and the dissociation of this compound at a higher temperature into nickel and carbon monoxide, which can be used again. The nickel-copper sulphide ores are first roasted to release sulphur and then smelted to a Ni-Cu-Fe matte, which is bessemerised to a matte of 75 to 80 per cent Cu-Ni. Some matte is used directly to make Monel metal, the rest being specially smelted to separate copper and nickel sulphides, the latter being roasted, reduced with carbon, and electrolytically refined to pure nickel.

Uses

The main use of nickel is in the form of alloys with other metals e.g. German silver (an alloy of copper, nickel and zinc), Monel metal (an alloy of copper and nickel), nickel brasses, nickel bronzes etc. The manufacture of nickel steels and nickel cast-iron absorbs the largest proportion of nickel produced. Nickel imparts to its alloys greater hardness, toughness, tensile strength, lightness and anti-corrosion, electrical and thermal qualities. It is, hence, preferred for many machines, tools, shafts, automobiles, aeroplanes, ships, railways, armour plates etc. It is also employed in coinage, electro-plating (nickel plating), and certain storage batteries. Several of its salts are used in chemical industry.

Mode of Occurrence and Origin

The nickel ores exploited till the end of 19th century were of sulphide type and of endogenous origin. A new type of deposit was discovered in New Caledonia in 1863 by J. Garnier : the nickel ore was associated with weathering mantle of ultramafic rocks (laterite), and the main minerals of the richest nickeliferous zones were green hydrated silicate minerals which were called garnierites. The deposit still represents the important reserves of nickel in the world. The nickeliferous leterite deposits are formed by in-situ weathering of ultramafic igneous rocks. Thus, the nickel deposits occur in two forms :

(i) as nickel-copper sulphide deposits formed by replacement or magmatic injection, and

(ii) as residual concentrations of nickel silicates from weathering of ultramafic igneous rocks.

These ores are found invariably with a few exceptions in deep seated magmatic rocks. That is, they are restricted to basic and ultrabasic rocks, though ore-veins may also be found in other rocks, situated nearby. Some nickel bearing rocks contain considerable quantities of sulphur, which gives rise to nickel sulphides. Where the nickeliferous magma contained no sulphur, nickel combined with silica, and when it congealed, a rock containing nickeliferous silicate minerals formed. Where a molten mass consists of various metals, sulphur and silica constitute not one but two hot immiscible liquids. It is, hence, the sulphide and the silicate ores of nickel are not found together. When the magma chamber cools, the molten sulphide mass begins to crystallize later than the molten silicate and, generally, appear as impregnations disseminated through the entire rock. Occasionally, the rock is so suffused with fine impregnations that, in places, they form solid sulphide pockets and patches. The size of sulphide bodies ranges from a few cu. cm to several tens of cubic metres. In shape they may be irregular bodies, elongated lenticular bodies and lastly, tabular sharply defined veins. Massive sulphides accumulate at the bottom of a magma chamber to form bottom deposit followed upward by disseminated ores, unless the sequence is otherwise distorted by orogenic forces.

The nickel deposits occurring in the weathered crust (laterite) are of the aerial, the linear fissure and the contact karst types. In the mantle of waste of aerial type, a gradual transition from the uppermost products of weathering to the primary rock can be traced and may be divided into three distinct zones :

(i) the uppermost ochre zone, derived from complete decomposition of ultramafites, does not contain any nickel;

(ii) the middle zone of brown and green clays, having a complex composition includes iron, magnesium, nickel and silica; they are generally yellow or green due to nickel and iron in the silica carrying solution; usually 1-6 metres thick, green nontronite clays containing veins of prase opal (nickel opal) and chrysoprase (nickel chalcedony) are good nickel zones; and

(iii) a zone of leached serpentine consisting of loose, partly weathered rock which has not yet been decomposed into clay; the leached serpentine is only nickeliferous in its upper part and further down where carbonates begin to appear, the amount of nickel decreases and finally it becomes barren.

The mineral goethite appears in the upper layer and the serpentine in the lower layer. In the intermediate transitional layer, smectite and chlorite are found. The nickel is concentrated with goethite in the upper layer; with serpentine in the lower layer and with smectite in the intermediate layer to form hydrous magnesium - nickel silicate mineral - garnierite. In the fissure type of crust of weathering one finds ochres, nontronite clays and leached serpentine, which do not follow a definite pattern, but are often intermixed and grade into each other. Nickel precipitates at a depth of not less than 2-4 metres :

(i) in ochres, green clays and ocherous porous leached serpentine;

(ii) in chalcedony stringers; and

(iii) as streaks of nickel minerals of the garnierite type.

In nickeliferous crust of weathering of the contact karst type, the nickel deposits are found at the contact of serpentine and limestone. The limestone has the property of precipitating nickel from its solution.

Distribution

The nickel in India is found associated with laterites of Orissa; copper-ores of Andhra Pradesh, Bihar, Himachal Pradesh, Manipur and Rajasthan; ultrabasic rocks of Andaman, Gujarat and Maharashtra; and sulphide-ores of J & K, Kerala, Karnataka and Tamil Nadu. It is also found in weathered talc-schist and serpentinite of Bhitardari region, Singhbhum district, Bihar. The details of some important occurrences are tabulated below.

Table 4.24—Details of Important Occurrences of Nickel Ores in India

State	Locality	Details of Mineralisation	Remarks
Bihar	Ukkam and Dhoba hills of Bhitardari regions, Singhbhum district	Nickel with cobalt occurs associated with talc-schist and serpentinite extending for about 12 km. The nickel percentage in the ore varies from 0.3 to 0.8.	The economic extraction of nickel is not feasible for the present. The area will assume importance with development of technology.
	Singhbhum Copper belt	It is found associated with copper-ore.	Hindustan Copper Limited produces nickel sulphate while refining for copper at Ghatsila Smelter, Singhbhum.
Kerala	Shreekandapuram area, Cannore district	Associated with partly weathered chlorite-tremolite schist. The highest nickel value recorded was 1,400 ppm, though the general nickel value in the zone ranges from 170 ppm to 300 ppm.	
Nagaland	Pokhpur, Tuensang district	Nickel shows maximum concentration upto 1.5% in the goethite matrix within a tabular body of magnetite, 1000 m x 300 m x 5-15 m in dimension, underlain by ultramafics and overlain by sedimentaries of probable Mio-Pliocene age. 4.45 million tonnes of nickel ores have been estimated with an average grade of 0.63% nickel.	

(Contd.)

State	Locality	Details of Mineralisation	Remarks
Orissa	Sukinda area, Cuttack district	The nickel-ores are of lateritic type related to intense weathering and limonitisation of silicified ultramafic rocks, intrusive into Iron Ore Group of metasediments. The ores are generally associated with chromite bodies and are classified as : (i) High grade ore with 0.9% nickel and above, (ii) Medium grade ore containing 0.7 to 0.89% nickel, and (iii) Low grade ore with 0.5 to 0.69% nickel. The over all reserves of 65 million tonnes of Ni-ores with an average nickel content of 0.85% have been estimated in the Kansa and Sarubil-Sukerangi sectors. The high grade ores with an average nickel content of 1.15% amount to 31 million tonnes.	
	Simlipal area, Mayurbhanj district	Nickel-ore is concentrated within weathered ultramafic rocks, in the laterites and soil cappings. The mineral occurs in silicate form i.e. garnierite. The nickel percentage in the rock varies from 0.55 to 1.1. Possible reserves of 9.7 million tonnes with an average grade of 0.97% Ni, and 17.2 million tonnes with an average grade of 0.79% Ni are found in Bhilapoga sector of Gurguria block. Prospective ore zones are also in Nawana block. The whole area covers 12 sq. km.	
	Kendujhar	Low grade nickel-ores associated with altered ultramafite with average grade of 0.2 to 0.7% have been recorded.	
Tamil Nadu	Chalk Hills, Salem district	Nickel is enriched in limonitized and serpentinised ultramafic rocks over an average area of 0.4 sq. km. In Sandalwood-Chimney area, eleven anomalous zones with 0.3 to 0.7% Ni have been demarcated.	

Reserves

A total reserve of 294 million tonnes nickel ores with 0.2 to 0.9% nickel has been estimated as on 1.4.90. Since the techno-feasibility of extracting nickel from these ores has not been established, the entire reserve has been classified as conditional (sub-marginal) resources. These conditional resources are distributed in Bihar (9 million tonnes) and Orissa (285 million tonnes).

Prospecting Guides

1. Sulphide nickel-ores occur in basic and ultrabasic rocks, though ore-vein may also be found in other rocks situated nearby.
2. Nickel silicate ores (Garnierite) are found associated with weathering mantle of ultramafic rocks (laterite).
3. Sulphide and silicate nickel ores are formed under entirely different natural conditions and never occur together. They form deposits of different types, which need separate prospecting and mining techniques.
4. The presence of nickel bloom (nickel green) on the surface of rocks points to the presence of nickeliferous minerals. These are hydrated and oxidized nickel minerals and are nickel indicators. Nickel green greatly resembles common copper green, the latter is, however, slightly bluish in colour. The nickel green always has a grass green or apple green tint and is never found with blue minerals unless copper minerals are associated.
5. The field test of identifying nickeliferous mineral is to grind a half tea-spoonful of specimen, to place in a glass test tube and a small quantity of aqua-regia poured in. Five to ten minutes later the solution is carefully shaken and 15-20 drops of ammonia and then 5-10 drops of 1% solution of dimethylglyoym added to it. The solution is brought to a boil in the flame of a burner or candle, after which it is allowed to cool. A caramine red precipitate will fall out of the solution, in case there is any nickel present.

8. COBALT

Cobalt is a metal of strategic importance due to its ability to impart hardness and corrosive resistance to alloys at high temperature. There is no domestic production of cobalt ore or metal and, hence, the demand is met through imports. The total world production of cobalt metal amounts to about 25 to 30 thousand tonnes a year of which Zaire is the largest producer, followed by Zambia, Canada, Finland and Japan. In 1989 the metal production was 25,700 tonnes against 30,100 tonnes in 1987. The world reserves of cobalt metal has been estimated as 8.34 million tonnes.

Mineralogy

Cobalt metal is grey with reddish tinge, malleable, sp. gr. 8.9, magnetic, has a high melting point between 1500°C and 1800°C, and can be produced by the reduction of its oxides by carbon or aluminium. It closely resembles nickel in appearance. It is also recovered as by-product of other ores, chiefly zinc, copper and silver. The commercial minerals of cobalt are tabulated below :

Table 4.25—Chief Minerals of Cobalt

Name	Chem. Comp.	Chief characters
Linnaite	Co_3S_4	Steel grey mineral, tarnishing coppery red, associated with pyrite, chalcopyrite etc.
Cobaltite	CoAsS	Silver white with a reddish tinge, greyish black streak, metallic, H = 5.5, sp. gr. = 6-6.3
Smaltite	$CoAs_2$	Tin white to steel grey, greyish black streak, metallic, H = 5.5-6, sp. gr. = 6.4
Absolite	Co-Oxide with oxide of Mn sometimes upto 50%	Amorphous, earthy with black or blue-black colour and streak, shining and resinous

The cobalt minerals weather on their exterior to pinkish cobalt "blooms" (Erythrite, hydrated arsenate of cobalt) - the cobalt indicators. The minerals are roasted and the residue is treated by a wet chemical process leaving a pure oxide of cobalt in the marketable form.

Uses

Cobalt is used as the major alloying element in making special steels and desired specially for magnet steel (35% of use), rustless and corrosion resisting steel, high speed steel, stellite steel for metal cutting, disc and valve steel, welding rods, temperature resisting and carbide-type alloys.

Beryllium-copper alloy containing cobalt has high electrical conductivity. Tungsten carbide used for many important operations like drilling and mining, are bonded by cobalt powder. The alloy stellite consists of 70 to 90% cobalt, 10-25% chromium, with small amounts of other metals. It is used in aerospace industry, because of its ability to maintain strength at high temperature. Cobalt is used to certain extent in electroplating, and finally, its compounds are extensively used in the mauufacture of pigments especially blues, employed in glass, enamel and pottery industries.

The conventional uses of cobalt are in the preparation of magnet and cobalt chemicals.

Mode of Occurrence and Origin

Cobalt deposits occur mainly as veins carrying smaltite and cobaltite, as cobaltiferous pyrrhotite, and as weathering product of basic and ultrabasic rocks. The last mentioned origin is similar to that of garnierite deposit of nickel (Singhbhum, Bihar). The sulphide and arsenide deposits are formed by replacement or magmatic injection. The oxides of cobalt occur as residual concentration from weathering of deep seated igneous rocks-basic and ultrabasic rocks.

Distribution

Cobalt occurrences are reported from several places in the country, the notable among them being Singhbhum district of Bihar, Kendujhar and Cuttack districts of Orissa, Jhunjhunu district of Rajasthan, Tuensang district of Nagaland, and Jhabua and Hoshangabad districts of Madhya Pradesh. Slag form of Ghatsila complex contains copper, cobalt and nickel, and attempts are being made to extract the metal. In the Ukkam and Dhoba hills of Bhitardari Singhbhum district, Bihar, cobalt associated with nickel has been found in talc-schist and serpentinite. In Orissa (Sukinda area) too, it is found associated in minor amounts with nickel in ultramafites. In Rajasthan it is associated with copper ore.

The conditional resources of cobalt estimated in Singhbhum district, Bihar alone are 9 million tonnes as on 1.4.1990.

Production

There is no production of cobalt for the present. A process has been developed by HZL for recovery of cobalt as cobalt sulphate and cobalt metal from B-cake, which is a waste-residue generated during zinc production. The process involves steps like leaching, roasting, solvent extraction, electro-winning etc. Based on above, cobalt recovery plant is being proposed to be set up. A hydrometallurgical method for the recovery of cobalt and nickel from ICC converter slag of Ghatsila Smelter containing 0.5 to 1% cobalt has been perfected by the HCL through bench scale tests. Erection of a dryer for processing leach cake was in progress.

9. TITANIUM

The titanium minerals of economic importance are ilmenite and rutile which constitute important exportable commodities. Japan, Germany, France, USA and UK remained the main importing countries. The world reserves of ilmenite are placed at 403 million tonnes with major reserves in Australia, Canada, India, Norway, Republic of South Africa, USA, Russia and China; while those of rutile have been estimated at 136 million tonnes. The world annual production of ilmenite and rutile is placed at 8.5 million tonnes and 0.457 million tonnes respectively (Production, 1989).

Mineralogy

Titanium has not been found in free state in nature. It is a greyish metal and resembles tin in its chemical properties. The important titanium minerals with their chief properties are listed in Table 4.26 below.

Table 4.26—Important Titanium Minerals and their Chief Characters

Minerals	Chem. Comp.	Chief Characters
Ilmenite	FeO TiO$_2$ (Hexagonal)	Iron black; black to brownish black streak; submetallic; conchoidal fracture; H = 5-6; sp. gr. = 4.5-5
Rutile	TiO$_2$ (Tetragonal)	Brownish red, yellowish or black; pale brown streak; sub admantine lustre; often knee shaped twins; H = 6-6.5; sp. gr. = 4.2
Anatase	TiO$_2$ (Tetragonal)	Brown, indigo-blue or black; streak colourless; perfect cleavage; admantine lustre; H = 5.5-6; sp. gr. = 3.8-3.95
Brookite	TiO$_2$ (Orthorhombic)	Hair brown, reddish or iron black; streak colourless; metallic, brittle; H = 5.5-6; sp. gr. = 4

Uses

The main uses of titanium ores are as follows :

1. **Alloys :** Certain special alloys with iron or with iron and carbon are produced. Ilmenite is smelted in the electro-furnace direct for making ferro-titanium alloys. Titanium carbide is a quite hard material and can be used for cutting tools. Ferro-carbon-titanium is alloyed with steel for high speed tools. It is also alloyed with chrome-steel.

2. **Pigments :** Titanium oxide, called titanium white, is used for imparting an ivory tint to artificial teeth, and as a yellow glaze in pottery manufacture. It has opacity, twice that of zinc-oxide and three times that of lead oxide. This can be mixed with other pigments without decreasing its opacity. Thus, tita-

nium paints are pure titanium oxide or combined with other paint compounds. These are also used in toilet articles, linoleum, artificial silk, white inks, coloured glass, and for dying leather and cloth.

3. **Electrodes/Synthetic rutile** : Ilmenite is also consumed in the manufacture of electrodes and synthetic rutile. The bulk of ilmenite is consumed in production of synthetic rutile.

4. **Other uses** : In the form of chloride, titanium is used for removing colours from cloths, and in the form of tetrachloride for making smoke screens and sky writings.

Mode of Occurrence and Origin

The titanium minerals, ilmenite and rutile, occur as accessory minerals in rocks or as magmatic segregations in veins. Ilmenite is also found in metamorphic rocks, in magmatic deposits associated with magnetite and minor amounts of hematite. Ilmenite and rutile may also occur as disseminated replacement deposits.

The above rocks upon weathering give rise to beach and stream placers, rich in titanium minerals i.e. ilmenite, rutile etc. Monazite and zircon are the common associates in the placer deposits. The exploitable deposits of both ilmenite and rutile are mostly of the detrital characters, such as beach sand.

Distribution and Reserves

The titanium minerals, ilmenite and rutile, constitute ingredients of beach sand deposits, found right from Ratnagiri in the west to Orissa coast in the east. But the concentrations of the above minerals are in the following three well defined zones :

(i) Over a stretch of 22 km between Neendakara and Kayamkulam, Quilon district, Kerala;

(ii) Over a stretch of 6 km from the mouth of Valliyar river to Colachel in Manavalakurichi and a little beyond in Kanyakumari district, Tamil Nadu;

(iii) On Chattarpur coast stretching for 18 km over an area of 26 sq. km in Ganjam district, Orissa.

Other small occurrences of ilmenite and rutile are the beach sands of Are-Kalbadevi areas of Ratnagiri coast, Maharashtra.

The Atomic Mineral Division of the Department of Atomic Energy has placed the total in-situ reserves of ilmenite and rutile at 146.31 million tonnes and 8.20 million tonnes respectively. The recoverable reserve figures of ilmenite and rutile up-dated by IBM as on 1.4.1990 are given as 86.75 and 5.18 million tonnes, respectively. The reserves of leucoxene are 74,000 tonnes and those of titaniferous magnetite about 10 million tonnes.

The conditional resources of ilmenite and rutile falling in probable category were of the order of 46.98 million tonnes and 2.56 million tonnes, respectively. The total titanium mineral resources, including those of anatase and titaniferous-magnetite, were 56.52 million tonnes.

Production

The mining of beach sands is being carried out by the Indian Rare Earths Ltd. (IRE), a Govt. of India Undertaking, and Kerala Minerals and Metals Limited (KMML), a Kerala Govt. Undertaking. The mining is both manual as well as by hydraulic dredging. The beach sand after sun-drying is fed to electromagnetic and electrostatic separators in a processing plant for physical separation of various mineral constituents. IRE has processing plants, one each in Kerala, Tamil Nadu and Orissa, while KMML has only one plant in Kerala. The details of the plants with the installed capacities are given below in the Table 4.27.

Table 4.27—Details of Processing Plants with Installed Capacity (in Tonnes)

Location of Plants	Installed Capacity in 1989-90	
	Ilmenite (tpy)	Rutile (typ)
IRE Plants		
Chavara, Quilon district, Kerala	80,000	9,000
Manavalakurichi, Kanyakumari district; Tamil Nadu	76,000	29,000
Orissa Sand Complex, Chhatarpur, Orissa	2,20,000	10,000
KMML Plant		
Chavara, Quilon district, Kerala	27,000	2,400

The ilmenite and rutile productions in 1989-90 were about 257 thousand and 9.6 thousand tonnes, respectively. Kerala continued to be the principal producing state (37% ilmenite and 58% rutile), followed by Orissa (34% ilmenite and 23% rutile) and Tamil Nadu (29% ilmenite and 19% rutile). The production in 1990-91 further rose to above 300 thousand tonnes of ilmenite and 13 thousand tonnes of rutile.

4.3 NON-FERROUS AND ALLIED METALS

1. COPPER

Introduction

The demand for copper metal has shown progressive increase under the Five Year Development plans of the country and the requirements are being met largely (about 70%) by exports from foreign countries. The world refinery production of primary copper metal was about 10.5 million tonnes in 1990, to which contribution of India was about 40,000 tonnes i.e. 0.4% with 30th position. The principal producers were Chile, USA, USSR, Canada, Zambia, Zaire, Poland and Peru which together accounted for more than 70% of world mine production. The world reserve of copper metal have been placed at 566 million tonnes.

Mineralogy

The chief ores of copper along with their chemical compositions, copper percentages and physical properties are given in the Table **4.28** below.

Table 4.28—Chief Ores of Copper

	Chem. Comp.	Copper %	Physical Properties
Sulphides			
1. Chalcopyrite	$CuFeS_2$	34.5	Brass yellow with greenish black streak, H = 3.5-4, sp. gr. = 4.1-4.3
2. Bornite (peacock ore)	$Cu_5 FeS_4$	63.3	Coppery red and iridescent with pale greyish black streak, metallic lustre, H = 3, sp. gr. = 4.9-5.4
3. Chalcocite	$Cu_2 S$	79.8	Blackish lead grey, metallic, H = 2.5-3, sp. gr. = 5.5-5.8
4. Covellite	CuS	66.4	Indigo blue, H = 1.5-2, sp. gr. = 4.6
5. Enargite	$Cu_3As S_4$	48.3	Greyish black to iron black, metallic, H = 3, sp. gr. = 4.4
6. Tetrahedrite (grey copper)	$Cu_8 Sb_2 S_7$	52.1	Steel grey to iron black, metallic, H = 3-4.5, sp. gr. = 4.5-5.1
Oxides			
7. Cuprite	Cu_2O	88.8	Different shades of red with shining brownish red streak, admantine or submetallic to earthy lustre, H = 3.5-4, sp. gr. = 5.8-6.15
8. Tenorite	CuO	79.8	Mostly in black powder, also in dull black masses, H = 3-4, sp. gr. = 6.25

(Contd.)

	Chem. Comp.	Copper %	Physical Properties
Carbonate			
9. Malachite	$CuCO_3 \, Cu(OH)_2$	57.3	Bright green encrustations, H = 3.5-4, sp. gr. = 3.9-4
10. Azurite	$2CuCO_3Cu(OH)_2$	55.1	Azure blue, H = 3.5-4, sp. gr. = 3.7-3.8
Silicate			
11. Chrysocolla	$CuSiO_3.2H_2O$	36.0	Bluish green or sky blue, white streak when pure, H = 2-4, sp. gr. = 2-2.2
Element			
12. Native copper	Cu	100.0	Copper red colour with metallic shining streak, H = 2.5-3, sp. gr. = 8.8

Uses

Copper is one of the most essential of all non-ferrous metals being used by man since before the "Bronze age". Its various uses due to its specific properties are summarised in the Table 4.29 below.

Table 4.29—Uses of Copper

	Specific properties
1. **Domestic use**	
Ornaments, vessels	Soft metal, can easily be shaped
2. **Industrial**	
(a) Electrical industry as conductors, windings and other components of chemical machines	High electrical conductivity.
(b) Steam engine automobiles, tractors, other machines	Unique technical properties not found in other metals
3. **Alloys**	
Bronze (88% Cu, 10% Sn and 2% Zn); Gun metal (90% Cu and 10% Sn); Duralumin (95% Al, 4% Cu and 1% Si + Mg)	Readily mixes with certain metals
4. **Chemicals**	
Chloride is used as disinfectant and in chemical operations; sulphate (blue vitreol) is employed in printing and dyeing textiles, for preventing rots in timber and fungicide; certain copper salts are utilised for colouring glasses.	Varied properties of copper salts

Mode of Occurrence and Origin

Copper occurs in a variety of ways - as magmatic segregatious, in disseminated forms, in veins and lodes, in contact metamorphic deposits, in bedded deposits etc. The occurrence of porphyry copper deposits is mostly associated with stocks or chonoliths intrusions of monzonite or diorite prophyry of early Tertiary age. These deposits are low grade and high volume ore, have greater horizontal than vertical dimension, are overtain by leached cappings and have been subjected to supergene sulphide enrichment. In India copper lodes occur mostly in veins, stringers, patches, disseminated forms, fracture and cleavage fillings etc., associated with different types of rocks, mostly Dharwar and Cuddapah ages. Associated with copper ores are frequently the other metals, namely lead, zinc, gold, silver, platinum, palladium, bismuth etc. which are recoverable at one or other stages of their treatment.

Copper deposits have largely originated from hydrothermal solutions, though various theories for the origin of copper ores have been propounded. The genesis of almost all known copper deposits is closely associated with igneous rocks, having invariably migrated and concentrated by uprising solutions penetrating the earth crust. The chemical solutions in their travel collected other metals and it is not uncommon to find gold, silver and other rare metals associated with copper ores.

Distribution

The geological and geographical distributions of copper ore in India are given in the Tables 4.30 and 4.31 below.

Table 4.30—Geological Distribution of Copper Ore in India

Formation	Deposits	Geological Details
PRECAMBRIAN		
Bihar		
Archaean quartz-chlorite-biotite schist, metamorphosed basic rock and soda-granite	(i) Singhbhum copper belt	It is localised in a shear-zone moulded along the northern and north-eastern margin of Singhbhum granite massif. The metalavas and quartzite of Dhanjori basin which overlie the granite and dip towards north and north-east, are overlain by high grade metamorphics - garnetiferous mica-schist and kyanite-quartz rock of Chaibasa stage.
Tremolite-actinolite-schist, calc-granulite and amphibolite	(ii) Hesatu-Belbathan belt	The mineralised rocks occur as lenses or pockets within Chotanagpur granite-gneiss.
Rajasthan		
Pre-Aravalli, Aravalli and Delhi groups of rocks	(i) Khetri belt (ii) Pur-Banera-Bhinder belt (iii) Other detached occurrences	The copper mineralisations are along favourable structural zones related to different orogenic movements.

(Contd.)

Formation	Deposits	Geological Details
Madhya Pradesh		
Quartz reefs in Bundelkhand Granite Complex, overlain by the Precambrian meta-sedimentaries of Chilpighat formation with an erosional unconformity	Malanjkhand copper deposit, Balaghat district	The copper mineralisation in steep dipping wide linear body of quartz reefs is controlled by structure like shear and fracture planes.
Karnataka		
Dharwar schist belt and occasionally associated with basic dykes and other intrusions in Peninsular Gneissic basement	Chitradurga and Hassan districts	In Chitradurga the mineralisation is along shear fractures in quartz-veins, traversing sheared greenstone.
Andhra Pradesh		
Archaean schists and gneisses	Mailaram, Khamman distt., and Gani area, Kurnool district	The mineralisation is in quartz-reef and quartz-chlorite-schist at junction of schists and granite.
Cuddapah Super Group, calcarious quartzite and dolomite of Cumbum formation	Agnigundala belt, Guntur district	It is located in the north-eastern corner of Cuddapah basin.

Table 4.31—Geographical Distribution of Copper Ore in India

State	Locality/Area	Remarks
Bihar	(i) Singhbhum copper belt : The potential areas are Pathargora, Rakha, Roam-Sidheshwar, Ram Chandar Pahar, Tamapahar and Turamdih.	The belt extends from Duarpuram (24°46' : 85°34') in the west to Barahagora (24°16' : 86°43') in the south east for over 128 km with a width ranging upto 5 km.
	(ii) Hesatu-Belbathan belt : There are four dozen occurrences covering the whole belt, the important of them are at Baraganda, Chandio, Jhalakdiha, Ganganpur, Damgi, Toolsitanr, Pirrah and Baghmari.	The belt extends in WNW-ESE direction for about 250 km having a width of about 50 km from Hesatu (Hazaribagh district) in West to Belbathan (Godda district) in the east.
Rajasthan	(i) Khetri copper belt, Jhunjhunu district : The potential areas are Madhan Kudan, Kalihan and Chanmari.	The Khetri belt runs for about 80 km in the north east part of the state.
	(ii) Pur-Banera-Bhinder belt : The potential deposit is at Dariba in Alwar district.	It runs for about 135 km in south-eastern part of the State and includes four smaller belts : Pur-Banera, Jasma-Rewara-Karo, Rajpura-Dariba-Bethuni and Wari-Bhopal-Sagar.

(Contd.)

State	Locality/Area	Remarks
	(iii) Detached occurrences, namely at Pali, Sirohi, Udaipur, Charu and Nagaur	
Madhya Pradesh	Malanjkhand, Balaghat district	It extends for 2.6 km in arcuate form.
Karnataka	Ingaldahu in Chitradurga district and Kalyadi in Hassan district	—
Andhra Pradesh	Mailaram, Khammam district; Gani, Kurnool district; and Bandalamottu, Nallakonda and Dhokonda, Agnigundela belt, Guntur district.	—

Reserves

The all-India recoverable reserves of copper-ore as on 1.4.1990 are placed at 324.79 million tonnes equivalent to 3.21 million tonnes of metal content. The reserves with 1% and above Cu as on 1.4.1990 in Bihar, Madhya Pradesh and Rajasthan, which are known to be rich in this ore, are about 110, 96 and 91 million tonnes, respectively. The all India conditional resources of copper-ore are 838.11 million tonnes (3.43 m. tonnes copper metal) and prospective resources are placed at 769.9 million tonnes of copper-ore.

An additional reserves of 434 million tonnes with 0.4-0.5% copper have been estimated at Ladera Sakkun, Jaipur district and Bedwelki-pal, Udaipur district, Rajasthan during 1990-91. In Hesatu-Belathan belt, Bihar further additional reserves of copper-ore along with zinc and lead (multimetal deposits) have been proved, and the work is still in progress.

Production and Consumption

The production of copper-ore, after maintaining an increasing trend upto 1987 (5.14 million tonnes) dropped slightly in 1988 when the production totalled 5.06 million tonnes. The production figures for 1985 and 1986 were 4.21 and 4.46 million tonnes, respectively, while those for 1989-90, 1990-91, 1991-92 and 1992-93 were 5.20, 5.26, 5.21 and 5.21 million tonnes, respectively.

Madhya Pradesh continued to be the leading producing state accounting for 36% of the total production (1988), followed by Rajasthan (35%) and Bihar (27%). Small amounts of ore are produced in Andhra Pradesh and Karnataka. The average grade of ore treated was 1.19% copper.

Copper metal is produced at two smelters of Hindustan Copper Ltd. (HCL) at Ghatsila in Bihar and Khetri in Rajasthan, having annual installed capacities of 16,500 and 31,000 tonnes of copper metal, respectively. Recovery of other metals like gold, silver, selenium, tellurium etc. is also made. Besides, there are plants for production of sulphuric acid and fertilizers (P_2O_5).

Mining and Metallurgy

Altogether ten mines are being operated at present by Hindustan Copper Ltd. These include five mines at Mosaboni, Pathergora, Surda, Kendadih and Rakha, respectively in Singhbhum copper belt; three mines at Khetri, Kolhan and Chandmari in Khetri copper belt; one mine at Dariba, Alwar district, Rajasthan and one mine at Malanjkhand, Balaghat district, M.P. Besides, a small output is reported from Chitradurga district, worked by Chitradurga Copper Co. Ltd., and Kalyadi in Hassan district, worked by Karnataka Copper Consortium Ltd. Productions at Mailaram in Khammam district, Andhra Pradesh; Rangpo in Sikkim and one mine in Sundergarh district, Orissa, though in small amounts, are also being made.

The copper ore is brought to size (about 10 cm) and then passed on a secondary crusher to reduce the size further to about 9 mm. It is finally ground in Ball mill, and ore goes to froth floatation machines where the ore is concentrated to 98%, carrying about 24% copper. After filtering and drying, the ore is roasted to remove sulphur and charged into reverberatory smelting furnace which produces a matte carrying 50% copper and a waste slag. The matte is treated in converters where the sulphur and iron is removed leaving a low grade or blister copper which then passes to a refining furnace for production of refined copper in ingot forms.

Prospecting Guides

The following few guides may be gainfully utilised in prospecting for copper :

(i) **Nature of host rocks :** Greyish-green compact sandstone, composed of oily quartz, grey or pink felspar and darker fragments of igneous rocks; secondary quartzite; skarn rocks; acid porphyries; basic plutonic rocks like diabase and spilite are favourable host rocks for copper mineralisation.

(ii) **Nature of alteration of host rocks :** Diabase altering to sericite and chlorite-schists, acid porphyries altering to sericitic quartzite or secondary quartzite, skarns showing alteration of its pyroxene to actinolitic amphibole, and sanstone showing enrichment of chlorite and not of hydrous iron-oxide are the usual sites.

(iii) Various alterations of copper-ores like iron-hat or gossan, presence of limonite, oxidised copper minerals like malachite and azurite, and quartz with inclusions of copper minerals are indicative of copper mineralisation.

2. LEAD AND ZINC

Lead and Zinc ores occur, generally, together in close association and so they are dealt here at one place.

India's present need for lead and zinc is met mostly (about 60%) by imports from foreign countries. Efforts by the concerned government and semi-government departments are being made since the beginning of independence to make the country self-sufficient in these important basic industrial minerals. With the growth of civilisation the demand of these metals is also increasing. The total world productions of lead and zinc metals in 1990 were about 5.7 and 7.4 million tonnes, respectively to which India's contributions were about 0.7% (0.040 m. tonnes) lead and about 1% (0.070 m tonnes) zinc. U.S.A., Russia, Japan, Germany, U.K., Canada, France, Australia, Belgium and China together produce about 70% of the world production of refined lead and zinc metals.

Mineralogy

Lead

Lead (Pb) occurs in native state, but it is quite rare. The metal is bluish grey in colour and shows on its fresh surface a bright metallic lustre which quickly oxidises on exposure to air. It is so soft that it can be scratched with finger-nail and shows a black streak on paper. It is, however, quite heavy, the sp. gr. being 11.34. The chief ores of lead with their principal characters are given below in Table 4.32.

Table 4.32—Chief Ores of Lead

Ores	Chem. Comp.	Characters
Galena	PbS	Massive, granular, cubes and octahedral forms, perfect cubic cleavage, lead grey colour with the same streak, metallic lustre, H = 2.5, sp. gr. = 7.4-7.6
Cerussite	$PbCO_3$	Prismatic, often cruciform twinned crystals, radiating, granular, massive, compact and sometimes stalactitic, white or greyish colour with colourless streak, admantine lustre, H = 3-3.5, sp. gr. = 6.55
Anglesite	$PbSO_4$	Prismatic, also massive, occasionally stalactitic, white, sometimes blue, grey, green or yellow tint, usually admantine lustre, H = 2.5-3, sp. gr. = 6.3-6.4

The lead ores were first partly roasted or calcined and then smelted in reverberatory or blast furnaces. Most lead ores contain silver, and this metal is obtained from lead by cupellation, repeated melting and crystallisation, alloying with zinc or by electrolytic process. The presence of zinc ores causes difficulties in smelting the lead ores and so mechanical separation by jigs floatation process etc. of the two minerals is resorted to.

Zinc

Zinc (Zn) is a bluish white brittle metal with a sp. gr. of about 7.15 and melting at 419°C. It may be rolled into sheets or drawn into wire between 100°C and 150°C, but reverts into a brittle condition at 300°C and may be readily powdered under hammer. It is soluble in dilute acids. The chief ores of zinc with their chemical composition and main characters are given in the Table 4.33.

Table 4.33—Chief Ores of Zinc

Ores	Chem. Comp.	Main Characters
Sphalerite (Zinc blende)	ZnS	Tetrahedra and rhombdodecahedra forms are common, also occurs as massive, perfect cleavage, usually black or brown in colour with white or reddish brown streak, resinous to admantine lustre, H = 3.5-4, sp. gr. = 3.9-4.2
Smithsonite	$ZnCO_3$	Commonly in massive, reniform, botryoidal or stalactitic, encrusting, granular or earthy forms, perfect rhombohedral cleavage, white, greyish, grenish or brownish white colour with white streak, vitreous lustre inclining to pearly, H = 5.5, sp. gr. = 4-4.5'

Zinc ores are roasted or calcined in admixture with coal or coke, and the zinc oxide produced is reduced to metal which being volatile, distils and is condensed. The sulphuric acid is produced as a by-product. Some amount of metallic zinc is obtained by electrolytic process. The association of its ores with galena is objectionable from the point of view of damage to zinc smelter and so the efforts are made to separate them before smelting.

Uses

The uses of lead and zinc and their compounds are manifold and rank next to copper as essential non-ferrous metals in the modern industry. The chief uses of both metals and their compounds are tabulated below.

Table 4.34—Uses of Lead and Zinc

Lead	Uses	Zinc	Uses
Metal	Accumulators (storage battery), sheeting and piping, cable cover (sheathing), ammunition foil etc., alloys (pewter, solder, babbit metal, type metal, bronzes, antifriction and fusible metal)	Metal	Galvanising (coating iron), dry battery, tubes such as those of tooth paste, alloys (brass, German silver, white metal)
Compounds		*Compounds*	
Oxides	Pigments, glass making,	Oxides and	
(Red lead,	flux and rubber industry	Sulphides	Pigments
litharge)		Chloride	Soldering, preventing
Carbonate	Pigment		decay in wood
(White lead)		Sulphate	Dying, glue making etc.
Nitrate	Calico-dying and printing		
Arsenate	Insecticide		
Acetate	Medicine		

Mode of Occurrence and Origin

Lead and zinc ores occur in a number of ways. The chief modes of occurrence are as lodes or veins, as metasomatic replacements and contact metamorphic deposits, as cavity or joint fillings or as disseminations. The ores of lead and zinc (galena and sphalerite) are found in close association, and are oxidised in their upper parts into oxy-salts, of which the most important economically are cerussite ($PbCO_3$), anglesite ($PbSO_4$), smithsonite ($ZnCO_3$), zincite (ZnO), willemite ($ZnSiO_4$) etc. Most deposits of lead and zinc are confined to limestone, dolomite and other calc-magnesium rich rocks. In Zawar, Rajasthan, they occur in fine-grained dolomite and gritty conglomeratic dolomite with subordinate interbedded phyllite and quartzite as sheeted zones, veins, stringers and disseminations, forming lenticular bodies arranged in overlapping en-echelon pattern; while in the Hesatu-Belbathan belt, Bihar, they are limited to calc-silicate rocks. The deposits may possibly have formed by the hydrothermal solutions of igneous derivation, by descending surface waters or by ascending artesian meteoric waters.

Distribution and Reserves

The important deposits of lead-zinc are in the states of Rajasthan, Andhra Pradesh, Gujarat, Bihar, Orissa and West Bengal. The area-wise distribution of lead-zinc deposits with reserves and grades are given below in the Table 4.35.

Table 4.35—Distribution and Reserves of Lead-Zinc Ores in India

State	District	Area	Total Reserves (ores in million tonnes)	Grade
Andhra Pradesh	Cuddapah	Zangamarajupal Gollapale, Verikunta	8.319	0.64 to 8.99% Lead, 1.08 to 4.21% Zinc
	Guntur	Bandalamottu Dhukonda Peddagavalakonda Karempudi etc. (Agnigundala belt)		
	Prakasam	Nallakonda		
Bihar	Godda, Banka, Jamui, Deogarh, Hazaribagh and Giridih	Hesatu-Belbathan belt	0.55	Av. 2 to 5.50% Lead and Zinc
Gujarat	Banaskantha	Ambamata Chitrasani	7.775	Av. 4.50% Lead and Zinc
	Vadodara	Khandia		

(Contd.)

State	District	Area	Total Reserves (ores in million tonnes)	Grade
Maharashtra	Nagpur	Kolari and Tambekhani	1.00	Av. 3% Zinc
	Bhandara	Mendki		
Meghalaya	Khasi and Jantia Hills	Umpyrtha	0.118	Av. 0.40% Lead, 2.83% Zinc and 1.35% Copper
Orissa	Sundargarh	Sargipalli	2.630	Av. 6.69% Lead, and 0.32% Copper
	Mayurbhanj	Kesarpur		
	Sambalpur	Karmali		
Rajasthan	Ajmer	Ghughra Lohakan Sawar Tikhi	332.716 + 17.16	0.51 to 5.49% Lead, 0.40% to 7.30% Zinc
	Bhilwara	Rampur-Agucha Pur Banera (Gaderiakhera, Malikhera, Devpura, Dedwas, Samodi etc.)		
	Sirohi	Deri (Basantgarh-Delri belt)		
	Udaipur	Zawar area (Mochia, Balaria, Zawarmala & Baroi) Rajpur-Dariba		
Sikkim	East Sikkim (Rangpo)	Bhotang and Dikchu	0.950	Av. 1.09% Lead, 2.59% Zinc and 1.08% Copper
Tamil Nadu	South Arcot	Mamandar	0.679	Av. 1.13 to 1.17% Lead, 5.20% to 5.40% Zinc and 0.53% to 0.67% Copper
Uttar Pradesh	Pithoragarh	Askot	1.030	Av. 2.64% Lead, 3.95% Zinc and 2.32% Copper
West Bengal	Darjeeling	Gorubathan	3.250	Av. 4.01% Lead, and 4.23% Zinc
Total	**All India**		**359.017**	**With about 5 million tonnes of Lead metal and 18.5 million tonnes of Zinc metal**

Reserves

The all-India recoverable reserves (all grades) of lead and zinc ores as on 1.4.1990 are placed at 167.56 million tonnes with 2.42 million tonnes of lead metal and 7.88 million tonnes of zinc metal. The all-India conditional resources of lead and zinc-ores are 177.51 million tonnes.

Reserves of about 27 million tonnes of ore with lead 2.13 to 5.21%, zinc 2 to 7.11% and copper about 0.44% have been estimated in the recent years in Rajpura - Dariba and Bamnia areas, Udaipur district, Badalikhera area, Bhilwara district and Madarpura area, Ajmer district of Rajasthan and Bandalmottu area, Guntur district of Andhra Pradesh. Further, 3 million tonnes of ore with 1.2% lead and 12% zinc in Kayar area, Ajmer district, and 14.16 million tonnes with 3.06 to 8.68% lead and zinc in north Sindesar Ridge, Udaipur district of Rajasthan have been estimated during 1990-91. In Pindara and other areas of Hesatu-Belbathan belt, Bihar, good prospect of lead and zinc along with copper (multi-metals) is indicated.

Production

The production of lead concentrates is mainly from the states of Andhra Pradesh, Orissa, Rajasthan and Sikkim. The last two states also produce zinc concentrates besides lead. The productions of lead and zinc ores for the years 1988-1989 and 1989-1990 were, 1728 and 1828 thousand tonnes respectively.

The all-India productions of lead and zinc metals for 1989-90 to 1992-93 are being given below in the Table 4.36.

Table 4.36—Production of Lead and Zinc Metals, 1989-90 to 1992-93 (in thousand tonnes)

Particulars	1989-90	1990-91	1991-92	1992-93
Lead (primary)	23	25	32	61 (conc.)
Lead (secondary)	13	15	15	
Zinc (ignot)	75	74	102	301 (conc.)

The public sector undertakings, namely Hindustan Zinc Ltd. (HZL), Sikkim Mining Corporation (SMC) and Gujarat Mineral Development Corporation (GMDC) have been producing and developing lead and zinc deposits. Of these the principal producer, HZL, has mining activities in Rajasthan, Orissa and Andhra Pradesh, while the latter two (SMC) and (GMDC) are mining in Sikkim and Gujarat, respectively.

There are four lead and zinc smelters in public sectors and three smelters in private sectors. The details about their locations and installed capacity are given below in the Table 4.37.

Table 4.37—Lead and Zinc Smelters in India

Smelters	Capacity in tonnes per year (typ)		
	Lead	Zinc	Remarks
Public Sector			
(Managed by HZL)			
1. Debari, Rajasthan	-	45,000	All the zinc smelters have facilities to recover sulphur and cadmium as by-products.
2. Chanderia (Chittaurgarh), Rajasthan	35,000	70,000	
3. Visakhapatnam, Andhra Pradesh	22,000	30,000	Based on imported concentrates to begin with.
4. Tundoo, Bihar	8,000	-	Silver is recovered as by-product.
Private Sector			
1. Binanipuram, Kerala (Managed by Cominco Benami Zinc Limited)	-	2,000	Based mainly on imported concentrates.
2. Thane (Maharashtra) and Kalipark (West Bengal). (Managed by Indian Lead Pvt. Ltd.)	22,500	-	Recovering lead metal from Lead scrap.

Sikkim Mining Corporation which has been mining the poly-metal ore deposit at Rangpo (Sikkim), treated the ore to produce copper, lead and zinc concentrates at its benefication plant at Rangpo. The copper concentrates are sold to HCL at Ghatshila smelter, Bihar and zinc concentrates to HZL's Vizag smelter, Andhra Pradesh. Lead concentrate remained unsold due to high bismuth content.

3. TIN

Introduction

Tin is known to mankind since the earliest times. It was possibly one of the first metal used by man as an ingredient of bronze (copper mixed with tin). Bronze objects with 10 to 14 per cent tin have been found in excavation of different ancient civilization. A bronze rod found in Egypt dates back 3700 B.C. The name has been derived from Anglo-Saxon word "Tin" or from Latin "stannum" means tin (Sn). This is a wide spread metal and the earth crust contains 0.0008 per cent or 8 grams per tonne of rock. The primary deposits are considered workable if they carry to a minimum of 0.3 per cent tin (3 kg to a tonnes of ore). It is mined principally in Malaysia, Indonesia, Thailand, Australia, Bolivia, Brazil and China. The Association of Tin Producing Countries (ATPC) was formed in 1983 with an aim of seeking fair and remunerative tin prices for its members and to intensify efforts towards research, development and marketing of tin. The world mine production of tin metal was 223 thousand tonnes. Brazil was the largest mine producer of tin, reaching a level above 50,000 tonnes. Other principal producers were China, Indonesia, Malaysia, Thailand, Bolivia and USSR. The total world reserves of tin metal have been placed at 4.41 million tonnes.

Mineralogy

The chief tin minerals with their chemical composition, tin per cent and chief physical properties are given in the Table 4.38 below.

Table 4.38—Chief Minerals of Tin

Name	Chem. Comp.	(Sn%)	Chief Physical Characters
Cassiterite (Tin stone)	SnO_2	78.6	Usually black or brown, white or pale grey to brownish streak, H = 6-7, sp. gr. = 6.8-7.1, occurs in tetragonal prisms, often with knee shaped twins in massive or fibrous forms or disseminated grains
Stannite (Tin pyrites)	Cu_2SnFeS	27.5	Steel grey to iron black in colour with blackish streak, metallic lustre, H = 4, sp. gr. = 4.4, usually massive, granular or disseminated form

Practically all the world's metallic tin comes from cassiterite.

Tin metal is silver white and malleable with low tenacity. It has a specific gravity of 7.3 and melts at 232°C. It is obtained from its ore by roasting to remove sulphur and arsenic and smelting with powdered anthracite in a reverberatory furnace. Pure cassiterite is easily smelted by adding coal, coke or charcoal, as oxygen is the only impurity. The impurities like iron and silica cause tin to go into slag by careful addition of fluxes and remelting. The molten tin is drawn off and purified by agitation with air or by electrolytic refining.

Uses

Tin is used in many forms including wrought and unwrought and as an alloy. A large part of tin is used for the manufacture of tin plate (steel sheet covered with a thin coat of tin) and corrosion resistant alloys such as bronze (copper and tin), brass (copper, zinc, lead and tin), solder (lead and tin) etc. Besides copper, zinc and lead it also forms alloys with magnesium, bismuth and cadmium. The addition of tin lowers the melting point of metal and imparts to them valuable properties which enable their wide use in industry. Tin alloys are used in the manufacture of bearings, machine parts (bushings), type metal etc. The electrical industry uses tin to make tin foil for condensers. In the food industry tin foil is used as a moisture proof packing material, and in pottery it is used as pigments, enamels and glazings.

Mode of Occurrence and origin

The tin ores occur mainly in veins, stockworks, disseminations, replacements and placers. Depending upon their modes of formation, they are classified as

    (i) tin-bearing pegmatites,

   (ii) cassiterite-quartz,

  (iii) cassiterite-sulphide ores, and

  (iv) tin-bearing sands.

(i) Pegmatite type

The ores of this type form irregular deposits, generally confined to the central section of granitic masses and occasionally in metamorphosed sandstones and schist. The cassiterite deposit of Koraput (Orissa) is in pegmatite-veins, intrusive into the Bengpal metasediments and is possibly related to nearby Paliam, Dariba granites in M.P.

(ii) Cassiterite-Quartz type

This type is widespread and is represented mainly by light grey and milky white quartz, books of light green mica and light grey felspars. The ore bodies occur in the granite masses or near their margin. The ores occur as veins and stockworks, which are ore-bodies characterised by thin irregular fissures filled with gangue and ore minerals. Ore-veins are of irregular size and usually persist to a great depth. Stockworks are large ore bodies, several hundreds of metres long and tens, and sometimes hundreds, of metres wide. The ore body consists of brecciated rock, intersected by a dense network of veinlets. The metal content is usually low but bigger deposit contains thousands of tonnes of metal. The Tosham cassiterite deposit of Haryana is an example of this type.

(iii) Cassiterite-sulphide type

It is characterised by the presence of green-blue chlorite, dark tourmaline, pyrite, pyrrhotite, galena, arsenopyrite and inclusions of fine crystalline cassiterite. This type differs from the ores of cassiterite-quartz type by its dark grey colour. Its tin content is high. The ore bodies occur as veins and stockworks in sandstone and schists, generally in the vicinity of granitoid masses away from the margin.

(iv) Tin bearing sands (placers)

These placers are formed as a result of destruction of cassiterite-quartz and, sometimes, pegmatitic and cassiteritic-sulphide deposits as well as rocks impregnated with cassiterite. They are the biggest source of tin. The placers may be eluvial, deluvial or alluvial, depending upon the conditions of their formation. The three elements distinguished in the structure of placers are top-soil, metal bearing gravels, and the bed rock. The important tin province of the world is the belt of placers found along the Malaya peninsula. In India the placer deposits are found at Tongpal-Leda-Kudripal and Pushpal areas of Madhya Pradesh.

The tin deposits (primary) contain high temperature minerals and are considered to have been derived by pneumatolytic action. It is commonly assumed that the tin was transported from the magma chamber as gaseous tin fluoride or tin chloride, which by reaction with water form cassiterite, releasing HF or HCl. The granite (wall rock) is generally altered to muscovite, quartz and topaz (greisen), presumably by attack of the acid gases. The associated fluorine-bearing minerals, topaz and fluorite, suggest this mode of origin. A hydrothermal origin has, however, been assigned to some of the tin deposits.

Distribution

The distribution of tin-ores in India is given in Table 4.39 below.

Table 4.39—Distribution of Tin Ores in India

State	Locality	Type of mineralization	Other geological details
Haryana	Tosham, Bhiwani district	Cassiterite-quartz and disseminated types	Established over a strike length of 1000 m on the western side of Tosham hill
Madhya Pradesh	Tongpal-Leda-Kudripal, Bastar district	Placers	
	Pushpal, Bastar district	Alluvial	
Orissa	Mandaguda-Salem, Koraput district	Pegmatite-veins	It is intrusive into Bengpal meta-sediments, and may be possibly related to nearby Paliam and Dariba granites of M.P.
MINOR OCCURRENCES			
Bihar	Nurungo, Hazaribagh district, Semritanr, Chappatanr and Pihra, Giridih district	Disseminated in granite, granulite lepidolite-granite	Associated with granite at Nurungo, granulite at Chappatanr and lepidolite granite at Pihra
	Chakrabandha, Aurangabad district	Cassiterite-Sulphide type	Associated with tungsten and copper minerals in garnet-mica schist and quartz-sericite-sillimanite schist
	Paharsingha, Ranchi district	Greisen type	Associated with greisen, intrusive into fine-grained quartzite and quartz-schist in the southern slope of Paharsingha hill
Gujarat	Hosainpura, Banaskantha district	Pegmatite type	Associated with tourmaline-pegmatite

(Contd.)

State	Locality	Type of mineralization	Other geological details
Rajasthan	Soniana, mica mine, Bhilwara district	Pegmatite type	Associated with albite and quartz of pegmatite, academic interest only
Tamil Nadu	Kadavur	Pegmatite type	Minor grains of cassiterite associated with wolframite

Reserves

The all-India recoverable reserves of tin-ore as on 1.4.1990 are placed at 29 million tonnes most of which are in M.P. (Bastar district). Recoverable reserves of tin-ore in Orissa are just about 13,000 tonnes, located in Koraput district. The reserves in Aurangabad district, Bihar; Bhilwara district, Rajasthan and other places have not been estimated. The all-India conditional resources are placed at about 30 million tonnes of which 8 million tonnes are in Tosham, Bhiwani district, Haryana, 1.85 million tonnes in Bastar district (M.P.) and the rest in Koraput district, Orissa.

In Bhiwani district, Haryana the mineralization is over a width of 76 m from 233 to 309 metres depth with an average of 0.42% tin, indicated by drilling.

Production

The production of tin concentrate was about 29,760 kg in 1988 compared to 29,040 kg in 1987. The production figures for tin metal for 1989-90, 1990-91 and 1991-92 were 10,158 kg, 29,459 kg and 27,493 kg, respectively, while its production in 1992-93 was 65,637 kg. The two tin mines, one each in M.P. and Orissa, are being operated by the public sectors, namely Madhya Pradesh State Mining Corporation Limited and Orissa Mining Corporation Limited, respectively. The tin plate industry is the largest consumer of tin metal. There are three tin plate units in the country, namely :

  (i) Rourkela Tin Plate Unit (Rourkela Steel Plant, Orissa),

  (ii) Tin Plate Company of India Limited at Gomuri, Jamshedpur (Bihar), and

  (iii) R.K. Steel Union at Kalwa, Thane (Maharashtra)

with installed capacities of 150,000, 90,000 and 60,000 tonnes per year, respectively.

A pilot tin smelter project has been set up by the M.P. Mining Corporation at Urla, 5 km from Raipur, close to Bhilai Steel Plant at a cost of Rs. 5 million. It is based on tin-ore deposits of Bastar district and is designed to process two tonnes of tin ore annually and is expected to be stepped upto 10-12 tpy.

Field Test for Identification

Finer grains of cassiterite are difficult to be recognised. The field test normally adopted are :

(i) Cassiterite is infusible ordinarily. When finally powdered mineral is treated with the blowpipe on charcoal with a mixture of sodium carbonate and charcoal powder, it gives a globule of tin with a coating of white tin-oxide.

(ii) The fine grains when put in a test tube and zinc foil or iron is added in a solution of cold dilute HCl or H_2SO_4, the hydrogen evolved as a result of the action of acid on zinc, reduces the stannic oxide leaving a thin film of metallic tin as a grey coating over the grains of the mineral.

(iii) *Microchemical test :* The mineral is fused in sodium carbonate bead in a loop of platinum wire and the residue is dissolved in small capsule containing a drop of 1 : 5 HCl. A drop of this solution is transferred to a slide by means of capillary tube and a fragment of rubidium chloride is added. On viewing through microscope, colourless octahedra of tin would be seen, if the mineral is cassiterite.

4. ANTIMONY

The annual domestic requirement of antimony is about 500 tonnes which is met by imports. There is no production of antimony in the country. The only antimony smelter in the country at Vikroli, Bombay, owned by Metal Distributors Ltd., has remained closed since 1978. China, Bolivia, Russia and S. Africa are the main countries producing antimony and are together accounting for over 70 per cent world production. In 1989 the world production was 68,362 tonnes to which China's contribution was 30,000 tonnes. The world's reserves of the antimony has been assessed at about 4.7 million tonnes.

Mineralogy

The important minerals of antimony are tabulated below giving their chief characters.

Table 4.40—Antimony Minerals and their Chief Characters

Minerals	Chem. Comp.	Chief Characters
Native Antimony	Sb	Rare occurrence, metallic, tin white, very brittle, $H = 3-3.5$, sp. gr. $= 6.6-6.7$, melts at 630°C
Sulphides		
Stibnite (Antimonite)	Sb_2S_3 (Sb = 71.7%)	Lead grey, metallic, $H = 2$, sp. gr. $= 4.5-4.6$

(Contd.)

Minerals	Chem. Comp.	Chief Characters
Jamesonite	$Pb_4Fe\ Sb_6S_{14}$ (Sb ± 29.5%)	Dark lead grey with greyish black streak, metallic, H = 2-3, sp. gr. = 5.5-6.0
Complex Sulphides		
Stephanite	Ag_5SbS_4	Iron black, H = 2-2.5, sp. gr. = 6.26
Pyrargyrite	Ag_3SbS_3	Black to cochineal red, H = 2-3, sp. gr. = 5.7-5.9
Prousite	$Ag_3As\ S_3$	Cochineal red, admantine, H = 2-2.5, sp. gr. = 5.55-5.64
Bournonite	$CuPb.Sb.S_3$	Steel grey to blackish, metallic, H = 2.5-3, sp. gr. = 5.7-5.9
Tetrahedrite	$(Cu, Fe)_{12}\ Sb_4S_{13}$	Steel grey to iron black, metallic, H = 3-4.5, sp. gr. = 4.5-5.1
Oxides		
Senarmontite	Sb_2O_3 (Cubic)	White or greyish, arising by oxidation of primary antimony - ores
Valentinite	Sb_2O_3 (Orthorhombic)	White, greyish or reddish, formed due to oxidation of antimony ores
Cervantite	$Sb_2O_3Sb_2O_5$	Acicular or powdery crusts of white or yellow colour, results from oxidation of primary Sb-ores

The chief ore of antimony metal is its sulphide antimonite (stibnite). For the production of the metal, the sulphide is freed from the gangue and then reduced in reverberatory furnace or the crude ore is volatilised in a blast furnace and the condensed fume reduced in reverberatory furnaces.

Uses

The metal has many indispensable and diversified uses in normal industries and in war. Its consumption in war time increases and so it is taken as a strategic metal. The metal expands upon cooling and this property makes it possible to have type metal alloy which when cast does not change size. It imparts hardness and stiffners to lead. Antimonical lead is used as an antifriction metal. Its major consumer is in automobile industry for making battery plates. The other uses of antimony lead alloy are in sheets and pipes, sheathing for electrical cables, collapsible tubes, foils, bullets, type metal, solder and anti-friction bearings. Antimony is also used in production of white metal alloys which include Britannia metal (Pb-Sb-Cu); pewter (Pb-Sn-Sb), Queen's metal (Sn-Sb-Cu-Zn), and sterline (Cu-Sb-Zn).

Metallic antimony is used in ornamental casting and brick-a-barc. The compounds of antimony are used for flame-proofing, pigments, enamels, safety matches, glass vulcanizing and medicines. Military uses consist of sharpnel balls and bullet cores, detonating caps and bursting charges.

Mode of Occurrence and Origin

Native antimony occurs very seldom in nature. Stibnite (Sb_2S_3) is widely distributed, but workable deposits are only a few. The antimony minerals occur both in deposits associated with volcanic rocks and also more deep seated veins formed under moderate to high temperatures and pressures. They, thus, occur with mercury ores but more commonly in veins associated with quartz, calcite, dolomite or barytes. They are also found associated with lead, zinc, silver, copper or arsenic minerals. Most antimony deposits may have originated due to hydrothermal solutions at low temperatures and shallow depths. Fissures, joints and rock-pores may have given access to these solutions. Irregular replacement deposits are also found. The primary antimony minerals (Stibnite, Jamesonite, etc.) are enriched in oxidised products with formation of antimony oxides and oxysulphides through residual weathering.

Distribution and Reserves

The occurrences of antimony are reported from Bara Shigri glacier area of Lahul-Spiti in Himachal Pradesh, Karimnagar in Andhra Pradesh, Hesatu (Hazaribagh) in Bihar, Chitradurga in Karnataka, Udaipur in Rajasthan and Chamoli in Uttar Pradesh. The reserves of antimony-ores are estimated only for the Bara-Shigri area and are tabulated below.

Table 4.41—Reserves of Antimony Ores in Bara Shigri Area, Himachal Pradesh (in tonnes)

State	District	Proved	Probable	Possible	Total	Types of Ores	Grade
						Primary	
Himachal Pradesh	Lahaul	-	-	3296	3296	Sb-ore	2.88% Sb
	& Spiti	-	-	7292	7292	Sb-Pb-Zn-ore	1.09% Sb, 1.4% Pb & 0.54% Zn
Total				10533	10533	Ores	
				171.19	174.19	Sb-metal	

5. BISMUTH

Bismuth (Bi) has some specific properties that make the metal quite important. The name bismuth was derived from German word "Weisse masse" meaning white mass, and later wismuth. The chief producing countries are USA, Peru, Canada, Mexico, Bolvia, Japan, Germany, Spain and Australia. The market for bismuth is very uncertain and statistics of production are incomplete.

Mineralogy

Bismuth is a white, brittle and crystalline metal with melting point 271°C and boiling point of 1450°C. It has diamagnetic property and unctuous feel. The chief bismuth minerals with their chemical compositions and characters are given below in the Table 4.42.

Table 4.42—Chief Bismuth Minerals and their Characters

Minerals	Chemical Comp.	Characters
Native Bismuth	Bi	Silver white with reddish tinge, metallic, silver white streak, H = 2-2.5, sp. gr. = 9.7-9.8
Bismuthinite (Bismuth glance)	Bi_2S_3	Lead grey, metallic, H = 2, sp. gr. = 6.4-6.5
Bismite (Bismuth ochre)	Bi_2O_3	A yellow earthy mineral, occurs as an alteration product of bismuth and bismuthinite
Bismutite	$Bi_2CO_5.H_2O$	A white grey or yellowish fibrous or earthy crusts, occurs as alteration of bismuth and bismuthinite

Bismuthinite is a common ore of bismuth, but a considerable portion of production now comes from anode slimes resulting from the electrolytic refining of copper and lead. It is also obtained from deposits of other metals such as tin, copper or silver. The extraction of bismuth from its ores needs their reduction with carbon or iron and soda ash in furnaces.

Uses

Bismuth alloys easily with lead, tin, cadmium, antimony etc., and the alloys so formed have low melting point e.g. wood's metal (Bi-Sn-Cd) melting at 60°C. As such these alloys are important in certain industrial processes like casting and for uses in electric fuses, boiler, safety plugs, hand grenades, automatic water sprinkler in fire protection etc. It forms anti-friction metal with brass and bronze. Bismuth wire is used in electric appliances and delicate measuring apparatus.

Bismuth has a great value in medical and cosmetic preparations. Its salts find use in wound dressing, bowel disorder, treatment of syphlis and other diseases, and x-ray examinations. Their smooth unctuous feel helps them to be used in cosmetics. The bismuth salts are also employed in printing fabrics, optical glass, enameling, putting glaze to porcelain and other purposes.

Mode of Occurrence and Origin

As mentioned above, bismuth occurs in free state in nature. The native bismuth and its sulphides and tellurides are of primary origin, while its oxides and carbonates occur as alteration products of primary minerals and are, thus, of secondary origin.

The bismuth minerals are found mostly in veins, associated commonly with ores of tin, silver, cobalt and nickel and also with pyrites, chalcopyrites, quartz etc. They may have originated from hydrothermal solutions as replacement deposits. Studies of trace-element dispersion haloes in the Singhbhum Copper belt, Bihar have indicated preferential enrichment of bismuth along with other elements of acid parentage like copper, nickel, cobalt, silver, zinc, molybdenum, lead etc. in the narrow central greisenised portion of the shear zone on the hanging wall side of the main ore-zone which supports its hydrothermal origin.

Distribution (Resources)

Stray occurrences of bismuth minerals have been noted at many places in the country. Two bismuth minerals, bismuthinite (Bi_2S_3) and bismutospaerite ($Bi_2(CO_3)_3$ $2Bi_2O_3$) were reported from Mathol (23°26' : 86°26'), Purulia district, West Bengal. They were found associated with barytes. A small lode of manganese- ore with bismuth at Sirhi (31°55' : 77°14') near Kulu, Himachal Pradesh has been recorded. A galena sample from south of Bankakacha (22°53'30" : 86°43'00") in Bankura district, West Bengal has analysed 1.98 per cent bismuth. Certain sulphide deposits, like those of lead, copper and tin, may prove to be the main source of bismuth in the country. Trace-element analyses of samples, carried out for the Singhbhum Shear Zone, have indicated presence of bismuth varying from 5 to 80 ppm, and this variation of 16 times is significant in the search of bismuth lode.

No production of bismuth in the country has been reported. However, it is recovered at various stages of refining of copper and lead-zinc smelting.

4.4 LIGHT METALS

1. LITHIUM

Lithium is the lightest of all metals and falls in Group IA of periodic table along with sodium and potassium. It is obtained from lithium bearing minerals in pegmatites, but recently brines of certain lakes are found to be the main source of lithium. USA, Canada, Australia, Russia, Germany, France, Czechoslovakia and Rhodesia are the chief producing countries of lithium. The total world resources of lithium have been estimated as about 11 million tonnes, both from pegmatites and brines.

Mineralogy

Lithium (Li) does not occur as native in nature. The common lithium minerals with their main characters are listed below in the Table 4.43.

Table 4.43—Common Lithium Minerals and their Chief Characters

Minerals	Chem. Comp.	Chief Characters
1. Lepidolite	$K\ Li_2Al(Si_4O_{10})\ (OH,F)_2$ (Lithium mica) upto 7.7% lithium	Generally in small scales, perfect basal cleavage, rose-red, lilac, violet, grey and white in colour, pearly lustre, H = 2.5-4, sp. gr. = 2.8-2.9
2. Zinnwaldite (lithium biotite)	$K(Mg,\ Li,\ Fe,\ Al,\ Si)_3O_{10}$ $(OH_2F)_2$ Lithium-iron mica	Violet, pale yellow to brown and grey in colour, micaceous, pearly lusture, H = 2.5-3, sp. gr. = 2.8-2.9
3. Spodumene (Lithium pyroxene)	$LiAl\ (SiO_3)_2$ upto 8% lithium	Greyish or greenish, pearly lustre, perfect cleavage, parting common, H = 6.5-7, sp. gr. = 3.13-3.20
4. Amblygonite	$Li\ (F,\ OH)\ Al\ PO_4$ (upto 10.2% lithium)	White to pale greenish, bluish, yellowish or greenish, perfect cleavage, pearly to vitreous lustre, streak white, H = 6, sp. gr. = 3.01-3.09
5. Petalite	$Li_2O.Al_2O_3.8SiO_2$ (upto 4.9% lithium)	White or greyish colour, cleavage perfect, uncoloured streak, lustre vitreous to pearly, H = 6-6.5, sp. gr. = 2.39-2.46

Heddenite (emerald green) and kunzite (lilac colour) are two transparent varieties of spodumene and are of gem quality.

Lithium is a white metal, which tarnishes slowly in moist air. It is the lightest alkali metal with specific gravity of 0.53. It has hardness of 0.60 on Mohs scale and can be rolled and welded or drawn into thin wire at ordinary temperature. It has melting point of 186°C and boiling point above 1220°C. It is extracted by electrolysis from its fused chloride.

Uses

The lithium metal imparts toughness, tensile strength and resistance to corrosion to the alloy. Small amounts of lithium have a hardening effect on lead alloys. The metal's strong affinity for oxygen enables it to be employed as "getter" to deoxidise copper and its alloys. Because of its lightness, it is used as an alloy of aluminium, magnesium and zinc for light aeroplane metals.

Lithium minerals like spodumene, lepidolite and amblygonite are used directly in ceramic and glass industries. They are used for colouration and for non-shattering glasses having least co-efficient of expansion. In ceramics they are introduced directly into the batch. Lepidolite with its fluorine and lithium, decreases expansion and increases the strength of ceramic bodies.

Hiddenite and kunzite, transparent varieties of spondumene, are used as gems.

The lithium compounds are used in various ways. Lithium-chloride and lithium-bromide serve as moisture absorbers in air-conditioning and refrigeration units. Thne Li-chloride is also used in industrial dyeing. Lithium fluoride crystals are efficient transmitter of ultraviolet and infra-red lights and so is used in superior achromatic lenses. Lithium-hydroxide is used for absorbing carbon-dioxide in submarines and as a constituent of respirators. It is also used with fatty acids to produce lithium soaps, employed in the manufacture of lubricating greases. Lithium-stearate is replacing zinc stearate in cosmetics, e.g. body and face powders. Lithium-carbonate is used as electrolytes of alkaline storage batteries, bearing and welding of aluminium, bleaching etc. Dry lithium-hydrochloride is used for chlorination of water. Lithium compounds also find use in photography, dental cement, curing meat, making ammonia, making rayons, reducing agents and purifying helium and other gases.

One of the important use of lithium compounds is as rocket propellants, and in nuclear reactors. Lithium isotope, Li_6, splits into helium and titrium when bombarded with neutrons. Titrium (extra heavy hydrogen) is a major constituent of the thermonuclear (hydrogen) bomb. Lithium deutride also finds similar use. Lithium-hydroxide and lithium-borohydride are claimed to be used for war purposes for producing gases.

Mode of Orrurrence and Origin

Lithium never occurs free in nature, but is widely distributed in traces in almost all igneous rocks, in water of many mineral springs, in plant ashes, soils, seaweed, milk, blood and muscle, lung tissues etc. The principal sources are, however, the lithium minerals, brines and volcanic ashes. The lithium minerals are found concentrated in pegmatites. Like other micas, lepidolite occurs near the peripheral zone of the pegmatites or in the zone of felspar and quartz intergrowth, and is usually associated with beryl, cleavalandite, fluorite, tourmaline and topaz. The other associates of lithium-bearing pegmatites are cassiterite, wolframite, molybdenite and columbite - tantalite. The association of lepidolite with cleavalandite in pegmatites is fairly well established. In brines it is associated with halite, potassium and boron, while in montmorillonite clay it has been found concentrated in crystal lattice in the form of its certain compounds.

In pegmatites the lithium minerals are formed during the late stages of consolidation of granitic magma. The association of beryl, fluorite, topaz, toumaline etc. with the lithium minerals in pegmatites is suggestive of their pneumatolytic origin. Larger crystals of lithium minerals may possibly have formed in the pegmatites during primary crystallization of highly fluid melts over narrow ranges of temperature with extremely rapid growth rates. The concentration of lithium in ferro-magnesian minerals (mica, pyroxene and amphibole) is indicative of relatively high temperature of formation. Here lithium appears to have entered in appreciable extent only when the ferro-magnesian minerals are formed in the late stage of temperature differentiation sequences, and the formation of lithium minerals takes place when the lithium concentration in the final fluid phase gains a significant value.

In brines of several closed basins, the lithium compounds are precipitated at late stage of evaporation. The brines of Searles lake, California and Silver Peak, Nevada, USA have yielded considerable amounts of lithium compounds.

Distribution

Lithium bearing minerals occur in rocks belonging to Precambrian to Tertiary in age. They have been found associated with Tertiary granites and their greisenised zones in many parts of the world. In India, however, the lithium bearing pegmatites are confined to Precambrian rocks, ranging in age from 1345 to 950 million years. The state and area-wise distribution of lithium mineralisation with geological details is given below in the Table 4.44.

Table 4.44—Distribution of Lithium Minerals in India

State	District	Area	Geological Details	Remarks
Bihar	Gaya	Pichli and Chatkari	The lepidolite bearing pegmaties are intrusive into biotite-granite and the associated minerals are beryl and columbite with smoky quartz, cleavalandite and tourmaline	Not of any economic significance
	Hazaribagh	Amsa, Charki, Jhamai, Kokaria, Nimadih, Dalchari Pahar, Barkola Pahar, Balbali, Lachchani Bathan, Ghutia, Nagaum, Manimahodar, Bedwa Pathal, Angar, Bhuladih, Pihra, Nagari, Basodih, Gawan-Tisri	The pegmatites extending for 150 m to 300 m in length with a max. width of 150 m are intrusive concordantly into mica schists. Lepidolite and amblygonite, associated with green tourmaline in many places (like Pihra), have been noted.	Li_2O content of lepidolite varies from 2 to 5.08 per cent, while amblygonite contains 6.6 per cent Li_2O

(Contd.)

State	District	Area	Geological Details	Remarks
	Monghyr	Bijaiya, Batia, Lamki etc.	The pegmatites striking for 150 to 170 m with width varying from 12 to 17 m are intrusive into mica-schists and and gneisses. Violet coloured lepidolite occurs in association with rubellite, pink beryl, columbite and tantalite.	Li_2O content of lepidolite varies between 3 and 4.4 per cent.
Karnataka	Hassan	Kabburu	The pegmatites intrude Dharwarian schists and gneisses and a few of them contain spodumene.	Not of much economic significance
		Oorangum Gold mine	Some spodumene is reported in pegmatites at deeper levels and contains 5.8 per cent Li_2O.	
Madhya Pradesh	Bastar	Tongpal, Bericupli, Govindpal, Chiurvada, Mundval, Chitalnar	Lepidolite-bearing pegmatites associated with cassiterite traverse basic stills, intrusive into Bengpal rocks. Lepidolite is accompanied by amblygonite in Govindpal and Bericupli areas. More than 50,000 tonnes of lepidolite with Li_2O more than 3% upto 5 m depth are estimated.	These may be potentital areas for lepidolite.
Maharashtra	Ratnagiri	Kodaval	Spodumene and lepidolite bearing pegmatite, 700 m in length and 325 m in max. width, intrudes granites and mica-schists and occurs as a domical body.	It appears to be the best occurrence of spodumene in India for the present.
Rajasthan	Ajmer	Jeevan mica mine, Rajgarh	The pegmatite, 120 m x 20 m, is emplaced in tremolite-bearing carbonate rock. Lepidolite occurs as flaky aggregates with smoky quartz and rubellite in fracture filling veins. Amblygonite and	It is the most important source of lepidolite in in Rajasthan.

(Contd.)

State	District	Area	Geological Details	Remarks
			rarely spodumene have been noted. The other associated minerals are beryl, tourmaline, apatite, garnet, columbite, tantalite etc. Li$_2$O contents range from 4-5 per cent in lepidolite and is upto 9.42 per cent in spodumene.	
	Bhilwara	Potlan-Bhaoli, Margi areas, Deora	Thin veins of pink and lilac coloured lepidolite occur in albite-zone of pegmatite. It is accompanied by rubellite (pink tourmaline).	

Systematic and detailed exploration is warranted in all the areas, particularly in Bihar, Rajasthan and M.P. where lithium mineralisation has been recorded, to assess its total reserves and to help its exploitation. Zones of albitisation accompanied by lepidolitisation, fluoritisation and greisenisation not only in the granitic terrains but also in satellite pegmatites should be looked for evaluating the resources of lithium in India.

At present there is no production of lithium in India and the country's need is met by imports from other countries.

2. MAGNESIUM

India is wholly self-sufficient in magnesite which is required as a raw material for metallic magnesium, magnesium compounds and is also used in its natural state. The world annual production of natural magnesite is about 12.00 million tonnes to which India's contribution is about 0.5 million tonnes (4.2%) ranking 8th in order of quantum of production (figure for 1989, I.B.M.). China, North Korea, USSR, Austria and Turkey continued to be the leading producers. There are about twenty plants throughout the world producing magnesia from brines with a total production capacity of 2.25 million tpy, the leading producing countries are USA, Japan, UK, Italy, Mexico, Ireland, Netherland and Israel. The world reserves of recoverable magnesite is estimated as 2800 million tonnes, to which more than 80% reserves exist in only four countries; China (820 million tonnes), Russia (720 million tonnes), North Korea (490 million tonnes) and India (233 million tonnes).

Mineralogy

The chief magnesium minerals of economic importance are its carbonates, sulphates, chlorides, aluminates and oxides. They are tabulated below giving their main characters.

Table 4.45—Chief Commercial Minerals of Magnesium

Name	Chem. Comp.	Diagnostic Characters
Carbonate		
Magnesite	$MgCO_3$	Crystalline and amorphous, perfect rhombohedral cleavage in crystals, white, greyish white, yellowish or brown, H = 3.5-4.5, sp. gr. 2.3-3
Dolomite	$MgCO_3.CaCO_3$	Massive granular or tri-rhombohedral crystals, perfect rhombohedral cleavage, white often tinged with yellow or brown, H = 3.5-4, sp. gr. 2.3-2.9, slightly higher R.I. than calcite and lower than magnesite
Sulphate		
Epsomite	$MgSO_4.7H_2O$	Commonly in fibrous crust or botryoidal, white in colour, bitter saline tastes, H = 2-25, sp. gr. 1.68
Kieserite	$MgSO_4.H_2O$	Massive granular or compact, white in colour as saline residue and salt deposits
Chloride		
Carnallite	$MgCl_2.HCl.6H_2O$	Usually massive and granular, pinkish or reddish-white with admixture of iron-oxide, shining greasy, bitter taste, soluble in water, H = 3, sp. gr. 1.6
Aluminate		
Spinel	$MgO.Al_2O_3$	Red, brown or black, sometimes green or blue crystals, commonly octahedra, H = 8, sp. gr. 3.5-4
Oxide		
Periclase	MgO	Dark green grains and octahedra, perfect cubic cleavage
Brucite	$MgO.H_2O$	White, often bluish, greyish and greenish, H = 2.5, sp. gr. 2.39, usually massive and foliaceous, sometimes fibrous

Magnesium metal is manufactured from natural brines, sea water, magnesite, carnallite, dolomite or brucite. It is a silver white metal, easily tarnishing to its oxide, and is one of the light metals known (sp. gr. 1.74).

Uses

The demand for magnesium is for light alloys that are used in aeroplanes and automobiles and many other materials requiring lightness. The chief alloying element is aluminium, generally with some zinc and manganese. Such alloys are non-corrosive, highly rigid and stronger than most aluminium alloys. They are also used in microscope mounting, field glasses, cameras, surveying instruments, artificial limbs, musical instruments etc. A little magnesium hardens lead for cable sheaths. A remarkably light alloy consists of magnesium and berylium. Other important uses for magnesium are for structural shapes and sheets and for deoxidising and desulphurising nickel, monel metal, brass and bronze. Because it burns at low temperature with a strong actinic light, it is used for flashlight, photography, fireworks and signal flares. Incendiary bombs are made of 93% Mg and 7% Al. The use of inportant minerals of magnesium are tabulated below.

Table 4.46—Use of Magnesium Minerals

Minerals	Uses
Magnesite	Refractory bricks, furnace lining and crucibles, in production of magnesium metal, magnesium salt and carbon dioxide, in special cement, in paper, ceramic, glass and sugar industries, chemical accelerator in rubber
Dolomite	Building material, cement, refractory, furnace lining, source of carbon dioxide
Epsomite	Medicine, tanning
Kieserite	Chemical manufacturing such as sodium-carbonate for glass making, soap making, tanning etc.
Carnallite	Source of magnesium metal, fertiliser
Spinel	As gemstone
Periclase	Source of Mg-metal, refractory material, drying agent at ordinary temperature
Brucite	Source of Mg-metal

Mode of Occurrence and Origin

Magnesium is estimated to constitute about 2.7 per cent of earth's crust. It does not occur in free native state, but enters into the composition of a large number of rock-forming silicates, such as biotite, pyroxene, amphibole, olivine etc. It also occurs in the form of its carbonates, sulphates, chlorides, aluminates and oxides.

The mode of occurrence and origin of chief non-silicate magnesium minerals are tabulated below.

Table 4.47—Mode of Occurrence and Origin of Mg-minerals

Minerals	Mode of Occurrence	Origin
Magnesite	1. Occurs as irregular veins with mainly hard amorphous magnesite and occupying fractures or crush zones in serpentinite or ultra-basic rocks. Most Indian occurrences fall in this category.	They result from the break down of serpentinite by hydro-thermal carbonate solution, accompanied by the release of silica which forms opal or chalcedony.
	2. As replacing dolomite and limestone forming bedded deposits, lens like or irregular in shape and large size. This is mainly of crystalline type.	Progressive replacement of limestone or dolomite by $MgCO_3$ through hydrothermal solution

(Contd.)

Minerals	Mode of occurrence	Origin
	3. As sedimentary bedded deposits in association with salt and gypsum or shales and limestones	They are interpreted as saline residues formed by deposition of magnesium carbonate brought about by chemical precipitation with subsequent dehydration.
Dolomite	1. As sedimentary beds	Many have formed from sea water deposition of calcium and magnesium carbonate, may also be due to sea-floor replacement of calcareous ooze
	2. As replacing limestone	Epigenetic replacement of limestone, calcite in limestone is replaced by dolomite.
	3. As veins	May be due to the action of hydrothermal solution in fractures, fissures, joints or any weak zone
Epsomite	1. Occurs in solution in sea water and mineral water	Deposited from water of saline lakes due to chemical precipitation
	2. As afflorescent crusts and masses in limestone caves, and also as encrusting serpentine and other rocks rich in magnesium	Solution and precipitation with subsequent dehydration
Karnallite	As saline residue and in salt dome	Drying up of salt water
Spinel	1. In crystalline limestone and schists	Regional and contact metamorphic origin
	2. In alluvial deposits	Formed due to degradation of parent rocks containing spinel
Periclase	Associated with crystalline limestone (marble) and dolomite	Results due to contact metamorphism of limestone
Brucite	Occurs with marble and also as veins traversing serpentinite	Contact metamorphism of impure limestone, secondary origin formed by hydration of periclase

Distribution and Reserves

The principal deposits of magnesite are in the states of Uttar Pradesh, Tamil Nadu, Karnataka and Jammu & Kashmir. Minor occurrences are in Himachal Pradesh, Kerala and Rajasthan. The total recoverable reserves of magnesite are about 233 m tonnes, of which 41 million tonnes are in the proved category. The state/district-wise distributions of magnesite with reserves are given in the Table 4.48 overleaf.

Table 4.48—Distribution and Recoverable Reserves of Magnesite as on 1.4.1990 (million tonnes)

State/District	Proved	Probable	Possible	Total
Himachal Pradesh (Chamba)	-	0.100	0.192	0.292
J & K (Ladakh, Udhampur)	2.500	-	0.112	2.612
Karnataka (Kodagu, Mysore)	0.128	0.283	0.762	1.173
Kerala (Palaghat)	-	-	0.038	0.038
Rajasthan (Ajmer, Pali, Udaipur)	1.336	1.353	2.798	5.487
Tamil Nadu (Coimbatore, Dharampuri, Nilgiri, N-Arcot, Periyar, Salem, Tiruchirapalli, Teruneveli)	18.673	12.114	12.349	43.137
Uttar Pradesh (Almora, Chamoli Pithoragarh)	18.413	128.175	34.005	180.593
Total - All India	**41.050**	**142.025**	**50.256**	**233.331**

Production

The Indian production of magnesite in 1989-90 was about 505 thousand tonnes maintaining the previous year level. The share of public sector in total production was 56% compared to 62% in the previous year 1988-89. Tamil Nadu continued to be the leading producer of magnesite accounting for about 80% of the production, followed by Uttar Pradesh and Karnataka. Rajasthan and Kerala also produce minor amounts of magnesite. The all-India production of magnesite for 1990-91 to 1992-93 were 529, 562 and 570 thousand tonnes, respectively.

Mining

Magnesite is worked by open cast method by developing benches. Major magnesite mines in Salem area belong to Tamil Nadu Magnesite Ltd. (State Govt. Undertaking), Dalmia Magnesite Corpn. (Private Sector Enterprise) and Burn Standard Company (Central Govt. Undertaking). These mines, are, by and large, mechanised and use compressors, wagon drills, jack hammers, power shovels, loaders, dumpers, dozers and pump. The blasted rock containing 25 to 30% magnesite, in situ, is subjected to manual sorting. The hand-picked crude magnesite is further subjected to sorting and dressing in the dressing yard to obtain useable magnesite containing less than 3% silica. The useable magnesite hardly constitutes 4 to 8% of the blasted rock, even though run-of-mine contains 20 to 35% magnesite. The magnesite mines in Karnataka are worked by TISCO. In Uttar Pradesh, Almora Magnesite Ltd. and Himalayan Magnesite Ltd. are two important producers, having mines in Almora and Pithoragarh districts, respectively.

Magnesite is, generally, marketed after calcination, that is after converting it into lightly calcined or caustic magnesite by heating the mineral to a temperature to 800°C to 1000°C and the dead burnt variety by heating to 1800°C.

3. BERYLLIUM

Beryllium (Be) is an alkaline earthmetal, having low degree of solubility as well as other earthy characteristics. Next to mica it is one of the most important accessory minerals in pegmatites. The new find that the beryllium makes a fatigue resistant alloy with copper has changed beryl, chief berryllium mineral, from a mineralogical curiosity to a quite valuable mineral. Much of the world's supply of beryl comes from Brazil and Southern Rhodesia. Argentina, India, U.S.A., South-West Africa and Russia are also important producers. Small reserves are known in Canada, Mexico and Madagascar.

Mineralogy

Pure beryllium is a silver-grey metal, quite hard and brittle, but can be rolled at high temperature. It is the least reactive of the alkaline earth metals and does not tarnish in air. Its specific gravity is 1.85 and is, thus, lighter than aluminium. Beryllium does not occur as native in nature. Its principal source is beryl. The other beryllium bearing mineral is chryso-beryl. The chief mineralogical characters of the above beryllium minerals are given below in the Table 4.49.

Table 4.49—Chief Beryllium Minerals and their Characters

Minerals	Chem. Comp.	Chief Characters
Beryl	$3BeO, Al_2O_3, 6SiO_2$ (BeO = 14%, Be = 5%)	Hexagonal crystals common, also massive, indistinct cleavage, emerald green to pale green, pale blue, yellowish white in colour, white streak, vitreous to resinous lustre, H = 7.5, sp. gr. 2.7
Chrysoberyl	$BeO.Al_2O_3$	Orthorhombic, prismatic crystals common, poor cleavage, shades of green, vitreous lustre, H = 8.5, sp. gr. 3.6-3.8

Beryllium is extracted from its ore by electrolysis in a fluoride bath containing sodium and barium. Its fused compounds are used for electrolysis. It may also be obtained by the reduction of its oxide or fluoride.

Uses

Beryllium metal is used in production of special alloys mainly with copper, and also with cobalt, nickel, aluminium and iron. The alloy makes the metal harder, tougher and more resistant to corrosion with very high tensile strength. A nickel alloy containing 2.2 per cent beryllium can be made even harder and stronger than beryllium bronze (beryllium alloyed with copper). It is used for parts that withstand tremendous stress at high temperature, such as aircraft engine parts and diamond core drills. The various beryllium alloys find use in various types of instrument springs, control parts, valves and aeroplane carburators. Since the alloy is non-sparking, it is used in explosive and petroleum industries.

Beryllium absorbs X-rays to a lesser extent than any other metal and so it is used in X-rays tubes. It is also employed in fluorescent lamps, neon signs, and cyclotrones. It works as a moderator in atomic reactor.

Beryllium oxide is used as a refractory.

Beryl is also used as gems. The gem varieties are emerald and aquamarine. Alexandrite is the gem variety of chrysoberyl.

Mode of Occurrence and Origin

The beryllium ores occur chiefly in mica-pegmatite. They also occur in alluvial deposits as placers after being derived from the host rock, pegmatite. As an accessory mineral they are found in acid igneous rocks and metamorphic rocks of various types. The albite and green and/or black spotted mica favour concentration of beryl. The Indian beryls are usually well developed hexagonal crystals, often tapering towards one end. Sometimes crystals appear to have been flattened resulting in two opposite prisms being wider than the remaining four. Instances of crystals broken at several points are not infrequent. Tourmaline, cleavalandite, columbite-tantalite, wolframite, cassiterite etc. are usual associate minerals.

Beryllium has a tendency to become enriched towards the late stage of pegmatite crystallisation. It is also present in hydrothermal solutions and may become incorporated in late hydrothermal minerals. Alkali felspars, nepheline, sodalite, micas, tourmaline, alkali amphibole, and alkali pyroxenes carry the bulk of Be present in the upper lithosphere. The immigration of the pegmatitic material in specific structural locale takes place when the pegmatitic fluids tend to be channelised rather than disbursed. High temperature is maintained over a period of time when intermittent shearing stress may possibly has resulted in the development of slip planes. These slip planes/weak zones form favourable structural locales into which alkali-alumina-silica fluids preferentially migrate and crystallise as pegmatites.

The mode of occurrence of large crystals of beryl in pegmatites, showing varying orientation, suggests their formation during primary crystallisation of highly fluid metals over narrow ranges of temperatures, with extremely rapid growth rates and the orientation in many cases may have been in response to temperature gradients. Replacement may possibly have played a small part, if any, in the formation of beryl.

Distribution

The important localities where pegmatites have yielded sizeable quantities of beryl are given in the Table 4.50.

Table 4.50—Occurrence of Beryl in India

State	District/Locality	Geological Details	Remarks
Andhra Pradesh	Nellore, Srikakulam and Visakhapatnam districts, (Kalichedu, Ellen, Pallimetta, Shah and Sankara are the main localities)	The rock types are mica schists and gneisses, calc-granulite, charnockite, khondalites and quartzites, associated with hornblende schist, amphiolite and talc-chlorite schist.	Beryl occurs sporadically.
Assam	Mikir hills	Beryl has been recorded in pegmatite veins traversing the gneissic rocks	It has no economic significance.
Bihar	Hazaribagh, Giridih, Monghyr, Gaya and Ranchi; (Ghortappe, Chakai, Tisri, Gawan, Dhab, Deghi, Charki, Bankhap, Pirha, Chhatkari, Bendi etc. are the well known areas for the incidence of beryl).	Pegmatite bodies in migmatitic gneiss, adjacent to hornblende-schist and amphibolite, are found to be favourable for beryl mineralisation	Giant crystals of beryl measuring 0.75 to 2 metres across and 2.5 m to 6 m in length have been found at Bendi, Chhatkari, Tisri and Hirankhan in Bihar.
Jammu & Kashmir	Ladakh district, Likche and Gayak areas	Pegmatites traversing Ladakh granites	Crystals measuring 2 mm to 3 cm long and 1 mm to 1 cm in diameter have been recorded.
Karnataka	Mysore, Hassan and Chikmanglur	Pegmatite traversing Dharwarian schists and gneisses is the host rock.	Not of economic importance
Kerala	Trivendrum, and Quilon	Minor impersistent and lenticular swarms of pegmatite-veins traverse Archaean gneisses. Also being won from river gravels	Chrysoberyl is found. Gem variety alexandrite has been obtained.
Madhya Pradesh	Bastar (Tongpal-Pushpal area in the southern part)	Pegmatite adjacent to hornblende-schist, amphibolite and basic-schist is favourable site	Associated with tin mineralisation
Mysore	Hassan amd Mysore districts (Doddakamer area)	Pegmatite traversing schists and gneisses	Not of much economic significance
Orissa	Koraput, Sambalpur and Ganjam	Pegmatites traversing schists and gneisses	-do-

(Contd.)

State	District/Locality	Geological Details	Remarks
Rajasthan	Ajmer, Bhilwara, Alwar Sikar and Udaipur - Rajasthan Pegmatite belt, Bisundi, Tiloli and Makrera are well noted areas	Mica-schists, hornblende schists, migmatites and gneisses are the host rocks for the pegmatites. The above rocks are intruded by granite plutons. Cleavalandite, tourmaline, lepidolite, columbite - tantalite etc. are usual associates.	
Tamil Nadu	Salem and Coimbatore (Idappadi, Kurumbapatti are the well noted areas)	It is found associated with mica within pegmatites, traversing granite suite of lower Palaeozoic.	Not of much economic significance
West Bengal	Purulia district (Sulung, Lohar and Belamu hill are the well known areas)	Pale green hexagonal crystals of beryl are reported from pegmatites, traversing schists and gneisses.	-do-

Bihar, Rajasthan and Andhra Pradesh are the main beryl producing states. Very small quantities of beryl have been recorded in the other states, noted above. Indian beryls analyse 10 to 13.5 per cent BeO. After independence the Atomic Minerals Department of Atomic Energy Commission was stock piling beryl. This department is the sole purchaser of beryl. No data of production is available.

4. ALUMINIUM

Introduction

Aluminium (Al) is the most abundant of metals. It is not found in free state, but in combination constitutes 8.07% of the earth crust. The element, Al, holds second position in abundance after silicon. The total world production of Al metal in 1990 was 17.9 million tonnes to which India's contribution (1990-91) was about 450 thousand tonnes (2.5%) with 10th ranking in the world production. The leading producers in the world were USA, USSR, Canada, Australia, Brazil and China with production figures of 4.03, 2.40, 1.56, 1.24, 0.94 and 0.83 million tonnes, respectively during 1989. The world production of aluminium mineral, bauxite, increased significantly to about 106 million tonnes (1989) in view of heavy demand for alumina.

Mineralogy

The chief industrial source of aluminium and its compounds is bauxite. Attempts have been made to extract the metal from andalusite, anorthosite, alunite and clays with little success. The other aluminium minerals commonly associated with bauxite are diaspore, boehmite and gibbsite. Their chemical composition, aluminium percentage and physical properties are tabulated overleaf.

Table 4.51—Chief Ores of Aluminium, their Chemical Composition and Physical Properties

Minerals	Chem. Comp.	$Al_2O_3\%$	Physical Properties
Diaspore	$Al_2O_3.H_2O$	85%	H = nearly 7, sp. gr. 3.5, scratch glass, pearly lustre, prismatic crystals, foliaceous and scaly forms, white colour.
Boehmite	$Al_2O_3.H_2O$		Microscopic minute rhombic or bladed crystals, lower sp. gr. than diaspore.
Bauxite	$Al_2O_3.2H_2O$	73.9%	Pisolitic and oolitic structure common, sp. gr. = 2.55, packed more or less densely within homogeneous non-crystalline matrix. The matrix differs from pisolites in colour, fracture and even chem. composition.
Gibbsite	$Al_2O_3.3H_2O$	65.4%	H = 2.5-3.5, sp. gr. 2.3-2.4, usually as concretions and accompanied by kaolin

Bauxite is no longer recognised as a mineral species, but is considered as a mixture of several hydrated aluminium oxides with considerable variations in alumina content. Iron-oxides, halloysite, kaolinite and nontronite are invariably present as impurities. A typical bauxite contains 55 to 65% Al_2O_3, 2 to 10% SiO_2, 2 to 20% Fe_2O_3, 1 to 3% TiO_2 and 10 to 30% combined water. For aluminium-ore, bauxite should contain atleast 50% Al_2O_3 and less than 6% SiO_2, 10% Fe_2O_3 and 4% TiO_2. For chemical industry, iron and titanium oxides should not exceed 3% each, and for abrasive, silica and ferric oxide should be less than 5% each.

Commercial bauxite occurs in three forms :

    (i) pisolites or oolites,

    (ii) porous sponge ore, and

    (iii) amorphous or clay ore.

Uses

The uses of aluminium in its different forms are tabulated below.

Table 4.52—Uses of Aluminium in Different Forms

Aluminium in different forms	Uses	Remarks
1. Aluminium metal	House hold utensils, wraping material, canning industry, electrical industry, aeroplane construction, automobile parts, alloys etc.	It finds manifold uses due to its lightness, durability and electrical conductivity.
2. Aluminium alloys	Its alloys with zinc, copper or magnesium are important. They are resistant and durable. Duralumin (95% Al, 4% Cu and 1% SiO_2 + Mg) is used in the manufacture of aeroplane.	It alloys with most metals and some nonmetals.

(Contd.)

Aluminium in different forms	Uses	Remarks
3. Al-salts	Al-salts are directly used for dyeing, tanning and printing. Al-chloride is used as disinfectant and for preservation of wood.	
4. Al-ore (Bauxite)	Manufacture of refractory bricks, alumina cement, chemical and metallurgical industry, purification of kerosene.	
5. Commercial products	'Alumino ferrite' and 'alferite' prepared by digesting crude bauxite with sulphuric acid and used in the manufacture of finest paper, precipitation of sewage and refuse liquid and purification of water supplies.	

Classification and Specifications

The classification of bauxite based on its chemical composition is given below in Table 4.53.

Table 4.53—Classification of Bauxite based on Chemical Composition

Types of Bauxite	Chemical Composition (% of weight)					Remarks
	Al_2O_3	Total impurities (%)	Sio_2 (%)	Fe_3O_4 (%)	TiO_2 (%)	
1. Normal Bauxite						
(a) High grade	> 60	< 20	< 5	< 5	< 5	Preferred for extra-
(b) Fair grade	55 to 60	< 20	< 5	< 5	< 5	ction of Al
2. White or siliceous bauxite	> 55	< 20	5 to 20	< 5	< 5	For chem. purposes, manufacture of alum & al-salts
3. Titaniferous bauxite	55 (Av.)	< 25	< 5	< 10	< 7	TiO_2 may be obtained as by-product. It is rare.
4. Ferruginous bauxite	52 (av.)	< 25	< 5	10-15	< 5	Generally used for metallurgical purposes

The specification of bauxite for different uses and those prescribed by ISI for metallurgical grade are tabulated in Tables 4.54 and 4.55 overleaf.

Table 4.54—Specifications of Bauxite for Different Uses

Uses	Chemical Composition of Bauxite (% by weight)			
	Al_2O_3	SiO_2	Fe_2O_3	TiO_2
Metallurgical	52	4.5	6.5	
Chemical	52	-	3.0	
Refractories	50	3-6	3-5	2.5-4
Abrasives	40-60	3-5	3-5	2.5-4

Table 4.55—ISI Specifications for Metallurgical Grade Bauxite

Constituents	Percent by weight			Temp. of drying : 105° ± 2°C (determined on dry basis)
	Gr-I	Gr-II	Gr-III	
Al_2O_3 (min.)	51	48	44	
SiO_2 (max.)	3.5	5	5	
P_2O_5 (max.)	0.2	0.2	0.2	
Fe_2O_3 + TiO_2 (max.)	30.0	30.0	30.0	
V_2O_5 (max.)	0.2	0.2	0.2	
Loss on ignition at 1100°C (min.)	20	20	20	

Mode of Occurrence and Origin

Bauxite deposits occurs as :

   (i) Blanket at or near the surface,

   (ii) Interstratified bedded deposits lying on erosional unconformities,

   (iii) Pocket deposits or irregular masses in limestone or dolomite, and

   (iv) Transported deposits.

(i) Blanket Deposits

They generally have some soil cover and are represented by the caps of the high plateaux, e.g. in western Chotanagpur and Rajmahal hills (Bihar), Amarkantak (M.P.), Kolhapur, Satara and Ratnagiri (Maharashtra), Belgaum (Karnataka) and other places. These may be the denuded remains of an original continuous sheet of Deccan trap which has been subjected to sub-aerial alteration over a very long time and this alteration in the tropical climate or alternating wet and dry seasons such as in India, has resulted in the formation of laterite, the greater portion of silica and other constituents having been removed in solution. The blanket deposits formed in-situ and the original textural features of the parent rocks viz. volcanic agglomerate, breccia, tuffs etc. are retained to some extent. Bauxite agglomerate of Kolodi and bauxite breccia of Netra, Kutch district and tuffaceous bauxite near Rann, Jamnagar district, Gujarat are examples.

(ii) Interstratified Deposits

The bauxite deposits in Saurastra and Kutch regions, Gujarat are in narrow zones intervening in between the Deccan trap and the Tertiary sediments and represent the interstratified type. The deposits of Guina and Arkansas lie on erosional surface beneath partly eroded Tertiary sands, clays and lignites, and those of France lie on erosional surfaces beneath folded upper Cretacous or Eocene limestone.

(iii) Pocket Deposits

Several isolated pocket deposits of bauxite with yellow fossiliferous limestone are noticed along the coast between Lamba and Miani, Saurastra region. The bauxite deposits in Ahmedabad region (Gujarat) occur as elongated and funnel-shaped pockets resting over uneven surfaces of limestone and show conspicuously inconsistent compositional variations. These deposits are formed by lateritisation and bauxitisation of limestone.

(iv) Transported Deposits

Bauxite may also be formed by reworking and moved from the site of formation and be redeposited in nearby sedimentary beds or as rubble accumulations, giving rise to transported deposits. The deposits of this type can be sub-divided as :

(i) those close to the source area, resulting in the pseudo-brecciated bauxites, and

(ii) those far from the source area as indicated by bauxite conglomerates and grits with well rounded pebbles and grains of bauxite.

The coastal bauxite in parts of Ratnagiri resting on conglomerate with pebbles of Deccan basalt represents the transported type. Bauxite conglomerate and grit deposits of Rataria, Cutch district are other examples.

Studying of sections across high plateaux with bauxite deposits in western Chotanagpur, Bihar reveals the presence of well defined layers of bauxite between the surface and the unaltered rock below.

The following succession from the top downward is indicated :

1. Red yellow thin layers of clay,
2. Hard ferrugious laterite,
3. Bauxite,
4. Soft porous laterite,

5. Laminated siliceous lithomarge,
6. Kaolinised trap,
7. Unaltered trap.

The above sequence is an ideal one and is discerned when the conditions of rock alteration, drainage etc. persist over a long period. Koalinisation takes places below the groundwater level. The more soluble constituents, such as silica, are carried downwards from above the groundwater level in solutions or colloids, and, thereby, leaving this zone richer in alumina. The iron-hydroxide is drawn to the surface by capillary action during dry seasons giving rise to a zone of bauxite by impoverishment of iron and automatic enrichment of alumina. The existence of two layers of laterite zones with a zone of bauxite in between is, thereby, explained. The alumina forms a gel and separates out in the form of pisolitic bauxite.

The fawn grey, massive and high grade bauxite is considered to have formed from original tuffs, subsequently lateritised. The gritty and conglomeratic types of bauxite may be the transported as reworked facies of earlier formed bauxite.

The geological and geographical distributions of bauxite are given in the Tables 4.56 and 4.57.

Table 4.56—Geological Distribution of Bauxite

Formation	Types of deposits	Occurrences and Geological details
Quaternary		
Associated with high level laterites	Blanket type	In M.P., Maharashtra and Karnataka overlying Deccan trap, in Ranchi-Palamau (Bihar) overlying granite gneiss, in Eastern ghats overlying Khondalite, in Jammu & Kashmir overlying Sirban limestone (Permo-Carbo), in Tamil Nadu overlying charnockite and in Kerala overlying Tertiary and Archaean formations.
Associated with low level laterites	Pockety type	Coastal deposits of Maharashtra overlying gneissic rocks, coastal deposits of Orissa overlying upper Gondwana sandstone, coastal deposits of Gujarat overlying limestone, and coastal areas of Karnataka, Goa, Daman and Diu overlying various rocks.
Associated with interstratified laterites	Interstratified type	In Kutch area between Mandir and SW Lakhpat, separating underlying trap and overlying Gaj beds (Tertiary) about 100 km x 1.5 km strip of bauxite with laterite is known.

It has been suggested that all the bauxite deposits in India including those on Precambrian charnockite in Tamil Nadu may possibly be related to world-wide formation in Eocene times.

Table 4.57—Geographical Distribution of Bauxite

State	District	Remarks
Bihar	Lohardaga, Ranchi, Gumla, Palamau, Rajmahal, Monghyr, Rohtas	All grades of bauxite are available.
Orissa	Koraput, Kalahandi, Bolangir, Sambalpur, Bandh/Khondmala, Keonjhar and Sundergarh	Metallurgical grade-I and chemical grade are rare.
Andhra Pradesh	East Godavari and Vishakhapatnam	Mostly metallurgical grades are available.
Madhya Pradesh	Surguja, Shahdol, Mandala, Bilaspur Balaghat, Bastar, Jabalpur, Rewa Satna, Raigarh, Durg, Sidhi and Katni	All grades available, Gibbsite and diasporic types have also been noted. Refractory grade is obtained from Katni and Bastar districts.
Maharashtra	Kolhapur, Kolaba, Sindhu, Durg Ratanagiri, Satara and Thane	All grades available.
Gujarat	Kutch, Saurastra, Jamnagar, Junagarh, Kheda, Bhavnagar, Valsad, Sabarkantha and Amreli	All grades available, Chemical grade is obtained from Saurastra and refractory grade from Jamnagar.
Karnataka	Belgaum, Chikmangalur, Uttar Kannad and Dakshin Kannad	All grades available.
Goa, Daman and Diu	—	Metallurgical grade and other mixed grades, excluding chemical grade are available.
Tamil Nadu	Salem, Nilgiri, Madurai	Metallurgical grades (I and III) and low grades are available.
Kerala	Cannore, Quilon and Trivandrum	Mixed grades, metallurgical, chemical and low grades are available.
U.P.	Banda, Lalitpur and Varanasi	Metallurgical, chemical (mixed) and low grades are available.
J & K	Udhampur	Diasporic and low grades are available.
Rajasthan	Kota	Metallurgical grade II is available.

Reserves

The total in-situ reserves of bauxite in the country is estimated to be 3037 million tonnes, out of which 2525 million tonnes are recoverable reserves. About 89% of the recoverable reserves constitute metallurgical grade. The availability of refractory and chemical grades is quite low, 1% and 0.44%, respectively.

Production and Consumption

The demand for aluminium metal exceeds the production level, and hence, metal is imported to bridge the gap. The production figures of bauxite and aluminium metal for the four years 1989-90, 1990-91, and 1991-92 and 1992-93 stood at 4.83 and 0.43, 4.99 and 0.45, 4.74 and 0.51 and 5.07 and 0.52 million tonnes, respectively.

There are five primary producers of aluminium ingots in the country. Their locations, sources of bauxite supply, and plant-wise installed capacities are given below in the Table 4.58.

Table 4.58—Details of Aluminium Producers in India

Producers	Plant location	Source of bauxite	Installed capacity (Al--ingots in tonnes per year)
Public Sector			
Bharat Aluminium Co. Ltd. (BALCO)	Korba, M.P.	Amarkantak, Phutka-Pahar, M.P.	1,00,000
Private Sector			
1. Indian Aluminium Co. Ltd. (INDALCO) three smelters (i)	Alupuram, Alwaye, Kerala	Chanda, Maharashtra	20,000
(ii)	Belgaum, Karnataka	(Local)	73,000
(iii)	Hirakund, Sambalpur, Orissa	Lohardaga, Bihar	24,000
2. Hindustan Aluminium Corp. (HINDALCO)	Renukoot, U.P.,	Richugula, Palamau district, Bihar; Amarkantak, Shahdol distt, M.P.	1,50,000
Madras Alumi-nium Co. Ltd. (MALCO)	Mettur, Tamil Nadu	Saurastra, Gujarat and Salem, Tamil Nadu	25,000
National Aluminium Co. Ltd. (NALCO)	Angul, Koraput, Orissa	Panchatmali, Koraput, Orissa	2,40,000
Total installed capacity			**6,32,000**

Alumina/aluminium industry continued to be largest consumer of bauxite, followed by refractory, cement, abrasive and chemicals. Minor quantity of alumina is used for manufacture of iron-free alum, synthetic cryolite and aluminium fluoride. About 50 per cent of aluminium produced in the country goes to the power sector towards the manufacturing of cables and conductors for which there are about 680 units. Inspite of all these the per capita consumption of aluminium is still at a low of 0.4 kg a year.

Mining

Some mines are systematic and fully mechanised like Bagru Hill mine of INDALCO, Phutkapahar and Amarkantak' mine of BALCO and Panchatmali (Orissa) of NALCO where compressed air powered drills are used for drilling blast holes, power shovel and excavator for handling and loading of ores, and aerial ropeway, trucks and dumpers for ore-transport. Separate benches are maintained for overburden and ore body.

In a number of mines, operations are semi-mechanised when handling of ore and over burden is normal. Whenever bauxite occurs as small lenses or pockets, it is difficult to mechanise the mining operations fully. Bauxite is hand-sorted to remove the low grade ore and waste. Majority of bauxite mines are small ones, producing less than 25,000 tpy and worked manually.

Prospecting Guides

(i) *Nature of host rock :* Bauxite can form from all types of rocks. But trap rock, limestone and mottled clayey formations are best suited to look for bauxite.

(ii) All red-brown and green-grey rocks with pisolitic structure require special attention to look for bauxite.

(iii) Bauxite can be distinguished from iron-ores by its hardness and streak, from sandstone by its regular shape of pisolites, from jasper like varieties by its perfect conchoidal and even fracture, from limestone by the greater sp. gr. and HCL test, from effusive rocks by its rounded densely coloured pisolites embedded in homogenous groundmass and from clayey schists by its greater hardness, greater sp. gr., soiling capacity, lack of plasticity, greasiness and presence of separate pisolites.

4.5 RADIO-ACTIVE METALS

1. URANIUM AND THORIUM

Introduction

In this new era of atomic energy for nuclear power generation, uranium and thorium constitute the potential atomic fuels. Uranium was discovered in 1789 by Klaproth in pitchblende (chief ore of uranium), while thorium was discovered in 1828 by Berzelius in thorite. Both the metals are extremely gas and air-sensitive and so their extraction has to be carried out under inactive gases, such as argon and helium. Pyrometallurgical reduction yields uranium in the form of a regulus (referred to as 'biscuits'), and thorium in the form of a powder. This is owing to differences in the melting points of the two metals.

Mineralogy

Uranium

Uranium is a radioactive element which breaks down at a steady rate and after passing through a number of intermediate stages, it is finally changed into a form of lead i.e. Pb^{206} that is not radioactive. Uranium as such is not found in nature in native state. The chief minerals of uranium with their characteristics are given in Table 4.59 below.

Table 4.59—Chief Minerals of Uranium

Name	Chem. Comp.	Chief Characters
1. Pitchblende (Uraninite)	$2UO_3.UO_2$	Colour and streak black, green brown at times, pitch like dull, greasy, sub-metallic lusture, H = 5.5, sp. gr. = 6.4-9.7.
2. Torbernite (Copper Uranite or Uran Mica)	$CU (UO)_2 P_2O_8$ $12H_2O$	Thin tabular or scaly form, colour and streak emerald or glass green, sub-adamantine lusture, H = 2-2.5, sp. gr. = 3.5
3. Autunite (Lime Uranite)	$Ca (UO_2)_2.$ $P_2O_8.8H_2O$	Form, lusture, hardness and specific gravity similar to torbernite, citron to sulphur yellow in colour and yellowish streak
4. Carnotite	$K_2O.2U_2O_3.$ $V_2O_5.2H_2O$	Powdery form and canary yellow colour
5. Uraconite (Zippeite)	Hydrated sulphate of uranium	Earthy or powdery crust, lemon yellow or orange or rosette colour

Uranium metal is white and hard with a specific gravity of 18.68.

Thorium

Thorium metal is grey (nickel colour), heavy (sp. gr. 11.3), hard to fuse, and emits radiations, similar to those of radium. Like uranium it is not found in nature in free state. The important thorium minerals with characters are given below in the Table 4.60.

Table 4.60—Chief Thorium Minerals

Name	Chem. Comp.	Characters
1. Monazite	Phosphate of cerium metals (Ce, La, Yt) PO_4, with thoria, THO_2, and silica, SiO_2	Rolled grains, or massive, colour pale yellow to dark red brown, white streak, resinous lustre, hardness H = 5.5 and sp. gr. = 5.27
2. Thorite	$Th.SiO_4$	Black or orange yellow in colour, dark brown streak, vitreous lusture, H = 4.5, sp. gr. = 5.3
3. Thorianite	$ThO_2.U_3O_8$	Colour and streak black, H = 6.5, sp. gr. = 9.3

Uses

The disintegration of uranium and uranium minerals gives rise to radium (Ra) which is important in the treatment of cancer, in certain X-ray apparatus and for luminous paint. The uranium salts were used to give yellow to brown colours for glass and glazes and for special alloys of steel, copper and nickel. Now, it is desired mostly for the atomic energy. The fission of uranium nuclei releases tremendous amount of energy. The fission of all atoms in a kilo of uranium releases as much energy as that obtained by 2,500 tonnes of coal or as much as is generated in ten days by a 100,000 KW power station. This makes it possible to use as nuclear fuel.

The radioactive isotopes of uranium and thorium which are products of fission reactions, are being increasingly used in research, industry, agriculture and medicine. In the iron and steel industry these find use to control and regulate production process, and thereby, to step up the output of open hearth and blast furnaces. They are widely used in many industries, and construction projects as a means of controlling quality. The chemical industry uses nuclear reaction products to speed up chemical process and to obtain new compounds such as plastics. The radioactive isotopes have helped scientists to ascertain the mechanism of the root and leaf nutrition of plants, the efficacy of various fertilizers, and to solve a number of fundamental problems of agriculture.

Mode of Occurrence and Origin

Uranium and thorium are widely disseminated in the earth's crust, but rarely form big deposits of rich ores. According to rough estimates, the upper layers of the earth's crust contain 100,000,000 million tonnes of uranium. Being highly active chemically, uranium and thorium never occur in native form. Both occur as an admixture in other minerals, and for the most part are disseminated as mobile compounds leached out with relative ease during weathering of igneous rocks. They appear as primary constituents of igneous rocks, granites and pegmatits, or in high temperature veins associated with tin, copper and lead minerals or accumulate in ores of hydrothermal deposits. Owing to their large ionic radii and other facts, they remain in solution until the very last stage of crystallisation. When their concentration in the residual fluids are quite high they may appear at times as disseminated grains in the interstitial spaces of mineral grains in a rock or form vein-type deposits in faults and shears or occur as breccia fillings. Practically all the uranium deposits are of epigenetic type.

Uranium and thorium also occur as secondary products associated with other minerals. They are found as closely associated with cerium mineral, monazite, which besides occuring as an accessory component of acid igneous rocks and large crystals and masses in pegmatites, is found as heavy residue in sediments. Monazite is obtained on a commercial scale from sands where natural concentration has gone on, such as a constituent of sea shore sands.

Primary uranium and thorium minerals in veins and igneous rock masses are converted in the surface oxidation zone into secondary minerals, which are easily soluble, and are carried by rivers into seas and lakes. These disseminate through the sedimentary rocks deposited on the floor of these seas and lakes or given favourable conditions, begin to reconcentrate in bedded deposits. A remarkable feature of the accumulation of these in sedimentary rocks is their invariable association with vanadium, phosphorus and molybdenum and also with selenium, lead and zinc, the contents of which are sometimes quite large.

Distribution

Uranium

The uranium deposits of India may be classified as :

(i) disseminated type,

(ii) vein type, and

(iii) infillings in fault breccia.

They all occur in the Dharwarian rocks and are generally genetically related to the post-Dharwar and post-Cuddapah granitic intrusions. While uranium is associated with lesser or greater amounts of copper mineralisation, not all copper deposits carry uranium.

Geographically, uranium ores are known to occur in several localities along the Singhbhum copper belt, Aravalli (Huronian) rocks at Umia and Udaisagar in Udaipur district, the Delhi formation (Algonkian) at Kho-Dariba in Alwar district and at Khetri in Jhunjhunu district of Rajasthan, and in the rhyolite (Archaean) outcrops near Baghnadi and Jangal in Durg district, M.P. and Parsoria in Bhandara district, Maharashtra. The indication of uranium mineralisation has been recorded in Parbati Valley, Kulu district, Himachal Pradesh, associated with quartzite of Dharwar age, and in the old copper workings in the Pokhri-Tunju tract in Chamoli district, Uttar Pradesh. Uranium bearing minerals also occur in the pegmatites in the Mica belt of Bihar. Besides, the inland placer deposits with columbite-tantalite, uraniferous allanite, cassiterite, barite etc. have been found over an extensive area in the older alluvium of sub-recent age in Ranchi, Purulia and other adjoining districts of Bihar and West Bengal.

Thorium

Thorium along with uranium occurs in the cerium mineral, monazite, which is a constituent of the heavy black sands on the sea coasts of India, from the Narmada estuary to Cape Camorin and from the latter locality to Orissa coast (Ganjam district). Promising areas of monazite occurrence are between Kumla and Mangalore in Cannore district, Varkallai, Chavara and other places in Quilon district and Vilinjam, Koralaim, Pachallur and Vehi in Trivandrum district of Kerala.

Production and Reserves

The Singhbhum copper belt in Bihar has so far remained the only place in the country where uranium ore is being produced. The uranium mineralisation occurs discontinuously in the meta-sediments, the important among them are at Jaduguda, Narwa Pahar, Bhalko, Kanyaluka, Tamadungri and Dhantuppa. At Jaduguda, the uranium ore occurs associated with the phyllites and schists. The grade of the ore, in general, increases with depth, the better grade of ore being found below 300 m depth. The indicated ore reserves at Jaduguda are estimated at 2.8 million tonnes with an average grade of about 0.08% uranium oxide. The deposit at Narwa Pahar is also another large occurrence among the presently known zones of uranium in Singhbhum district. The tentative reserves of uranium ore in Singhbhum belt have been placed at 20-21 million tonnes.

As regards monazite in the coastal sand, reserves of some two million tonnes have been estimated, the total thorium content of which is between 150-180 thousand tonnes and that of uranium between 6-7 thousand tonnes. The copper tailings of the copper mines of Singhbhum, Bihar, and Rajasthan also yield uranium.

4.6 RARE METALS

1. TANTALUM AND COLUMBIUM

Tantalum (Ta) and Columbium (Niobium - Nb) are rare metals and invariably occur together. The world production of their ores is only a few thousand tonnes, but during war there is significant increase in their production for use in aeroplane radiosets and special steels. The main producers of these ores were Western Australia, Rhodesia, Nigeria, South Dakota, Russia and USA.

Mineralogy

The occurrence of tantalum in nature has been recorded, though very rare. The only tantalum-columbium (niobium) mineral of economic importance is tantalite-columbite (Fe, Mn) (Nb, Ta)$_2$O$_6$. The nearly pure tantalate is named tantalite, while pure niobate is columbite, and they form isomorphous series. The mineral is grey, black or brown and frequently iridescent. Its streak is dark red to black, and displays sub-metallic to sub-resinous lustre. Its hardness is 6, while sp. gr. is 5.3-7.3 increasing with tantalum content. It is distinguished from wolframite by much less distinct cleavage.

The tantalum metal is hard, white and ductile, and has a great tensile strength. It is extremely resistant to corrosion. Its melting point is 2850°C. The metal is produced in electric furnace.

Uses

Tantalum being highly resistant to corrosion, is used for acid resisting chemical ware, in absorption systems for making hydrochloric acid in certain electrical processes. Tantalum and columbium, having great affinity for gases, are used in vacuum tubes. They are also employed in making synthetic rubber. Tantalum was widely used for electric filaments, but this is now replaced by tungsten. Tantalum is used for making special steels employed in dental and surgical instruments. It makes extremely hard alloys for abrasive and cutting purposes. Tantalum carbide is almost as hard as the diamond and is used in making tools, gun barrels, saws etc. Tantalum wire is stronger than steel and allows electric current to flow in one direction only. Columbium bearing stainless steels are very resistant to high temperature and corrosion.

Mode of Occurrence and Origin

It is widely known that pegmatites are mineralogically complex with high concentration of rare metals. Tantalite and columbite occur in granitic pegmatites where they are associated with wolframite, cassiterite, samarskite, tourmaline, lepidolite, beryl, cleavalandite etc. Several small production units operate on deposits of this nature. They also occur in alluvial deposits derived from above occurrences.

Field evidences suggest considerable interaction between the silicate fluids of granitic magma and the invaded basic rocks and the localisation of tantalite-columbite mineralisation along interaction zones with older basic rocks. Gabbros and amphibolites are reported to be favourable rocks for rare metal mineralisation through process of magmatic interaction with the silicate fluids of the granitic magma *(Solodov N.A., 1964 - Rep-XIIth Session, IGC, Part-VI, pp 193-212)*. Tantalite-columbite are typical for albite-pegmatite, though tantalum-rich varieties are known in microcline-albite-pegmatites and columbites is very characteristic for albite-spodumene and microcline-pegmatites. The type of pegmatite is also found to be one of the chief factors to control rare metal mineralisation.

Distribution

Tantalite and columbite have been reported from many places in India. The details are being given below in the Table 4.61.

Table 4.61—Distribution of Tantalite and Columbite in India

State	District/Area	Rock types/Association	Remarks
1. **Andhra Pradesh**	Dhenkanal-Visakha-patnam Mica-belt and Nellore Mica belt	The major rock formations in Dhenkanal-Visakhapatnam belt are mica-schists, calc-granulites, charnockites and khondalites, associated with amphibolite and talc-chlorite schist, while in Nellore belt the rock types are mica-schists and quartzites, associated with hornblende-schist and amphibolite. The pegmatites are emplaced in the above rocks and function as host for the the tantalite and columbite.	Pegmatitic pneumatolytic deposit.
2. Bihar	Koderma, Jhajha and other areas in Bihar Mica-belt	Pegmatitic bodies in migmatitic mica-gneisses, adjacent to basic rocks (hornblende-schist, amphibolites etc.) control localisation of the rare metals. Cleavalandite (albite), tourmaline, beryl etc. are the common associates of columbite and tantalite.	An aqueous phase as well as water saturated silicate melt appear to have participated in this complex evolution of the rare metals.
	Abrakhi Pahar near Singar (24°34' : 85°30') in Gaya district	The host rock is mica pegmatites within migmatitic schists and gneisses, and associated rare minerals are pitchblende, monazite and triplite.	The occurrence is only of mineralogical interest.
3. **Jammu & Kashmir**	Ladakh	Beryl is the associated rare mineral in pegmatitic bodies traversing Ladakh granites.	-do-

(Contd.)

State	District/Area	Rock types/Association	Remarks
4. **Karnataka**	Mysore, Hassan and Chikmagalur	Pegmatites within schist and gneisses of Dharwar age.	Only of minera-logical interest
5. **Madhya Pradesh**	Southern part of Bastar district (Tongpal-Pushpal area)	Associated with tin minera-lisation which is generally confined to the pegmatites emplaced in basic rocks.	
6. **Maharashtra**	Bhandara, Pipalgaon (20°46' : 79°55')	Granulated Quartz-veins and pegmatites within meta-sedi-ments and closely associated with basic meta-lavas. Tanta-lite and columbite upto 800 ppm are found associated with tungsten ores. No mineral of Tantalite-columbite has been noticed so far.	Only of academic interest
7. **Rajasthan**	Rajasthan Pegma-tite belt (Ajmer, Bhilwara, Alwar, Sikar and Udaipur)	The main formations hosting pegmatites are largely mica-schists, hornblende-schists, migmatites and gneisses in-truded by plutons of granitoids. Cleavalandite, tourmaline, beryl and lepidolite are the usual associates.	
8. Tamil Nadu	Kadavur (10°36' : 78°12'), Tiruchirapalle district	Associated with mica minerali-sation in pegmatites within granite suite of lower Palae-zoic age (Ca 500 Ma).	Only of academic interest

No data on production of this rare metal is available. Atomic Mineral Division of the Department of Atomic Energy was stock piling columbite-tantalite along with beryl after independence.

2. CADMIUM

Cadmium (Cd) is finding undreamt of applications in this era of nuclear science, electronics and space. It is a by-product metal recovered during zinc smelting. Russia, Japan, USA, Canada, Australia, Belgium, Germany and Mexico are the main producers. The present world production of cadmium is above 21,000 tonnes annually. In 1989 the total world production was 21,300 tonnes, decreased by 3% from that in 1988. Japan was the largest producer contributing 13% to the world production, followed by USSR (12%), Belgium (8%), and Canada (7%). Certain restrictions have been imposed in some countries in the use of cadmium to safeguard environmental pollution and this may cause relative fall in its consumption.

Mineralogy

The cadmium minerals are of rare occurrence. The only mineral of importance is its sulphide, though its oxide and carbonate are also known. They are described below in the Table 4.62.

Table 4.62—Minerals of Cadmium

Minerals	Chem. Comp.	Characteristics
Greenockite	Cds	Orange yellow with orange yellow to brick-red streak, adamantine to resinous lustre, H = 3-3.5, sp. gr. 4.9-5, usually with sphalerite as earthy coatings
Cadmium Oxide	CdO	Forms a thin black coating, brilliant metallic lustre
Otavite	Cadmium carbonate of uncertain composition	White to reddish in colour, shows minute rhombohedral crystals in crusts

There is no deposit worked for cadmium alone, and is found in association with zinc-ores. It is volatilized from zinc-ore (sphalerite) or is recovered from dust chamber and electrolytic slimes. In the extraction of zinc, it comes off first and condenses as brown oxide. It is then reduced with carbon.

Cadmium is a soft, bluish-white metal having a brilliant lustre, tarnishes in air and forms oxide when heated. It has a low melting point (320°C) and a boiling point of 767°C. It is ductile, softer than zinc, and emits a peculiar cry when bent. Its specific gravity is 8.6.

Uses

The main use of cadmium is in various alloys, especially in low melting and bearing alloys. It alloys readily with other metals. With nickel or silver and copper, it forms the best high pressure antifriction bearing metal for automobile bearings. It alloys with bismuth and tin to form an alloy known as wood's metal (Bi-Sn-Cd) with melting point of 60°C, which is used in automatic water sprinklers for fire protection. It is also employed in electroplating and metal spraying and in manufacture of electric transmission wires. In cadmium plating, particularly on iron, it forms thin rustless surface alloy.

The main consuming industries of cadmium metal in the country are paint, pigment, glass, battery and chemicals. Cadmium compounds are also used in photographic materials, rubbers, soaps, fireworks and textile printing. The manufacture of nickel-cadmium battery is increasing all over the world.

A special cadmium-bearing fluoride glass which is ten times more resistant to water corrosion has been developed in USA by the Corning Glass Works. This is used in infra-red optics and in optical windows.

Cadmium serves to absorb atomic radiations, particularly neutrons. Cadmium rods are employed to control the rate of fission in atomic piles. Boeing Aerospace company reported the manufacture of solar cells with high energy conversion from copper-indium diselenide/zinc cadmium sulphides which can be utilised in terrestrial solar power applications and potential uses in space.

Mode of Occurrence and Origin

Cadmium invariably occurs with the zinc ores, on which it forms coating. A little of cadmium is often present in the composition of sphalerite and, sometimes, in galena. In the north Sindesar Ridge block, Udaipur, Rajasthan, 15 to 30 m wide ore zone with 4-10% Pb and Zn, and 400 to 500 ppm Cd has been located. Cadmium never occurs uncombined in nature. In its mode of occurrence and origin it is closely allied to sphalerite or galena.

The mineralisation is generally found in limestone and dolomite where the ores may have formed by replacement processes. The cadmium ore with lower melting point than that of zinc ore may have formed from the residual liquid at the later stage. The early formed zinc mineral, say sphalerite, may possibly have been replaced to some extent by cadmium sulphide, greenockite. These ore minerals may have been formed from hydrothermal solution of igneous origin or ascending artesian meteoric water or descending surface water. The occurrence of cadmium minerals as surface coatings or as minute crystals in crusts supports its formation by deposition on the earlier formed sphalerite surface through hydrothermal solution. The outer margins of the earlier sphalerite might have also been attacked by the residual solution rich in cadmium which with fall of temperature deposited as crusts.

Production

It has already been mentioned that cadmium is obtained as a by-product during smelting of zinc. Indian production of cadmium during 1990-91 and 1991-92 were 263 and 285 tonnes, respectively, while during 1985 it was only 194 tonnes. The details of various plants and their locations with installed capacities are tabulated below.

Table 4.63—Details of Various Plants Producing Cadmium

Unit	Location	Installed Capacity in tonnes per year
1. Hindustan Zinc Ltd.		
(a) Debari Zinc smelter	Debari, district Udaipur, Rajasthan	250
(b) Vizag Zinc smelter	Visakhapatnam, Andhra Pradesh	115
2. Cominco Binani Zinc Ltd.		
Alwaye Zinc smelter	Benanipuram; district Ernakulam, Kerala	36

Hindustan Zinc Ltd. has installed a lead and zinc smelting complex near Chanderiya, Chittaurgarh, Rajasthan which will have facility to recover 375 tonnes of by-product cadmium from Rampur-Agucha lead-zinc deposit, Bhilwara district, Rajasthan. The plant has now come in its production in September 1991.

3. MERCURY

Mercury (Hg) is called as 'hydrargyrum' (water silver) by Greeks and from this the chemical symbol Hg is derived. The equivalent of this in English is 'quick silver'. These names described accurately the metal that is liquid at room temperature. The old literatures bearing the name of this metal, indicate that the metal was known to ancients as early as 300 B.C. The main mercury producing countries are Spain, Italy, USA and Mexico. Some quantities of the metal come from Canada, China, Japan, Russia, S. Africa, Chile and Peru. Mercury is sold in flasks, each flask containing 75 pound (34 kg).

Mineralogy

Mercury is a silver white, shining metal with melting point -39°C, boiling point 357°C and specific gravity 13.59. It is the heaviest metal known. It does not wet glass and is slightly volatile. Its vapour is highly poisonous. It combines with most metals to form alloys called amalgams, and these decompose on heating with volatilisation of metallic mercury. The main mercury minerals with their characters are given below in the Table 4.64.

Table 4.64—Some Mercury Minerals and their Characters

Minerals	Chem. Comp.	Characters
1. Native Mercury	Hg	Rare, occurs as small fluid globules disseminated through its gangue, tin white with brilliant metallic lustre, sp. gr. 13.59, freezes at −38.87°C.
2. Native Amalgam	Ag.Hg	Silver white, metallic, H = 3-3.5, sp. gr. 10.5-14
3. Cinnabar	HgS	Usually massive, conchineal red colour, streak vermilion, H = 2-2.5, sp. gr. 8.09
4. Colomel (Horn quick Silver)	Hg_2Cl_2	Whitish, greyish or brownish, H = 1-2, sp. gr. 6.48, found associated with cinnabar

The chief source of mercury is cinnabar. The tenor of the ore is 0.3 per cent. The mercury is readily extracted by volatilisation of its ores and the vapour is condensed to a liquid in cooling tubes and drained to collecting tank. Because of the simplicity in extraction the low grade ores can also be extracted.

Uses

The chief uses of mercury are in instrument industry in the manufacture of barometers, hydrometers and vacuum gauges for measuring fluid pressures. It is used in thermometer, since it expands like a typical metal when heated.

Mercury has the highest electric resistivity of all metals and its ionised vapour is a much better conductor than the liquid. Hence, both are used in electrical industries for high power electronic rectifiers, automatic switches, mercury vapour lamps etc.

Mercury dissolves many metals at room temperature, yielding alloys known as amalgams. This process is employed in the extraction of silver and gold, in the preparation of certain kinds of tooth fillings and in the silvering of mirror.

It is also used for the manufacture of drugs and chemicals, vermillion pigments, anti-fouling paints and explosive salts used in detonators etc.

Mode of Occurrence and Origin

Mercury ores occur as disseminations, impregnations, stockworks and veins in many kinds of rocks. They may occur in fractured and sheared rocks, permitting ingress of solution.

The mercury deposits may have originated from hydrothermal solutions at relatively low temperature and are classed as replacement deposits, fissure veins, breccia fillings, stockwork etc. In many cases they are the result of volcanic activity and are deposited by certain hot springs in volcanic areas. The common associates are chalcopyrites, pyrite, realgar, stibnite, quartz and opal, chalcocite and often bitumen.

There is so far no known mercury deposit in India. Intensive search is required to locate such deposits in the possible areas.

4. PALLADIUM, SELENIUM AND TELLURIUM

Palladium, Selenium and Tellurium are grouped under rare metals which are obtained mainly as by-products during copper, lead, zinc, gold and platinum-ore processing. A brief description of each of the metals is given below.

PALLADIUM

Palladium (Pd) is one of the minerals of Platinum group. The platinum group of minerals is mined in several countries, but major production over 95% comes from South Africa, Russia and Canada. The USA and Japan also share the production. The reported world production of platinum group metals in 1989 reached a level of 286 tonnes showing a marginal increase of 1 tonne over the previous year. The world resources of platinum group of metals are estimated at about 66 million kgs. (66,000 tonnes). South Africa and Russia reported to have a major share of reserves. The other countries being the USA, Canada, Australia, Ethopia and Finland.

Mineralogy

Palladium is a silver white hard metal, but not so ductile as platinum. It oxidises more readily than platinum. It has H = 4.5, sp. gr. = 11.3-12, and melting point = 1546°C.

Uses

It is used in dental alloys, for coating the surfaces of silver reflectors, in searchlights, in the construction of delicate graduated scales, jewellery, chemicals etc. Telephone, jewellery, chemical and automobile industries are the main consumers of palladium.

Mode of Occurrence and Origin

Palladium occurs native in crude petroleum, and in small quantities in cupriferous pyrites, especially those containing nickel and pyrrhotite. In India it is recovered as by-product from sludge during copper refining at Hindustan Copper Limited's plant at Ghatsila, Bihar. For the first time in 1980, HCL produced a few kgs of palladium on bench scale through its R & D efforts.

Distribution

Search for platinum group of metals in Sukinda-Nausahi and Simlipal areas, Cuttack, Keonjhar and Mayurbhanj districts, Orissa indicated a distinctive platinum bearing zone with Pt 29 to 100 ppb in Kaliapani. In Nausahi area presence of Pt, Pd and Ir values in the range of 2 ppb to 12 ppb were indicated. There is no reported commercial deposit in India.

SELENIUM

Selenium (Se) belongs to the same group as sulphur and tellurium. The world reserve base of selenium is about 130,000 thousand tonnes. Though annual world production is about 1400 to 1500 tonnes, Japan continued to be the major producer contributing 33.6 per cent. World production in 1989 at 1400 tonnes decreased by about 7% over the previous year.

Mineralogy

The chief minerals of selenium are clausthalite (PbSe), berzelianite (Cu_2Se), tiemannite (HgSe), and naumanite (Ag_2Se). They occur in native sulphur and in all pyritic ores, though often in negligible traces. Selenium bearing sulphur is selen-sulphur. It substitutes for sulphur in sulphide minerals of copper.

Uses

Manufacturers of photo-copiers, electronic products, pigments, metals, glass and chemicals are the main consumers of selenium. It is used in the production of red glass, enamels and glazes, in rubber manufacture and in production of special steels. Its demand is increasing with advancement in television sets and photo-electric apparatus.

Mode of Occurrence, Origin and Production

Large reserves of by-product selenium exist in copper and other metal deposits. Selenium is obtained from the anode mud of slime, produced in electrolytic refining of copper, lead, zinc, silver etc. and matte, and from deposits in sulphuric acid chambers. Coal contains an average of 1.5 ppm of selenium, but its recovery from coal apears unlikely in the foreseable future. The output of selenium is dependent on copper-ore production. In USA on an average 0.25 kg selenium exists in per tonnes of copper ore.

A selenium extraction plant was established in 1973 by HCl at Ghatsila which reported its production in 1984 and 1985 at 4.19 and 4.85 tonnes valued at Rs. 1.44 million and Rs. 2.84 million, respectively. The production of selenium metal in 1989-90, 1990-91 and 1991-92 were 4000 kg, 3052 kg, and 8250 kg, respectively.

TELLURIUM

Tellurium (Te) is closely related to selenium, and is one of the few elements that combines with gold. The metal is marketed as (-) 200 mesh powder, one pound ingots or five pounds slabs. Tellurium dioxide is sold as powder ranging from (-) 40 mesh to (-) 200 mesh containing a minimum of 75% of tellurium. The reported world production of tellurium in 1989 was about 144 tonnes based on available data. It was same as in the previous year. It is recovered mainly from copper anode slimes, a by-product of electrolytic copper refining. The major production comes from Japan, Canada, USA and Peru. The world reserves of tellurium are estimated at 38,000 thousand tonnes.

Mineralogy

Tellurium is found in small quantities free in nature in sulphur and pyrite. The chief tellurium minerals with their characters are given below in the Table 4.65.

Table 4.65—Chief Tellurium Minerals and their Characters

Minerals	Chem. Comp.	Diagnostic Characters
Native Tellurium	Te	Nearly pure with a little gold and iron, tin white in colour and streak, metallic, brittle, H = 2-2.5, sp. gr. 6.1-6.3
Tellurite	TeO_2	White or yellowish, results from oxidation of tellurium or tellurides
Tellurides		
Tetradymite	$Bi_2(Te, S)_3$	Pale steel grey, metallic, splendent, H = 1.5-2, marks paper, sp. gr. 7.2-7.6
Fessite	Ag_2Te	Lead grey, metallic, H = 2.5, sectile, sp. gr. 8.4
Sylvenite (Graphic Tellurium)	(Au, Ag) Te_2 with 24.5% gold	Steel grey to silver white, sometimes yellowish, metallic, brittle, perfect cleavage, H = 1.5-2, sp. gr. 8-8.2

(Contd.)

Minerals	Chem. Comp.	Diagnostic Characters
Calaverite	(Au, Ag)Te$_2$ with gold pre-dominant	Pale yellow, H = 2.5, sp. gr. 9
Petzite	(Ag, Au)$_2$.Te	Steel-grey to iron black, H = 2.5-3, sp. gr. 8.7-9
Negyagite (Black tellurium)	Sulpho-telluride of Pb and Au with little Sb	Usually massive or foliaceous, dark lead grey, metallic, sectile, perfect cleavage, H = 1-1.5, sp. gr. 6.8-7.2

Tellurium is obtained with selenium from the anode slime of electrolytic copper refineries. When pure, it has a greyish white colour and a metallic lustre. It has a specific gravity of 6.3 and melts at 450°C and boils at 1400°C. The gold bearing tellurides are important gold ores.

Use

Tellurium is employed to produce tough hard rubber for sheathing hose and cables, as an alloy metal, for adding blue and brown colour to glass, to make lead harder and less corrosive and to improve the creep of tin. It is also used in the purification of zinc and in certain processes of electro-deposition on magnesium.

The principal use of tellurium is as an alloying metal in the production of free machining steels. The addition upto 0.1% tellurium improves the machinability of steel. Similarly, the addition of tellurium improves the machining characteristics and corrosion resistance of copper alloys. Photo-conductive mercury-cadmium-telluride is the most widely used infra-red sensing material for thermal imaging device used in military applications, such as night vision and navigation systems, and space systems. Of the total demands the 55% is in metallurgy, 25% in chemicals, 15% in electricals and 5% in other applications like pigment, in blasting caps etc.

Mode of Occurrence, Origin and Production

Though tellurium occurs native in nature, it mostly occurs combined with metals as tellurides, such metals being gold, silver, bismuth and lead. It is found associated also with selenium. Tellurides occur mainly in veins and replacement deposits in which they are associated with pyrite and other sulphides. The tellurides of the upper part of the veins are decomposed and most of the tellurium is removed in solution.

The extraction of tellurium in India on laboratory scale was initiated by HCL in 1978. The existing tellurium recovery process has been modified. Presently, tellurium is being recovered from deselenised slimes instead of 'Dorrfurnace' slag. Productions of tellurium in 1984 and 1985 were 310 kgs and 50 kgs respectively.

5. RARE EARTH GROUP OF METALS

A group of 17 metallic elements, consisting of lanthanides which comprise 15 chemically similar elements, scandium and yttrium, are the rare earths. The fifteen lanthamides are lanthanum, cerium, praseodymium, neodymium, promethium, samarium, europium, gadolinium, terbium, dysprosium, holmium, erbium, thulium, ytterbium and lutetium. They have atomic numbers ranging from 57 to 71, with lanthanum having 57 and lutatium 71. Since scandium and yttrium have chemical properties similar to lanthanides, they are also included under rare earth metals. Being quite uncommon in the earth's crust, they are called rare earth metals.

The world production of rare earth metals in 1989 in terms of their oxide contents was estimated as 56.7 thousand tonnes. China emerged as the largest producer with about 18.7 thousand tonnes in 1988 and 20 thousand tonnes in 1989 (World mineral statistics 1985-89). The world reserves in terms of rare earth oxides are placed at about 48 million tonnes, of which China is holding about 36 million tonnes, followed by USA (6.5 million tonnes), India (1.9 million tonnes), Australia (750 thousand tonnes) and other countries (USBM, Mineral Commodity Summaries).

Mineralogy

The rare earth metals are characterised by high density, high melting point, and high thermal and electrical conductance. For the sake of convenience, they are divided into two sub-groups :

   (i) light or cerium sub-group, and
   (ii) heavy or yttrium sub-group,

named after the most abundant element in the sub-group. The cerium sub-group consists of the first seven lanthanides named above, while the yttrium sub-group comprises the rest of lanthanides and yttrium. There are a number of rare earth minerals, of which a few important with their chemical composition and characters are included below in the Table 4.66.

Table 4.66—Chief Rare Earth Minerals and their Characters

Minerals	Chemical Composition	Characters
Phosphate		
Monazite	(Ce, La, Di) PO_4 with thorium oxides and silicates in varying proportions	Massive or rolled grains, light yellow to clove brown in colour, white streak, resinous lustre, H = 5.5, sp. gr. 4.9-5.3
Xenotime	$Y_2O_3.P_2O_5$, may contain erbium, cerium, thorium and silicon	Mainly in rolled grains, pale-yellow to reddish brown in colour, pale brown to yellowish streak, resinous to vitreous lustre, H = 4-5, sp. gr. 4.5
Carbonate		
Bastnasite	(RF) CO_3 R means Ce, La, Nd, Pr (Rare earths)	Wax-yellow to reddish brown in colour, H = 4.5, sp. gr. 4.95

(Contd.)

Minerals	Chemical Composition	Characters
Oxide		
Samarskite	$(R^{II}_3 \, R^{III}_2 \, Nb.Ta)_6 O_{21}$ R^{II}_3 = Fe, Ca, VO_2 etc. R^{III}_2 = Ce, Y, etc.	Velvet black in colour, dark reddish brown streak, H = 5-6, sp. gr. 5.6-5.8
Silicate		
Allanite	(Ca, Fe)$_2$ (Al, Fe, Ce)$_3$ (SiO$_4$)$_3$.(OH). Cerium-bearing epidote	Brown to black in colour, submetallic to resinous lustre, H = 5.5-6, sp. gr. 3-4.2
Gaddinite (yttrium earth)	Be Fe Y$_2$ Si$_2$ O$_{10}$ a complex group containing considerable cerium, scandium etc.	Black to greenish black or brown in colour, vitreous to greasy lustre, H = 6-7, sp. gr. 4-4.5

Monazite is the main source of rare earths in India and abroad. Bastnasite is mined extensively in China and USA. Xenotime is mined in Malaysia, Thailand and Australia.

Uses

The rare earths find use in petroleum refining, iron and steel, glass and other industries. Iron-cerium (30 : 70) alloy is used for producing spark in lighters. Misch metal, a complex alloy of rare earths, is employed as tracer bullet in projectiles and this with smarium is used in making permanent magnet. Smarium-cobalt magnet had long been in use and is now replaced by neodymium-iron-boron magnet. The presence of dysprosium improves the thermal stability. Rare earth polishers are used in optical lenses, television tubes and other glass products. The rare earths are also employed in making gas mantles, cracking catalyst, signalling, photography etc.

Cerium mineral, monazite, is used as a source of thorium besides that of cerium. It contains thorium oxide varying from 0.001% to as much as 31.5%, the worldwide average being 7.2%.

Thorite and thorianite though contain more thorium than that in monazite, these minerals are not preferred to monazite, as source of thorium.

Mode of Occurrence and Origin

Rare earth minerals occur associated with alkalic plutons and in placers derived from them. The alkali plutons include acid igneous rocks and pegmatites wherein they occur as accesory constituents. In pegmatites they form larger masses or crystals. Commonly cerium or uttrium group lanthanide elements partially or wholly replace calcium in minerals such as apatite and fluorite, and thus give rise to rare earth minerals.

The commercial source of rare earth minerals is the placer deposits in most of the cases. In India beach sands of both the eastern and the western coasts form the source for the rare earths. Cerium mineral, monazite, occurs in appreciable quantity in the beach sands. This is associated with other heavy minerals like garnet, magnetite, rutile, ilmenite, zircon and sillimanite. The parent source of monazite and other heavies are naturally the neighbouring monazite-bearing alkali plutons where natural concentration has gone on.

Distribution

The commercial source of rare earths in India is monazite mineral which is concentrated in beach sands. The state-wise distributions of exploitable zones of monazite are given below in the Table 4.67.

Table 4.67—Distribution of Monazite Concentrate in Beach Sands of India

State	District/Area	Other details
Andhra Pradesh	Bhimunipatnam, Visakhapatnam district	Reserves of 1000 cu.m of recoverable monazite have been estimated
Kerala	Quilon district	The average monazite content in beach sand is 0.5%
Orissa	Chhatarpur, Ganjam district	The average monazite content in beach sand is 0.4%
Tamil Nadu	Manavalakurichi, Kanyakumari district	The average monazite content in beach sand is 2%

The primary occurrences of rare earth minerals, found distributed in pegmatites and acid igneous rocks as accessory constituents, are not of commercial importance. Some of the recorded occurrences are given below in the Table 4.68.

Table 4.68—Primary Occurrences of Monazite in India

State	District/Area	Geological Details
Andhra Pradesh	Nellore Mica belt	The mica pegmatite contains sparsely distributed rare minerals like allanite, triplite, torbernite, samarskite, columbite, tantalite etc.
Bihar	Mica Pegmalite belt (Hazaribagh, Giridih, Monghyr districts)	-do-
	In Gaya district at Abrakhi Pahar near Singar (24°34':35°30;)	The presence of monazite with triplite and pitch blende has been recorded in mica pegmatite, emplaced within gneisses and migmatites. The occurrence is of mineralogical interest
	In Singhbhum district at Kanyaluka (22°29' : 36°31')	Monazite and xenotime associated with apatite have been recovered
Rajasthan	Mica pegmatite belt (Ajmer, Bhilwara, Alwar, Sikar, Udaipur)	The presence of rare earth minerals like allanite, samarskite, columbite-tantalite etc. are occasionally seen in mica-pegmatite.
West Bengal	Purulia district	Monazite and allanite associated with columbite and pitchblende have been noted.

Reserves

The Indian reserves of rare earths are predominantly of monazite ore. Monazite contains about 60% of rare earths of cerium group expressed as oxide (M_2O_3) plus an average of 7.2% thorium and minor yttrium. The monazite sands of Kerala and Tamil Nadu contain 0.2 to 0.4% uranium oxide (U_3O_8) and 5 to 11% thorium oxide (ThO_2) besides rare earth oxides - CeO_3 La_2O_3, Pr_3O_3, Nd_2O_3, Y_2O_3 etc. The total Indian reserves of rare earth metals in terms of oxides have been placed at about 1.9 million tonnes (USBM, Mineral Commodity Summaries, 1990).

Production

The present annual production of monazite is about 5,565 tonnes (1989), used mainly in the manufacture of rare earth chloride. The Indian Rare Earths Ltd. (IREL) reported the production from its deposits from Manavalakurichi in Tamil Nadu (4784 tonnes), Orissa Sand Complex in Ganjam district, Orissa (588 tonnes) and Quilon district, Kerala (193 tonnes). Kerala Minerals and Metals Ltd. (KMML) also produces a small quantity of monazite. The total production of monazite during 1987 and 1988 stood at 3746 and 5661 tonnes, respectively.

Mining and Processing

Mining of beach sand is being carried out by the Indian Rare Earths Ltd., a Govt. of India Undertaking, and Kerala Minerals & Metals Ltd., a Kerala Govt. Undertaking. It is mined both manually and by mechanical means using hydraulic dredging. After sun-drying the beach sand or the concentrate is fed to electromagnetic or electrostatic separators in a processing plant for physical separation of various mineral constituents including monazite. The locations of the different processing plants with their installed capacities are given below in the Table 4.69.

Table 4.69—Plants Producing Monazite from Beach Sand

Producer	Location of the Plant	Annual installed capacity
Indian Rare Earths Ltd. (IREL)	(i) Manavalarakurichi, Kanyakumari district, Kerala	4800 tpy
	(ii) Orissa Sand Complex, Malikhalo, Ganjam district, Orissa	1500 tpy
	(iii) Chavara, Quilon district, Kerala	200 tpy
Kerala Minerals & Metals Ltd.	Chavara, Quilon district, Kerala	N.A.

5

Non-Metallic Minerals

5.1 MINERAL FUELS

1. PETROLEUM AND NATURAL GAS

Introduction

Petroleum has been called 'liquid gold' because of its value in our modern civilization, and possession and lack of it can virtually affect the fortunes of a nation, both in peace and war. The worldwide demand for petroleum has grown up enormously within the present century, especially since petroleum products began to be used as fuel for power in gasoline, diesel and other types of engines. India's annual consumption of petroleum product was about 54 million tonnes in 1989-90, while it was about 43 million tonnes in 1985, about 37 million tonnes in 1983 and about 12 million tonnes in 1963. During the last 30 years the consumption has increased about fourfold and is increasing continuously at least by 10-12% a year. If the world's annual average per capita consumption of 0.4 tonnes oil is applied to India, then in future its requirement will amount to 150-200 million tonnes per year. The gas consumption will also increase correspondingly.

The present worldwide production of crude petroleum is about 3035 million tonnes to which India's contribution is about 33 million tonnes i.e. about 1.1% (Production Figure, 1990). The USSR continued to be the chief producer of crude oil constituting about 21%. With the break up of USSR, America has now become the leading producer. The other important producers are Canada, Mexico, Venezuela, UK, Iran, Iraq, Saudi Arabia, Indonesia, Nigeria and China.

Constituents and Properties

Crude petroleum is dark and sticky substance which was known as 'Pitch' in good olden days. It is a very complex substance, consisting of a large number of hydrocarbons, all mixed together, with minor oxygen, nitrogen and a little sulphur. The numerous members of the different CH series have differing properties. At normal temperatures some are gases, some are liquids and some are solid waxes. Since proportion of these varies in different oil-fields, no two oils are alike in their properties or constituents. Even the smell of crude petroleum is not alike, some has odour of foul eggs, while others smell as petrol. A crude oil may contain CH members that yield the oil a high gasoline content, or certain ones may be absent and the oil will contain little or no lubricants. Thus, oil may be referred to as paraffin base, or asphaltic and naphthene, or mixed base. The paraffin base oils have low sp. gravity and yield good lubricants while asphaltic base oil are heavy and may be usable only for fuel oil. The variations in these hydrocarbons are responsible for the differing characteristics of the substances, produced by the refineries.

Uses

Petroleum is the key fuel of modern times and is of immense value in the development/progress of a country. Our agriculture, industry, transportation and communication depend upon petroleum in hundreds of ways. The chief use of petroleum is to produce energy for power or heat and for lubricants. In the remote past Indians used to rub their bodies with this mineral oil in crude form believing that it toned up their muscles and made them active and quick. It may, however, be used in crude state for fuel oil or road oil. But most of it is refined into its component parts. The refining consists of heating crude oils and driving off as vapour first the most volatile parts, followed by the less volatile. These are condensed and fractionated to fluids, such as benzene, gasoline, distillate and kerosene. The residium is treated for lubricants and other constituents. Crude petroleum is a priceless raw material, from which are derived paints, perfumes, explosives, dyes, greases, asphalt, plastics, soaps and many other products. In addition, several organic compounds made from petroleum compounds are used for chemicals, medicines, solvents, textiles, resins, saccharine, antiseptics, rubber etc.

Origin

The organic origin of petroleum has been postulated by most of the workers in this field. It is believed that the organic remains, such as remains of plant and animal organism, accumulate in the bottom muds of lagoons or in depressions on the floor of shallow sea and become incorporated in the accumulating sediment. The bacterial actions that take place, decompose the organic matter slowly into the mother material of oil by removing oxygen and nitrogen and giving rise to other changes. The lower planktonic organisms, such as diatoms and algae, that thrive near the surface of the sea, are considered the most probable and important source materials. According to Trask the composition of average plankton is 24 per cent protein, 3 per cent fat and 73 per cent carbohydrate. The complex protein and carbohydrates yield oil after slow oxygen-free decomposition. Further, post-consolidation changes take place during migration. Pressure exerts some influence, but temperatures could not have been high because certain compounds escape at 140° to 300°C. During conversion of organic matter into petroleum, natural gas which is dominantly methane is also formed.

The organic theory of oil genesis appeared for the first time in the form of Lomonosov's distillation hypothesis (1763), according to which oil is considered to be a product of sub-surface distillation of coals. Different views were expressed as regards specific conditions required for the accumulation and transformation of source organic matter of oil. A new theory was by Gubkin (1932) that dispersed oil covers huge areas of the earth and is found in tremendous quantities. This developed into the concept of micro-oil present widely spread in stratisphere due to dispersed hydrocarbons. At the stage of katagenesis micro-oil becomes mature and acquires additional features similar to oil which is found in oil pools. The initial transportation of oil occurs in the form of solution in groundwater and gases. The subsidence of rocks containing micro-oil to zones of increased pressure and temperature is required for the development of the primary migration of micro-oil.

Formation of Oil Pools

The dispersed droplets of oil ultimately go to form oil pools under the following necessary conditions :

1. Migration and accumulation,
2. Suitable reservoir and cap rocks,
3. Suitable traps/structures.

The migration of dispersed droplets of oil generated within the source rock is caused due to :

(a) compaction,
(b) capillarity,

(d) its lower specific gravity than water, and
(e) currents of sub-surface waters.

The migration of oil leads to its accumulation in suitable reservoir rocks which must be both porous and permeable. The clay having high porosity but small permeability is not suited to the accumulation of oil. The most suitable reservoir rocks are sands and sandstones. Porous and covernous limestone and dolomite and other fissured and jointed rocks may also form suitable reservoir rocks. The greater the porosity, the greater the amount of oil that the reservoir rock can contain, and the larger the pore size, the greater the amount of oil it will yield, since rocks with small grains and small pores create more frictional resistance to flow. Further, a good impervious cap rock is necessary to retain oil in the reservoir rock. Clay and shale are the most common cap rocks. Some migration of petroleum could have occurred in a gaseous form along the highly impermeable micro-fractures and micropores towards the low pressure areas.

The migrating oil required to be arrested in some suitable traps where it accumulates to form an oil pool. Anticlines and domes are the commonest types of traps where the up-dip migration of oil is arrested when it reaches the top of the arch. There are many kinds of oil and gas traps which may be divided into structural and stratigraphic traps as given in the Table 5.1 below.

Table 5.1—Important Oil Traps

Structural Traps		Stratigraphic Traps	
Kinds	Details	Kinds	Details

1. Folds
  (a) Anticlines
  (b) Dome
  (c) Monocline
  (d) Syncline
	Upfolds giving rise to anticlines and domes, are the main source of oil so far produced.
	Folds with gentle dips (Monocline) offer larger areas for oil accumulation. Free gas is at the top, oil lies beneath and water below, e.g. Ventura, California. In deep wells where pressure is high, gas is contained with the oil and is liberated only upon drilling. Synclince in a few cases serves as oil trap where water is absent.
		1. Unconformities	Underlying titled sand beds may be sealed at the unconformity by overlying beds to form oil trap. Angular unfonformities are more effective e.g. Oklahoma oil field.

		2. Ancient shore-line	Sands deposited on the low-lying submerged coastal plains, tapering in shore and deeper off-shore may be covered by clays and form suitable oil reservoir.
2. Faults	In inclined beds an impervious shale may be faulted against the up-dip continuation of an oil sand causing an effective seal and permitting oil accumulations beneath it. A thick fault gouge may also serve as an effective upward seal to an oil sand.	3. Shoestring sand	These are bodies of sand enclosed within shaly beds.
		4. Sandstone lenses	These form oil traps when enclosed within shaly beds.
		5. Up-dip wedging of sand and up-dip porosity dimunition	These become fine grained impervious offshore or grade into shale and form an oil reservoir.
3. Salt domes	The up-thrust plastic salt plugs that pierce the overlying beds create excellent traps for oil where upturned edges of sandstone beds are sealed against the salt, e.g. Gulf coast of Texas and Luisiana. In some cases arched beds above salt dome or porous cap rocks also serve as reservoir.	6. Overlaps	These form the oil-seal.
		7. Reflected buried hills	Buried granite ridge of Texas where projected parts received a mantle of sediment, thick on flanks and thin out on the top, simulating anticline and forming overlap in the form of oil trap.
4. Terraces, fissures, igneous intrusions etc.	These also form oil traps. In Mexico igneous intrusions in the form of volcanic necks, dikes and sills have similarly upturned and sealed petroliferous beds to give rise to oil traps.	8. Buried coral reefs	These are found to from excellent oil reservoir in the offshore region.

It sometimes occurs near the surface of the earth, but most of it is stored at greater depths. Many deeper wells are known in California and Gulf coast exceeding 4500 metres in depth.

Distribution

Geologic

Petroleum is found almost invariably in marine strata or associated freshwater beds of sedimentary origin. The host rocks of commercial pools are sand, sandstone, conglomerate, porous limestone and dolomite and rarely fissured shale of all ages from upper Cambrian to Pliocene, the length of time being about 550 million years. But it is prolifically found in younger rocks i.e. Tertiary rocks. In India exploration of oil remained concentrated in the above Tertiary sediments.

A very thick Mesozoic and Tertiary sediments in Indus valley, which are considered a suitable geological regime for petroleum, wedge out towards Kutch and Cambay basin in Gujarat and Rajasthan shelf. A thick Tertiary development is also found in the foothill parts of Punjab and Ganga basins underlain by Mesozoic beds with a well marked unconformity. The marine and non-marine sedimentary formations of Tertiary age belonging to the geological province of Arakan - Naga hills extend in Assam towards north, while its southern extension is found in Andaman and south-western extension in the east coast. A small thickness of marine Miocene sediments are developed in west coast.

Geographic

Petroleum is found to some extent in every continent, in many islands of sea, and even beneath the oceans. In India natural oil seepage have been known for quite longtime, though their first description was made in 1825. Until 1956 only three oil-gas fields were known, all in Assam, namely Digboi (1889), Naharkatiya (1953) and Moran (1956). With the formation of Oil and Natural Gas Commission in 1956, extensive geological and geophysical exploration work in all the prospective areas of the country was taken up. With the result several oil and gas fields have been discovered like those at Ankaleshwar, Gujarat (1960), Bombay High (1974) and other places. The geographical distribution of oil and gas field in India is given in the Table 5.2 below.

Table 5.2—Geographical Distribution of Oil and Gas-Fields in India

State	Location	Remarks
Andhra Pradesh	Kaikalur, Chintalapalli and Mandepetta, Off-shore areas of Krishna-Godavari.	Natural Gas field
	Narsapur and Kaza structure in the off-shore area of Krishna Godavari	Oil and gas field
Arunachal Pradesh	Ningru and Dum Duma	Oil and gas field
Assam	Digboi, Moran, Dudrasagar, Changaigaon, Lakwa and Sonari	Oil and gas field

(Contd.)

State	Location	Remarks
	Adamtilla in Cachar district	Gas field
Gujarat	Ankaleshwar, Kalol, Navgaon, Balal, Dahej, Cambay basin (Nada), Kutchch off-shore	Oil and gas field
Rajasthan	Jaiselmer basin	Gas field
Tamil Nadu	Off-shore areas of Cauvery at Kovil-Kallapali and Nariman	Oil and gas field
Bombay high	Several structures like in B-74, D-18, B-174, Panna east etc. accounts for the major supply of oil and gas	Oil and gas field

Besides the above oil has been struck in the off-shore areas of Andaman islands, Py-3 and Daman.

Reserves

The all-India reserves for both on-shore and off-shore areas have been placed at 806.15 million tonnes of crude oil and 729.79 billion cubic metres of natural gas (as on 1.1.1991). The recoverable reserves of both crude oil and natural gas, statewise, as on 1.1.1991 are given in the Table 5.3 below.

Table 5.3—Recoverable Reserves as on 1.1.1991 (Proved and indicated balance recoverable reserves)

	Petroleum (in million tonnes)	Natural gas (in billion cu. m.)	
A. On-shore			
Assam (includes oil reserves in Nagaland, Tamil Nadu and Andhra Pradesh, and Natural gas reserves in Tripura, Nagaland, Tamil Nadu, Arunachal Pradesh and Andhra Pradesh)	156.22	151.68	
Gujarat	158.26	93.99	
Rajasthan		1.22	
B. Off-shore			
Bombay high	491.67	483.50*	
Total India	806.16	729.79	* Reserves of other off-shore areas are not available.

Production

Domestic production of crude petroleum, while maintaining a rising trend, reached the highest level of about 34 million tonnes in 1989-90 from about 25 million tonnes in 1983 and 30 million tonnes in 1985. The productions made in 1990-91, 1991-92 and 1992-93 were about 33 million, 30 million and 27 million tonnes, respectively. Bombay high continued to be the largest producer of petroleum (crude) in 1992-93 with 59% of the total output, followed by Gujarat (21%) and Assam (18.5%). The remaining small quantity of petroleum crude were reported from Tamil Nadu and Arunachal Pradesh. The output of natural gas (utilised) were about 13 thousand, 12 thousand and 14 thousand million cu. m. during 1990-91, 1991-92 and 1992-93, respectively.

During 1992-93 about 79% of the production of natural gas came from Bombay high, followed by Gujarat and Assam. Tamil Nadu and Tripura also reported nominal production of natural gas.

Petroleum Refining

The installed refinery capacity in the country in 1989-90 was 49.85 million tonnes of crude oil per year spread over 12 refineries. The location and capacity of each of the above refineries are given below in Table 5.4.

Table 5.4—Details of Petroleum Refineries in India

Location	Name of Company	Capacity (million tonnes per year)
1. Barauni	Indian Oil Corpn. Ltd. (IOC)	3.30
2. Digboi	Indian Oil Corpn. Ltd. (IOC)	0.85
3. Gauhati	Indian Oil Corpn. Ltd. (IOC)	0.50
4. Koyali	Indian Oil Corpn. Ltd. (IOC)	9.50
5. Haldia	Indian Oil Corpn. Ltd. (IOC)	2.75
6. Mathura	Indian Oil Corpn. Ltd. (IOC)	7.50
7. Madras	Madras Refineries Ltd. (MRL)	5.60
8. Cochin	Cochin Refineries Ltd. (CRL)	4.50
9. Bombay	Bharat Petroleum Corpn. Ltd. (BPCL)	6.00
10. Bombay	Hindustan Petroleum Corpn. Ltd. (HPCL)	3.50
11. Visakhapatnam	Hindustan Petroleum Corpn. Ltd. (HPCL)	4.50
12. Bongaigaon	Bongaigaon Refinery and Petro-Chemicals Ltd.	1.35
Total		**49.85**

In addition to above, there is a swing capacity of 2 million tonnes per year in HPCL at Bombay.

The total production of petroleum products in 1989-90 was 48.69 million tonnes.

It has been decided by the government to set up a 6 million tpy and a 3 million tpy joint sector refineries at Karnal in Haryana and Mangalore in Karnataka, respectively. It has also been decided to set up a 2 million tpy refinery in Assam, having a provision of increasing its capacity to 3 million tpy.

2. COAL AND LIGNITE

Introduction

Coal meets a large part of our energy needs and plays a very important role in mineral industry. It outranks all other sources of utilizable energy, like petroleum and natural gas, wood, water, power and solar and atomic energy. Though the occurrence of coal is widely distributed, 4/5 of the reserves (80%) are in only three countries, Russia, USA and China. The total world reserves of coal and lignite is estimated at about 13,800 billion tonnes which will last hundreds of years to come. India's industrial development has been founded on the country's coal resources. A number of steps, taken to step up coal and lignite production, have resulted in continuously upward trend in production touching 226 million tonnes in 1990-91 against world production of 4738 million tonnes in 1990.

Physical and Chemical Constituents

Physical

The physical (visible) components of coal are the following :

(i) **Vitrain :** It is made up of bright, glassy looking, jet like coal band with conchoidal fracture. Woody structure is not visible. Vitrain imparts coking qualities to the coal.

(ii) **Clarain :** It is characterised by bright colour and silky lustre, and is composed of transluscent material.

(iii) **Durain :** It is dull coal band with earthy appearance, black to lead-grey in colour, hard, consists of cuticles and spores, and formed in water less toxic than for vitrain.

(iv) **Fusain :** It is carbonised wood, resembling charcoal and is believed to have been derived from burnt wood. It is high in ash content.

(v) **Attritus :** It is finely divided plant residue, composed of more resistant plant products.

Chemical

Coal is composed chiefly of carbon with varying proportions of hydrogen, oxygen, nitrogen and impurities, like sulphur, phosphorus, silt, clay etc. The fuel ratio (fixed carbon/volatile matter) which is the main factor in determining the rank of coal, is high in anthracite and low in lignite. The carbon is present as fixed carbon is the source at heat. The presence of sulphur is deleterious and more than 1.5% sulphur excludes coal for making gas or coke. Ash comes from the included silt, clay, silica or other substances in the coal and high content of it makes the better quality of coal to lower rank. The chemical composition of coal is determined by proximate analyses i.e. by analysing its moisture content, volatile matter, ash, fixed carbon and calorific value; or by ultimate analyses wherein carbon, nitrogen, hydrogen, oxygen, sulphur and phosphorus contents are found out.

The broad classification of different types of coal depends mainly on their carbon contents and may be put in the following ascending order :

Wood - 50% carbon, Peat - 52 to 60%, Lignite and Brown coal - 55 to 65%, Sub-bituminous - 65 to 70%, Bituminous - 70 to 85%, Cannel coal - 75 to 80% and Anthracite - 75 to 95%.

Above this the substance becomes graphite or almost pure carbon, and is unburnable.

Uses

Sizeable quantities of coal are consumed in railways, brick manufacturing and domestic fuel. However, its major use is in thermal power plants, iron and steel and cement industries. It finds its use in fertilizer, chemical and paper industries besides foundry, glass, textile, ferro-alloys, rubber, sugar, alloy steel, vanaspati, electrical, paint, abrasive, ceramic, refractory, cosmetics, asbestos products, insecticides, charge-chrome etc. Apart from direct use as fuel for power purposes, the coal is converted into hard coke for metallurgical purposes. Since the reserves of the coking coal is limited, it is blended with semi-coking and weakly coking coals before converting it into hard coke. During the manufacture of hard coke, several by-products are formed, such as volatile oil, tar ammonia, nepthalene etc.

Classes and Grades of Coal

Department of Coal, Ministry of Energy, Government of India, has laid down a new classification of various types of coal in 1979 which continued to be in force and is given below in Table 5.5.

Table 5.5—Classes and Grades of various types of Indian Coals

Sl.No.	Class	Grade	Specification
1.	Coking coal	Steel, Gr-I	Ash content not exceeding 15%
		Steel, Gr-II	Ash content between 15 and 18%
		Washery, Gr-I	Ash content between 18 and 21%
		Washery, Gr-II	Ash content between 21 and 24%
		Washery, Gr-III	Ash content between 24 and 28%
		Washery, Gr-IV	Ash content between 28 and 35%
2.	Semi-coking and Weakly coking	Semi-coking-I	Ash plus moisture content not exceeding 19%
		Semi-coking-II	Ash plus moisture content between 19 and 24%

(Contd.)

Sl. Class	Grade	Specification
3. Non-coking coal produced	A.	Useful heat value exceeding 6200 kilo cal/kg
in all states other than	B.	Useful heat value between 5600 and 6200 kilo cal/kg
Andhra Pradesh, Assam,	C.	Useful heat value between 4940 and 5600 kilo cal/kg
Arunachal Pradesh,	D.	Useful heat value between 4200 and 4940 kilo cal/kg
Meghalaya and Nagaland	E.	Useful heat value between 3360 and 4200 kilo cal/kg
	F.	Useful heat value between 2400 and 3360 kilo cal/kg
	G.	Useful heat value between 1300 and 2400 kilo cal/kg
4. Non-coking coal produced in Andhra Pradesh, Assam, Arunachal Pradesh, Meghalaya and Nagaland	Not graded	

Mode of Occurrence and Origin

Coal occurs in three forms :
 (i) horizontal and low dipping platforms,
 (ii) confined to major folded structures of simple composition, sometimes slightly dislocated, and
 (iii) associated with complicated folded structures, dislocated with sharp changes in attitude.

It has a vegetable origin and seems to have originated from the accumulation and wreckage of ancient forest floor. These forests were swamped under later sediments. The strata were folded, broken up, sunk under ancient seas and raised again by the movements of earth's crust. Tremendous pressure, heat from depths and exclusion of air have resulted in the formation of coal. This process is like that where wood is transformed into charcoal in the kiln without access to open air. The varying composition of coal depends on the nature of vegetable matter, of which it is formed of, and the conditions under which it has been formed, that is conditions of place, time, pressure and influence of heat and of chemical changes. The physical and chemical changes that take place in progressive change from peat to anthracite are summarised below in Table 5.6.

Table 5.6—Physical and Chemical Changes in progressive change from Peat to Anthracite

Physical	Chemical
1. Consolidation and stratification with drying and hardening	Progressive water loss upto Anthracite
2. Development of cleavage, schistosity and joints	Progressive removal of oxygen and preservation of hydrogen upto graphite stage
3. Reconstruction	Successive increase of ulmin and loss of bitumins
4. Colour change and increase of density	Manifestation of heavy hydro-carbons
5. Change in lustre and optical characters	Large loss of hydrogen in anthracite
6. Change in fracture-from bedding cleavage to conchoidal fracture	Increase of resistance to oxidation, heat and solvents

The successive higher ranks of coal are formed due to the above changes which are dependent largely on pressure, temperature and time.

Distribution

Coal occurs in post Devonian periods - Carboniferous to Miocene. The Carboniferous received its name because of its worldwide inclusion of coal. Permian coals are less wide-spread but are abundantly found in India and to some extent in China, Russia, South Africa, and Australia. The Triassic coals are found in Australia, Central Europe and eastern Asia, while the Jurassic in India, Alaska, China and Austria. Cretaceous contains extensive beds of coal in western North America and central Europe. The Tertiary yields most of the lignite in the world, though high rank Tertiary coals occur in Alaska and elsewhere. In India Tertiary coalfields are found in the north-eastern states and Jammu & Kashmir. In Antarctica, Miocene coal is found.

In India coal occurs mainly in the Gondwana Super-Group of rocks. Both the Barakar and Raniganj formations of lower Gondwana group (Permian) are quite rich in coal, though the Rajmahal, Kota, Chikiala, Jabalpur and Umia formations of upper Gondwana (Jurassic to lower Cretaceons) also contain some coal. The important coalfields in India are Jharia, Bokaro, Giridih, Karanpura, Ramgarh, Auranga, Hutar, Daltonganj, Deoghar and Rajmahal (Bihar), Raniganj, Barjora and Darjeeling (West Bengal), Godavari valley (Andhra Pradesh), Singrauli, Korba, Chirimiri, Sohagpur, Pench-Kanhan-Tawa valley, Patheskhera, Sonhat, Umaria, Korar, Sendugarh, Hasdo-Arand, Mand-Raigar, Johilla, Bisrampur, Jhagrakhand, Jhilmili, Lakhanpur and Mohpani (Madhya Pradesh), Chanda, Kamte, Umrer and Bander (Maharashtra), and Talcher and IB - river (Orissa). The Tertiary coalfields are Namchik-Namphuk of Arunachal Pradesh, Makum, Delli-Jeypore and Mikir hills of Assam, Khasi, Jaintia and Garo-hills of Meghalaya, Borjan of Nagaland and Jigni-Mitka-Jangalgali and Kalakot of Jammu (J & K state).

Jharia is the most important coal-field in India, being responsible for about 35 per cent of Indian production and the storehouse of the best coking coals. This coal-field is located at about 275 km west of Calcutta, falling in Dhanbad district of Bihar, and lies between longitudes 86°06' and 86°30' and latitudes 23°38' and 23°52', covering an area of about 450 sq km. This field is roughly sickle-shaped, being about 19 km N-S and 38 km E-W. The Talchir, Barakar, Barren Measures and Raniganj formations are exposed. It is faulted against Archaeans along the southern boundary. The beds dip inward towards the centre of the field. Two small horsts of gneiss are found, one in the north-eastern and other in the north-western part. There are also faults on the other sides as well as cross faults within the field, the most of them being of the 'sag fault' type. A large number of dikes and mica-peridotite traverse the field. The coal seams occur in both Barakar and Raniganj formations, the Barakars are by far the most important and contains atleast 25 seams of over 1 m in thickness. The seams are numbered I to XVIII from below upwards, the seams X and above being of good quality. Out of the total reserves of about 20,000 million tonnes of coal upto a maximum depth of 1200 metres, about 8,000 million tonnes are coking coals.

The occurrences of lignite are known in the Neyveli area, South Arcot district, Tamil Nadu; at a number of places in Pondichery; Cutch, Bharuch, Surat, Bhavnagar and Bhuj, Gujarat; Bikaner, Nagar and Jaisalmer districts of Rajasthan; and Baramula district, J & K. They all belong to Tertiary formations. The major deposits of lignite are in the Neyveli area, South Arcot district, Tamil Nadu, where it is interbedded with Cuddalor sandstone (Miocene) and occupies about 300 sq km of area.

Reserves

Total known reserves of all types of coal in Gondwana and Tertiary coal-fields of India, estimated for seams having 0.9 m and more thickness and to various depths upto 1200 m, stood at about 192 billion tonnes as on 1.1.1991. Out of these, reserves of prime coking coal have been placed at 5.3 billion tonnes. Statewise break-up of reserves are given in the Table 5.7 below.

Table 5.7—Reserves of all types of Coal as on 1.1.1991
(in million tonnes)

(For seams 0.9 m and above in thickness and upto max. depth of 1200 metres)

Field	Proved	Indicated	Inferred	Total
A. Gondwana Coal Fields				
1. Andhra Pradesh (Godavari valley coalfield)	5278.30	1650.50	3842.55	10771.35
2. Assam (Singrimari coalfields)	-	2.79	-	2.79
3. Bihar (Jharia, E & W Bokaro, Ramgarh, N & S Karanpura)	27787.14	27477.64	6820.07	62084.85
4. Madhya Pradesh	8566.91	18843.20	9643.07	37053.18
5. Maharashtra	2891.16	1307.02	1873.00	6071.18
6. Orissa	4826.79	20621.05	18856.59	44304.43
7. Uttar Pradesh (Part Singrauli)	662.21	400.00	-	1062.21
8. West Bengal	8714.00	13573.17	7860.00	30147.17
Total	**58726.51**	**83875.37**	**43895.28**	**191497.16**

(Contd.)

Field	Proved	Indicated	Inferred	Total
B. Tertiary Coal Fields				
1. Arunachal Pradesh (Namchik coalfield)	31.23	11.04	47.96	90.23
2. Assam	133.38	64.38	94.63	292.39
3. Meghalaya (Garo, Khasi and Jaintia hills)	88.99	69.73	300.71	459.43
4. Nagaland	3.43	1.35	15.16	19.94
Total	**257.03**	**146.50**	**458.46**	**861.99**
Grand Total of Gondwana and Tertiary coal-fields	**58983.54**	**84021.87**	**49353.74**	**192359.15**

Total known reserves of lignite stand at about 6.5 billion tonnes as on 1.1.1991. Statewise reserves are shown below in the Table 5.8.

Table 5.8—Reserves of Lignite (in million tonnes)

	Field	Total Reserves (million tonnes)
1.	Gujarat	383.00
2.	J & K	90.00
3.	Kerala	100.00
4.	Pondicherry	580.00
5.	Rajasthan	870.00
6.	Tamil Nadu (Neyveli)	4450.00
	Total (known reserves of Lignite in India)	**6473.00**

Mining

The mining of coal is carried out by the public sector companies except TISCO which obtains coking coal from its captive mines. There were six coal-producing companies in the public sectors. Four of them, namely BCCL (Bharat Coking Coal Ltd.), CCL (Central Coal Ltd.), WCL (Western Coal Ltd.) and ECL (Eastern Coal Limited), were subsidiaries of a holding company, CIL (Coal India Ltd.). The SCCL (Singareni Collieries Co. Ltd.) is a joint undertaking of the Govt. of India and the Government of Andhra Pradesh. The sixth is an undertaking of the Government of J & K. The coal mines in Assam and Meghalaya are operated by the C.I.L. directly through its division, North Eastern Coal.

Coal is produced from both open cast and underground mines. Thrust is now given to increase the production from the opencast mines. Many opencast mines are of 10 million capacity. In so far as coal distribution is concerned, allocations of coking coal to the steel plants is made by Coal controller, while there is no statutory control for the distribution of non-coking coal.

Lignite is mined at present in three public sector mines, one in Tamil Nadu (Neyveli) and two in Gujarat (Umerser and Akhrimota).

Washeries

Coal washeries are operated by different companies. At present 18 washeries are being operated. Six washeries, namely Kargil, Kathra, Sawang, Gidi, Basara I and West Bokaro, beneficated medium-coking coal, while one i.e. Nandan in Nowrozabad in M.P. was set up in 1984-85 for washing non-coking coal. The rest eleven, namely Dugda-I, Dugda-II, Bhojudih, Patherdih Sudamdih, Lodna, Jamadoba, Chasnala, Durgapur (DPL), Durgapur (DSP) and Moonidih beneficiated prime-coking coal. The middlings generated in the washeries are used in the steel plants, thermal plants, loco manufacturing units etc.

The raw coal used by the washeries contains 24 to 33% ash content. The ash contents in the washed coal and middlings, produced by the washeries, vary from 19 to 22% and 35 to 40%, respectively. The rejects in most washeries contain over 50% ash.

Production

The production of coal is gradually having a rising trend. About 150 million tonnes was produced in 1985, 177 million tonnes in 1987, 188 million tonnes in 1988 and 201 million tonnes in 1989-90. Bihar continued to be the leading producer accounting for about 30% output (1992-93), followed by Madhya Pradesh (29.8%), Andhra Pradesh, Orissa, Maharashtra, West Bengal and Uttar Pradesh. Small quantities of coal are also produced in Assam and J & K. Bihar, West Bengal, M.P. and Andhra Pradesh together account for about 90% of the total mines in the country. The production of coal during 1990-91, 1991-92 and 1992-93 were about 212 million, 229 million and 233 million tonnes respectively.

5.2 GEMSTONES

Diamond, being the most valuable gem, is dealt with separately, while all other gems, precious and semi-precious, are discussed, hereunder, the heading "Gemstones (General)".

1. GEMSTONES (GENERAL)

The minerals with special beauty and physical properties which attract attention to possess them for personal adornment and decorative purposes are gemstones. Beauty, rarity and durability are the main virtues of a gemstone. They must be hard enough to resist the mechanical and chemical actions of every day's life. The cutstones are known as gems, while the uncut-ones are gemstones. The typical characteristics of gems are sheen, chatoyancy, opalescence, iridescence, and dichroism. The gemstones may be grouped into :

(i) precious stones, namely, diamond, emerald, sapphire, ruby, precious opal and pearl; and

(ii) semi-precious stones which include aquamarine (beryl), moonstone (felspar), ametbyst (quartz), peridote (olivine) and a good number of minerals.

The standard international weight of gems are in metric carat which is 1/5 of gramme (200 mg). For pearls, the unit of weight is a pearl grain (1/4 carat or 50 mg).

India is the largest exporter of cut and polished gems (70% of world's sale), but continued to depend on imported raw materials. The Diamond and Gem Development Corporation of India have set up Diamond and Gem parks for cutting and polishing of gems at Sachin near Surat (Gujarat), Jaipur (Rajasthan), Haldwani (U.P.), Tiruchi (Tamil Nadu) and Govindpur, near Jehanabad (Bihar).

Classification and Identification

The gems may be classified based on their colour, transparency, chemical composition, specific gravity and host rock. The properties like hardness, refractive index, dichroism and ultraviolet tests besides specific gravity are employed in identification of gems.

Colour

Colour is not a reliable character in classification and identification of gems and gemstones, since there are different colours for the same specie and nearly same tint of colour for different mineral species. The natural gems may, however, be classified as :

(i) idiochromatic minerals in which the colour is due to the element which is the essential part of composition like copper in malachite imparting green colour to it, and

(ii) allochromatic minerals in which colouring element exists as an accidental impurity as iron in sapphires gives it one of the three colours, yellow, green or blue. Due to mixed impurities and inclusions translucent or opaque gems are formed.

Transparency

Based on transparency the gems may be grouped into :

(i) transparent gems which may be coloured, tinted or colourless, and

(ii) translucent or opaque.

The former is comparatively costly and preferred.

Chemical Composition

Natural gems can be grouped into pure element (diamond), oxides (ruby, sapphire, amethyst etc.), silicates (emerald, garnet, tourmaline etc.), sulphides (pyrite, blende), carbonates (malachite, pearl, calcite), phosphates (apatite) and fluorides (flourspar). The majority of gems belong to the silicate group.

Specific Gravity

The classification of gems based on specific gravity may be reliable. Most of the synthetic unmineralised products used as semi-precious gems like opal, amber, erinoid etc. have sp. gr. between 1 and 2. The specific gravity of synthetic emerald is 2.65, while that of natural one is 2.71. The diamond and corundum (ruby and sapphire) have specific gravities 3.52 and 3.99, respectively.

Host Rock

Certain gemstones are associated with a particular type of rock, like diamond and pyrope with periodotite (kimberlite), tourmaline, spodumene, emerald, aquamarine, amazonstone, etc. with granite - pegmatites, corundum (ruby, sapphire), cancrinite, etc. with syenite, and quartz, opal, zeolite, chert etc. with volcanic rocks.

Identification of gems may be made determining specific gravity by using Walker' Steel Yard (for big pieces), density bottles and by a set of heavy liquids (sinking and floating test). Clereci solution (density 4.25), methylene iodide (density 3.32) and liquid bromoform (density 2.88) are the usual heavy liquids employed for this test. The methylene iodide is diluted with benzal (density 0.88) or bromoform to have liquids of desired densities.

The hardness test can be carried out by a set of standard specimens of gypsum (2), copper coin (3), pen knife (5.5), quartz (7), topaz (8) and corundum (9). Absolute hardness determination is laborious. The refractive index test is quite reliable, since each gem has a constant value. If a grain is immersed in a liquid of known R.I., it is possible to determine whether it has a higher or lower R.I. than the liquid as examined under the microscope. The property of dichroism is taken advantage of in identification of gems. Some gems show no dichroism, some weak and others

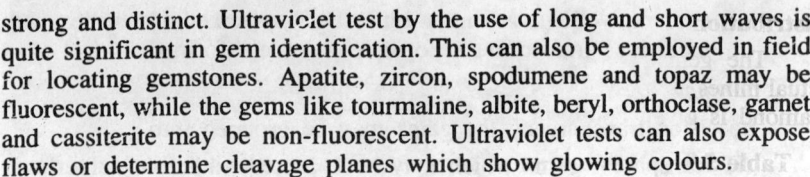

strong and distinct. Ultraviolet test by the use of long and short waves is quite significant in gem identification. This can also be employed in field for locating gemstones. Apatite, zircon, spodumene and topaz may be fluorescent, while the gems like tourmaline, albite, beryl, orthoclase, garnet and cassiterite may be non-fluorescent. Ultraviolet tests can also expose flaws or determine cleavage planes which show glowing colours.

Uses

The gemstones find wide application in jewellery. The semi-precious stones apart from jewellery are used as ornamental stones for wall decoration, floors of building, vases, statuettes etc. Some of which are neither useful for gems nor for ornamental purposes, are used as abrasives where hardness is taken into account. Diamond being the monarch of all gems fetches fabulous price. Some other gems like sapphire, ruby and emerald are also very costly. The precious stones are considered as a concentrated wealth.

Diamond which is primarily a gemstone, is also used as industrial stone. The industrial diamond is badly coloured or flawed stone in small fragments. Rubies and sapphire are used as watch jewels. Quartz and tourmaline are used in optical instruments. The balance bearings and the mortars and the pestles are made of agate.

The natural gems are being replaced by synthetic stones and imitations. The synthetic stones may be manufactured by fusing pure alumina. For colour a metallic oxide is added while fusing.

Mode of Occurrence and Origin

Gemstones are found dominantly in igneous rocks, granitic pegmatites and as detrital grains in alluvial. They occur as veins (quartz), in cavities (opal, agate, zeolites etc.), in volcanic pipes (diamond) and in a variety of ways.

During crystallisation of magma the magnesium olivine (peridot) and the ultrabasics are first to separate out. Some of these ultrabasics occupy volcanic vents as pipes or dykes giving rise to diamondiferous kimberlites with crystallisation of diamond, pyrope and olivine. The remaining magma becomes enriched in silica with lithium, beryllium, boron and volatile constituents like fluorine, chlorine and hydroxyl, and result into pegmatites with large crystals of quartz, felspar and mica with gems like tourmaline, beryl, topaz, spodumene, cassiterite etc. Ruby and spinel are formed due to contact metamorphism of impure limestone. Nephrite, staurolite, kyanite and sillimanite result under regional metamorphism. Lepis-lazuli, serpentine and thomsonite are formed by the action of volatile constituents on intruded rocks. Hydrothermal solutions yield opal and agate, while supergene processes give rise to turquois.

The gemstones get concentrated in the alluvial and massive placers by weathering and transportation. These are quite important source of gemstones from economic point of view, since it avoids the extraction cost from the hard rock.

Distribution

The gemstones are collected from various localities in India, but actual mines are only a very few. The distribution of important ones except diamond is given below in the Table 5.9.

Table 5.9—Distribution of some important Gemstones in India

Gemstone	State	District	Locality	Other details
Emerald (Transparent pale green to sea green beryl)	Rajasthan	Udaipur	Kaliguma and Tikhi	Emerald found at the contact of pegmatites intruding schists, mainly amphibole-schists.
Aquamarine (Transparent pale blue beryl)	Tamil Nadu	Coimbatore	Kangayam Chennimalai	It has been reported from pegmatites intruding Archaean schists and gneisses.
Sapphire (Transparent blue corundum)	J & K	Doda	Paddar	Sapphire occurs in syenite and pegmatite, intruding actionlite-tremolite-schists. Nearby alluvial placers have been intermittently worked for sapphire. The mine remains closed since 1973.
Ruby (Transparent red-corundum)	Andhra Pradesh	Khammam	Gobbagurti	The country rocks are sillimanite corundum rocks-metamorphised schists and and pegmatites. Detrital ruby is got in soil covered areas.
	Karnataka	Bangalore, Chikmagalur, Mysore, Kolar	Several localities in the districts	Sporadic occurrences of ruby at the contact of basic and ultra-basic rocks with the country rocks
	M.P.	Bastar	Bhopalpatn Kuchnor	semi-gem variety ruby found in schist and gneisses, intruded by plagio-granite and metabasic dykes (Bengal group of rocks).
	Tamil Nadu	Salem Dharampuri	Sitampundi Palcode and Popparapalli	Ruby is occasionally picked along with corundum. Minor occurrence of gem quality was noticed.
Garnet	Andhra Pradesh	Khammam		Opaque to translucent varieties were mined in kyanite-schist.
		Nellore		Translucent purple coloured garnet*are found in mica-pegmatite.
	Bihar	Hazaribagh		It is reported associated with mica-pegmatites in several localities.

(Contd.)

Gemstone	State	District	Locality	Other details
	Karnataka	Bangalore, Hassan, and Mysore		The country rock is mica-schist. Pink transparent varieties have been found at several localities. Green garnet has been found at Mandhya. Stream sands have yielded reddish pink garnet.
	Rajasthan	Bhilwara, Ajmer, Jodhpur, and Tonk	Several localities	Working mines on small scales exist. Gem variety of garnet has estimated reserves of about 3400 tonnes in Bhilwara and Tonk districts. About 2000 kg, 1400 kg and 3200 kg of gem garnets were mined in 1987, 1988 and 1989, respectively.
Quartz and its different forms	Gujarat Maharashtra & Andhra Pradesh		Deccan Trap region	They have been collected from several localities.
(Amethyst cetrine, rock crystal, agate, chalcedony, jasper, chert, opal)	Bihar, A.P. and Rajasthan		Mica Pegmatites	-do-
	Madhya Pradesh	Bastar,	Govindpal	Smoky quartz with cassiterite in pegmatite is known to occur.
	Andhra Pradesh	Rajah-mundry		In Godavari river bed agate and jasper are found.
		Kurnool		Red and green jasper occurs in abundance in Bangapalie and Tadpatri shales.
Topaz	Bihar	Singhbhum		Occurs in thin veins and patches in kyanite rock of Lapsaburu, Ghagidih etc.

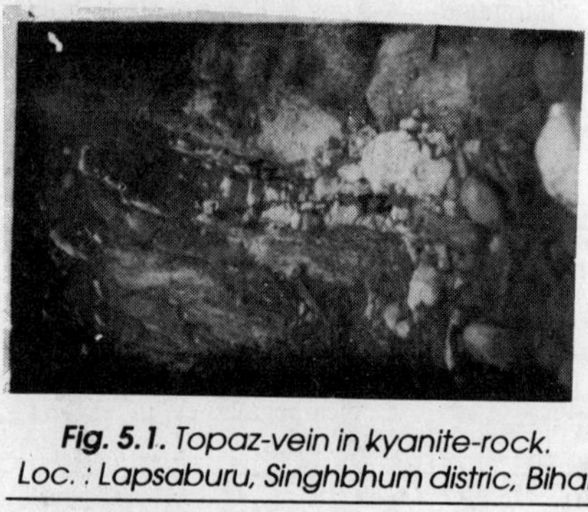

Fig. 5.1. *Topaz-vein in kyanite-rock. Loc. : Lapsaburu, Singhbhum distric, Bihar*

(Contd.)

Gemstone	State	District	Locality	Other details
Tourmaline	Bihar	Hazaribagh		Transparent green variety is recovered from Pilura.
		Singhbhum		Blue and brown tourmaline occurs near Lapsaburu.
Apatite	Rajasthan,	Ajmer,		Sea-green apatite is reported.
	Tamil Nadu	Devada		-do-
Kyanite	Kashmir			Gem quality kyanite is reported.
	Andhra Pradesh	Nellore Mica-belt		Blue to greenish kyanite in kyanite-schists near Shaw mine is reported.
	Karnataka	Chikmagalur		Transparent blue kyanite is reported at Kadmane and Melkoppa.
Cordierite	Karnataka	Bangalore		Semi-precious variety is found in Channapalma taluk.
Amazonstone	Andhra Pradesh	Nellore Mica belt		Small quantity is reported.
Lepidolite	Madhya Pradesh	Bastar	Govindpur	It occurs in good quantity.

Resources and Production

Mica-pegmatites and metamorphic suite of rocks with kyanite, sillimanite, corundum, garnet etc. like schists, khondalites and equivalent formations are the main sources for gemstones. Some of the semi-precious stones occur in trap rocks. But no systematic detailed evaluation of gemstones has so far been made in these rocks. Except for diamond, emerald and sapphire which fetch high price, low incidence of gemstones in host rocks would not be a good prospect. The alluvial placers formed by marginal streams bordering the granite masses with pegmatites and the metamorphosed sediments are considered most viable sites for the gem-stones. The important gemstone areas in India are shown in the map appended.

The so far known reserves of garnet, and productions of some gemstones are given below in the Tables 5.10 and 5.11, respectively.

Table 5.10—Reserves of Garnet in India as on 1.4.90

Gemstone	Proved	Probable	Possible	Total
Garnet (in million tonnes)	0.01	0.04	0.09	0.14

Table 5.11—Productions of some Gemstones in India

Gemstones	1990-91	1991-92	1992-93
Agate (in tonnes)	589	648	601
Sapphire	Only mine at Paddar, J & K state, remains closed since 1973		
Ruby	It is picked up occasionally along with corundum in Andhra Pradesh and Karnataka		
Jasper (in thousand tonnes)	5	4.7	4.3
Garnet (in thousand tonnes)	1.64	1.01	0.54
Fuchsite quartzite (in tonnes)	951	2120	

A brief description of some gems is given below.

(i) **Sapphire and ruby :** They are gem varieties of corundum and are respectively blue and red in colour. The other gem types of corundum are oriental amethyst, oriental emerald and oriental topaz which are purple, green and yellow, respectively.

(ii) **Emerald :** Emerald is transparent pale-green to sea-green variety of beryl.

(iii) **Semi-precious silica group of minerals :** Agate, amethyst, fuchite and jasper are the semi-precious minerals, belonging to silica group. They are described below.

Agate

Agate is a variety of chalcedony, showing regular or irregular bands with pleasant colours. Agate pieces after cutting and polishing are sold as semi-precious stones. There is a thriving cutting and polishing industry for agate pieces in Bharuch and Cambay. Studs, buttons, stone for ear-ring and curios are prepared from agate. Big pieces are used in making mortars and pestles for laboratory use. Agate pieces cut into requisite shapes are sold as fulcra of scientific balances and making edges, planes and bearings of precision instruments. The selected cut pieces find good export market in the European countries as decoration stones. Chalcedony pebbles are used as balls in the ball mills for crushing and grinding felspar, calcite and barytes.

It occurs mainly as fillings in the voids in the Deccan Trap rocks. The most important occurrences are in Rajpipla area and further west between the mouths of river Tapti and Narmada in Bharuch district, Gujarat, where it is found as pebbles in varying sizes associated with clay washed down with river flood. Other occurrences of economic importance are known at Amravati, Aurangabad, Buldhana, Chandrapur, Jalna, Nasik and Pune districts in Maharashtra, beds of Krishna and Godavari rivers in Andhra Pradesh, Rajmahal and Sahebganj districts in Bihar, Dhar, Mandsaur, Sehore and Shahdol districts in Madhya Pradesh, and Kuchch and Surat districts in Gujarat. The main production is from Bharuch district, Gujarat.

Amethyst

It is used as semi-precious stone and in making curios. It is mined in Penimalai R.F. area in Madurai district, Tamil Nadu.

Jasper

Jasper, a chalcedonic variety of silica, is usually of a deep red and brownish red colour and is used for making decorative stone. Jasper mining is being carried out mainly in Jodhpur district of Rajasthan.

Fuchsite Quartzite

Fuchsite (green quartzite) is utilised for making curios, paper weights, hand cuffs, studs and similar other artifacts. It is mined in Chitradurga district of Karnataka and Anandpur district of Andhra Pradesh.

2. DIAMOND

Introduction

The first diamond was found in India about 800 B.C. This country was a great producer of large sized diamonds such as Kohinoor and others, and has been home for diamond until the discovery of the Brazalian fields in the eighteenth century (1720). Any diamond not from India was regarded with suspicion. India which produced several thousand carats, now produces less than 2000 carats a year.

The present total world production of natural diamond is about 100 million carats (figure, 1990) to which India's contribution is only about 15,000 carats (0.015%). Australia as per production figure of 1990 is the world's largest producer (about 35 million carats), followed by Zaire (19 million carats), Botswana (17 m. carats), USSR (15 m. carats) and South Africa (8.7 m. carats). India covers about 70% of world sales inspite of having very little local production of natural diamonds. India's diamond industry concentrates on import of natural diamonds and re-export of cut and polished diamonds.

Properties

Diamond is a natural crystalline form of the chemical element 'carbon' and is the only gemstone made up of a single element. Crystallized under tremendous pressure, it is the hardest substance known. Chemically, it is identical with graphite, but in view of their different internal structures the physical properties of both differ radically. Diamond crystallises in the cubic system and the crystals are mostly octahedral or dodecahedral. It has a perfect cleavage and can be nicked, cracked or fractured by a sharp blow, provided the blow is in the cleavage direction. It has a hackly or more rarely . conchoidal fracture. Its sp. gr. varies from 3.47 to 3.56, but drops to 3.2 in case of carborando. Diamond is quite durable and its high index of refraction (2.41 to 2.43) and dispersion give it unexcelled brilliance and 'fire'. It is generally colourless, but some are yellow, pink, red, orange, green, blue, brown and black. The colour can be changed in the atom smasher, known as the cyclotron and the diamonds coloured by this method are now being sold quite widely.

The hardness of diamond differs in different directions and is lower on the octahedral faces than on the cubic faces. This property is used in cutting the diamonds themselves and in making diamond tipped tools.

It has a strong mirror like lustre and emits a distinct blue, green or yellow phosphorescene (luminescene) after being exposed to X-rays, cathod or ultraviolet rays which is utilised in extracting them from washing concentrates. It is transparent to X-rays which property is also used to identify it. It is non-magnetic, repels water and adheres well to certain greases and so is used for recovering diamonds from concentrates. It is highly resistant to chemical reagents. It gradually burns away at a temperature of 700°C (1200° to 1300°F) in an oxyhydrogen blowpipe flame, giving off carbon dioxide and leaving a small residue due to minute inclusions of pyrope, chrome-diopside, olivine, chromite, magnetite etc.

Types

Based on their crystallisation, the transparency of crystals and the presence of inclusions or flaws, the following four varieties of diamonds are distinguished :

  (i) **Diamond proper :** Well formed, perfectly transparent, colourless or uniformly tinted crystals without inclusions, cracks or other flaws, which may be used as gems,

 (ii) **Bort :** Opaque or slightly translucent or cracked crystals or crystal fragments with inclusions and mechanical flaws,

(iii) **Balas :** Spheroid aggregate with radiated structure, enclosed in a denser and harder surface layer, and

 (iv) **Carborando :** Fine-grained and compact, occassionally slightly porous, dark grey or brown aggregates, usually occurring as irregular fragments.

Uses

Diamond is the most valued gemstone. It also finds numerous industrial and technical applications because of its hardness, tremendous resistance to abrasion, great mechanical strength and chemical stability. It is widely used in engineering, machine tools and instrument making industries, in hard rock boring and in the ceramic, glass, watch, abrasive and other industries. Diamond tipped tools greatly increase efficiency and improve working condition. They are invaluable for precision work when tools made of the hardest alloys prove to be not strong enough. Stone cutting industry is its recent application. Diamond bits are used for drilling holes. The estimated apparent consumption of industrial diamond in India is about 2.5 million carats a year.

Industrial diamonds account for nearly 80% of the world output. However, value of gemstones which make up about 20% of the total world output by weight, exceeds the value of industrial varieties more than three-fold.

Mode of Occurrence

Diamond occurs in two forms :

(i) Primary deposits, associated with kimberlite pipes or other forms of ultrabasic intrusives, and

(ii) Secondary (Placer) deposits, associated with stream and beach gravels.

Primary Deposits

The kimberlite pipes are tabular or funnel shaped, vertical or steeply dipping with a width of tens of metres and, sometimes, several hundred metres. These pipes diminish in size with depth and gradually pass into veins. They are filled with explosive breccia, consisting of fragments of wall rocks and other rocks entrained by the magma when it rose from the deep seated layers of the earth crust, cemented by kimberlite, a typical ultrabasic effusive rock with high alkaline content. Diamond occurs as crystals sparsely disseminated in the kimberlite, while a few occur in eclogite (garnet-pyroxene rock) inclusions. The kimberlite is made up of Mg-Olivine, pyroxene and phlogopite-mica with minor amounts of picroilmenite (Mg-ilmentite), pyrope garnet, chrome-diopside, chromite, magnetite, perovskite, apatite, rutile etc. Pyrope, chrome-diopside and picroilmenite serve as guides to diamond deposits.

Secondary (Placer) Deposits

The placers are the most common type of diamond deposits. They are derived from the destruction and reworking of primary diamond bearing rocks by various surface processes.

Origin

Primary deposits

Diamond is suposed to have formed in gas-saturated ultrabasic magma. This magma is capable of dissolving great quantities of carbon, which is the source from which diamonds crystallise. When the magma welling upto upper layers of the earth's crust cooled, the pressure of the gases increased, and a number of successive explosions took place at a certain depth. Barrel shaped holes, called pipes, were blown out of the gases. The diamond bearing magma together with fragments of the wall rocks and of the rubbles entrained by the magma when rose from its chamber, congealed in these pipes. Of all the various theories of the formation of primary diamond deposits, this seems to be the most possible one. There are three views as to the place where diamonds are formed. They are :

    (i) the diamonds crystallised in situ,

    (ii) the diamonds were originally contained in the underlying layers of eclogite, which was fused by upwelling kimberlite that carried along the released diamonds, and

    (iii) the diamonds crystallised in the original magma chamber and were carried up as crystals in kimberlite to the present site of cooling.

Most scientists believe that the main diamonds crystallise in the magma chamber before the eruption begins and only a small quantity of diamonds mostly small crystals, form during the rise of magma.

Secondary (Placer) Deposits

Depending on the mode of their formation the following gentic types of diamond bearing placers are, generally, recognised :

    (i) alluvial,

    (ii) sea-beach,

    (iii) aeolian, and

    (iv) mixed origin.

The alluvial placers are the most common and are derived from the denudation of primary deposits, and diamond bearing placers of other types by the action of running water, e.g. High Plateau placers of Brazil and Lichtenburg placers of South Africa. The sea beach placers are confined to beach barriers and terraces, and were formed through the reworking and reposition of diamond-bearing bed by surf and coastal currents, e.g. placers of Namaqualand (South Africa). Aeolian placers are formed in regions with desert climate through the agency of wind as some placers in south-west Africa. Placers of a mixed origin are quite widely distributed. The most typical of these are gulch placers which occur in valleys of small creeks and gulches with an irregular flow.

There is no general rule governing the distribution of diamonds in placers. In most case the diamond content tends to increase towards the lower layer of the gravels as well as in the irregularities of the bed rock. Most of the placers being worked contain a minimum of 0.5 to 1.0 carat diamond in per cubic metre of gravels.

Distribution

Diamond deposits in India occur in three types of geological settings :

 (i) kimberlite pipes,

 (ii) conglomerate beds, and

 (iii) alluvial gravels.

The main diamond bearing areas are the Wajrakarur kimberlite pipes in Anantpur district, Andhra Pradesh, Panna belt (conglomerates separating different formations of upper Vindhyan, kimberlite pipes and gravel beds), Madhya Pradesh, Ramallakota-Banganapalli conglomerate (Lower Vindyan) in Kurnool district, and the gravels of Krishna river basin, Andhra Pradesh. The original home for diamond, whence they reached conglomerates, is considered to be the basic dykes of Bijawar (Cuddapah) which traverse the local crystalline rocks at many places. Stray occurrences of diamond are also recorded in other localised areas of Uttar Pradesh, Bihar, Orissa and Maharashtra. The sands of the North Koel river near Sima (23°35' : 84°17') in Palamau district and of the Sankh river near Rajadera (23°16' : 84°17') in Ranchi district are reported to be worked for diamond.

A number of new diamondiferous pipe rocks in addition to several conglomerate and gravel beds have been located in Andhra Pradesh and Madhya Pradesh, most of which are at the various stages of assessment.

Reserves

Reserves have been assessed for the Panna belt, M.P. In addition to recoverable reserves of about a million carats, there are conditional sub-marginal resources of 79 thousand carats. The reserves of diamond as on 1.4.1990 are given in Table 5.12 below.

Table 5.12—Recoverable Reserves of Diamond as on 1.4.1990
(in carats)

	Proved	Probable	Possible	Total	
India : Total	1260785	—	209381	1470166	(Grade : Unclassified)
Madhya Pradesh (Panna only)	1065795	—	130359	1196154	
Conditional (sub-marginal resources)	—	—	79022	79022	
Andhra Pradesh	194990	—	—	194990	

Production

Madhya Pradesh is at present the lone producer of diamond in the country. It produces about 18,000 to 19,000 carats a year. The productions for 1989-90 to 1992-93 are given below in the Table 5.13.

Table 5.13—Production of Diamond - 1989-90 to 1992-93 (in carats)

1989-90	1990-91	1991-92	1992-93
16,473	17976	18247	18752

Of the total productions (1989-90), about 48% was of gem variety, 35% off colour and 17% industrial.

The National Mineral Development Corporation Ltd. (NMDC) and the Madhya Pradesh Directorate of Geology and Mining are the two producers of diamonds. Majhagawan diamond mine (Panna district) is operated by NMDC. It is the only mine in the country which is more or less continuously under operation since a long time. The mine is worked by open-pit in tuff rock and by underground method in the main pipe rock. Diamonds are also recovered from conglomerate and gravel beds at shallow depths (not allowed to go beyond 3 m) by private small operators on the basis of annual permit granted by the Govt. of Madhya Pradesh.

Industrial Diamond

Though India is the largest exporter of cut and polished diamonds in the world, it continued to depend on imported raw material. The cutting activity is highly labour-intensive. Nearly 3,50,000 people are employed in the Industry. The cutting and polishing centres are mostly located in Surat, Navasari, Palanpur, Bhavnagar, Bombay and Cambay. Another significant development in this direction is the setting up of

5.3 ABRASIVE MINERALS

The materials used in cutting, crushing, grinding, abrading and polishing are abrasives. These are of two types, natural or mineral abrasives and artificial abrasives. The artificial abrasives are also manufactured from the mineral products e.g. carborundum (silicon carbide), alundum (fused alumina), norbide (boron carbide) etc. The capability of the material to cut, crush, grind, scour and polish is dependent upon its hardness which is the essential property of an abrasive.

The mineral abrasives may be classified into :

1. High grade natural abrasives, consisting of diamond, corundum, emery and garnet,
2. Siliceous abrasives, including various forms of natural silica and siliceous rocks, like quartz, sand, diatomite, tripoli, pumice and pumicite, flint, chert, sandstone, quartzite, quartz-mica-schist etc.
3. Soft abrasives, comprising felspar, dolomite, magnesite, bauxite, limestone, chalk, clays, fuller's earth, talc, rouge etc.

These abrasives may be used in natural form, after shaping or after being ground into grains or powders. They may be made up into wheels, paper or cloth. There is wide use of abrasives in various industries. The largest consumer is the automobile industry. The aeroplane industry demands industrial diamonds to a great extent. Metal polishes require emery, pumice, diatomite, silica, tripoli, chalk, fuller's earth, clay, bauxite, magnesite and oxides of metals. The soap industry needs abrasives like pumice, felspar, diatomite, silica, chalk, bentonite and talc.

The important abrasive are dealt hereunder.

1. High Grade Natural Abrasives

(a) Industrial Diamonds

There are two types of industrial diamonds;

(i) carbonado, which is black, hard and tough, and is employed for diamond drill bits, dies in wire drawing, tools used in boring hard substances (metals), dressing and truing abrasive wheels and in the auto and electrical industries, and

(ii) bort which is opaque or slightly translucent crystal fragments with mechanical flaws and inclusions, is mainly utilised in the manufacture of aeroplane and motor car engines for boring and abrasion of surfaces. Because of its lower cost it substitutes carbonados for many industrial purposes. Diamond dust prepared by grinding bort is used in metal bond for cutting wheel.

For mode of occurrence, origin, distribution, reserves, production and other details consult chapter on 'Diamond', dealt separately under gemstones.

(b) Corundum

Corundum (Al_2O_3) is the hardest mineral next to diamond. The common variety is grey, greenish or reddish and dull, and is used in abrasive. Under corundum only ordinary types and not of gem quality are included. It occurs mostly in barrel shaped crystals though massive and granular forms are also met. It is utilised as loose grains in optical glass grinding, as wheels in metal cutting and lens polishing, as papers and cloths in rubbing metals, hardwood and optical glass.

The abrasive wheels are made with the crushed corundum and the binding material, such as shellac. Artificial corundum is known as alundum which is fused bauxite or alumina.

It occurs as result of contact metamorphism of shale (silica poor hornfelses) or limestone. It is found associated with sillimanite or kyanite as observed in Meghalaya, Assam and Singhbhum district, Bihar. It also occurs in veins and magmatic segregations associated with periodotite and silica deficient rocks such as nepheline syenite as in Khammam district of Andhra Pradesh. It is found as residual deposits, derived from gneisses and pegmatites as found in Bastar and Morena districts, Madhya Pradesh.

The occurrences of corundum are found in Hassan, Bellary, Tumkur, Bangalore, Mysore and Chikmagalur districts of Karnataka, Tonk district of Rajasthan, Khammam district of Andhra Pradesh, Bhandara district of Maharashtra and Sidhi and Baster districts of Madhya Pradesh. Occurrences are also known in Salem and Dharampuri districts of Tamil Nadu, Singhbhum district of Bihar and in Assam and Meghalaya. The all India estimated reserves of corundum is about 141 thousand tonnes with about 94 thousand tonnes probable and 47 thousand tonnes possible reserves (as on 1.1.1985). Karnataka is the richest in corundum with about 136 thousand tonnes. Maharashtra happened to be the leading producing state accounting for about 90% of the total production, followed by Karnataka and Madhya Pradesh.

The mineral is marketed after grinding to different mesh sizes for use as abrasive. About 90% consumption is in abrasives, while the rest is in foundry and refractory industries. The abrasive factories which consume corundum, are located at Hanumana, Rewa district, Madhya Pradesh and at Bangalore, Karnataka. Synthetic corundum is produced by Emery (India) Pvt. Ltd. at Badreshwar, Jamnagar district, Gujarat.

(c) Emery

Emery is a greyish black variety of corundum, containing admixed magnetite, hematite and spinel. Three commercial types are known :

 (i) Greek which contains considerable corundum and is the hardest,
 (ii) Turkish is next in hardness, and
 (iii) American is the softer and contains considerable spinel in place of corundum.

The quality of emery depends upon the amount of corundum present. It is coarse or fine grained and tough, and is able to withstand intense heat. It is used as grinding wheels for snagging metals, as emery cloth and paper for grinding and rubbing metals and hard wood and as loose grains and flour for glass grinding and polishing metals.

It occurs in irregular bodies in crystalline limestone, basic igneous rocks, chlorite-amphibole-schist, and sillimanite-quartz-mica-schist. It is sometimes, found in veins. It is formed mainly by contact metasomatism. There is no mine of emery in India. It is, however, produced by Emery (India) Pvt. Ltd., Badreshwar, Jamnagar district, Gujarat by melting calcined alumina mixed with coke and iron filings in an electric furnace at about 2200°C.

(d) Garnet

The common garnet, almandite (Fe-Al garnet), is generally utilised as an abrasive, though, other varieties, andradite and pyrobe, are also sometimes used. Hardness, toughness and fracture (sharp angular) are necessary properties for its use in abrasive. It is employed in loose grains and finely ground powder for glass and optical lens grinding and surfacing ornamental stones and in paper and cloth for rubbing hardwood, automobile bodies, copper and brass, removing paints and varnishes, finishing hard rubber, celluloid, leather, felt and silk hats etc. It is also consumed in grinding wheels and grinding stones for metal polishing.

Garnet is widely distributed, but commercial deposits are only a few. It occurs in two forms, in-situ deposits and placers. As in-situ deposits, it occurs in metamorphic rocks, such as gneisses and schists of argillaceous parentage, crystalline limestone and metamorphosed basic and other igenous rocks. The occurrences are seen in many parts of the country, such as in Khammam district, Andhra Pradesh, Singhbhum district, Bihar, Hassan, Bangalore and Mysore districts, Karnataka, Ajmer, Bhilwara, Jhunjhunu, Sikar, Sirohi and Tonk districts, Rajasthan and several other places. It occurs as primary mineral in igneous rocks such as mica-pegmatites of Bihar, Andhra Pradesh and Rajasthan. It is also found as heavy detrital residues in sediments, e.g. placer deposits in beach sands of Kerala, Tamil Nadu and Orissa.

The total recoverable reserves of garnet have been placed at about 50 million tonnes (as on 1.1.1985), out of which the placers contain over 90% of the reserves. The existing mines of abrasive are located in Kanyakumari district, Tamil Nadu, Bhilwara and Tonk districts, Rajasthan and Khammam district, Andhra Pradesh. About 6000 tonnes of abrasive garnets are produced annually. The beach sand of Tamil Nadu contains 7% garnet in addition to other heavy minerals. The different minerals are separated on the basis of their magnetic property, electrical susceptibility and specific gravity.

2. SILICEOUS ABRASIVES

(a) Quartz and Abrasive Sand

Quartz is ground, graded and utilised in sandpapers. It is used in harsher metal polishes, metallurgical works and in several scouring compounds. It is also employed as burrstone for grinding paints, fertilizers, flour etc., and as pebbles in rotating ball mills for grinding metallic ores, ceramic minerals, cement, gypsum, powder abrasives etc. Abrasive sand must be fairly uniform in size, angular and clean. These may be utilised in scouring stone, sand blasting, grinding and surfacing glass and for sand paper. As abrasives they are of three categories, namely;

    (a) blasting sand,

    (b) glass grinding sand, and

    (c) stone sawing and rubbing sand.

The occurrences of quartz and abrasive sand are in almost all states of India. Their total reserves (all grades) have been placed over 500 million tonnes. Besides abrasives these find use in various other industries, like glass, foundry, ferro-alloys, cement, ceramic, refractory etc. Andhra Pradesh is the leading producer in the country contributing about 45%, followed by Rajasthan about 20%.

(b) Diatomite

It is made up chiefly of silica with a little of alumina, iron-oxides, alkalies and some water, and look like clay or chalk. Diatomaceous earth, diatomaceous silica, kieselguhr etc. are the different names given to it. It is porous and friable with specific gravity of 0.45 in the form of dried powder. It is soluble in alkalies and insoluble in acids. Its porosity, fineness of pores, absorptive power, light weight and low heat conductivity are the important properties which create its demand in market. It is utilised as an abrasive in metal polishes, automobile polishes, dental powder and paste and as an abrading agent in match heads and box sides. It is found in lake and swamp bottoms and its origin may be fresh water or marine. It is quarried or dredged and removed by scrappers or mechanical shovel. After drying, it is milled and utilised in the form of powder, aggregate or brick.

(c) Tripoli

It is a light, soft, and porous earthy material, consisting almost pure silica which results from residual weathering of chert, cherty and siliceous limestone or other siliceous rocks. The siliceous material, concentrated after removal of included calcium carbonate and other impurities, converts to tripoli in blocky or friable massive form. It is utilised in powder form in metal buffing. It is also employed in rubbing auto-bodies and painted surface and in scouring powders and soaps.

(d) Pumice and Pumicite

Pumice is the natural volcanic product or siliceous lava, and pumicite is volcanic ash. They contain thin sharp glass shards which make them excellent cleanser. They are ground and employed in preparation of scouring soap and various polishing compounds. Lump pumice is utilised in rubbing, cleaning and polishing metal surfaces, lithographic stones, fine tools and instruments.

(e) Sandstone and Quartzite

The sandstone and quartzite of fine grained, even texture are used as grinding stones, sharpening hones, millstones and pulpstones. They must be sufficiently cemented to ensure tenacity. They are found in almost all states of India. Vindhyan sandstone and quartzite of Rohtas and Palamu districts of Bihar and Mirzapur district of Uttar Pradesh are used for this purpose. The quartzites of Chotanagpur, Bihar and other places are also suitable and readily available. The total all India reserves of all grades quartzite (refractory, flux, ceramic, abrasive and others) have been placed at about 216 million tonnes with Bihar having about 177 million tonnes. Orissa is emerging as leading producer of quartzite (above 60%), followed by M.P. (28%) and Bihar (12%).

3. Soft Abrasives

(a) Felspar

It is used in ground-form as scouring powder and soap for cleaning porcelain, glass and enamelled surfaces. It is also utilised as bond for cementing vitrified abrasive wheels. It is widespread in occurrence. Mica pegmatites of Bihar, Rajasthan and Andhra Pradesh yield good amount of felspar. The deposits also exist in Karnataka, Madhya Pradesh, Maharashtra, Meghalaya, Tamil Nadu, West Bengal and other places. The total reserves have been placed at about 12 million tonnes, while the annual production was about 50 to 60 thousand tonnes. Rajasthan was the principal producer accounting for about 70%, followed by Andhra Pradesh and Bihar. Its main consumption is in ceramic, glass and refractory industries. Only two to three hundred tonnes are required for use in abrasives.

(b) Dolomite

It is calcined to unhydrated oxide of calcium and magnesium, and is used as buffer for various metals, pearl and celluloid. It is greatly used in polishing nickel plate. Dolomite occurrences are in almost all parts of the country with total reserves (all grades) of about 4,400 million tonnes. The major reserves are in Madhya Pradesh, Orissa, Arunachal Pradesh and Karnataka. The annual output of dolomite is about 2.5 million tonnes. Its consumption in abrasive is insignificant, the main consumer being iron and steel, and refractory.

(c) Magnesite

It is utilised in ground condition as soft polisher for metal and mineral surfaces. The use of magnesite in abrasive is only about 3 to 4 hundred tonnes annually. The deposits are in Uttar Pradesh (Almora and Pithoragarh), Tamil Nadu (Salem), Jammu and Kashmir (Udhampur) and Karnataka (Mysore).

(d) Bauxite

It is, generally, employed in powdered form for putting a finishing surface on wood and metallic instruments. It is also used in such domestic purposes as scouring and cleaning. India is endowed with extensive bauxite deposits. The important deposits occur in Orissa, Andhra Pradesh, Madhya Pradesh, Maharashtra, Gujarat, Bihar, Karnataka, Goa, Tamil Nadu and Uttar Pradesh. The consumption of bauxite in abrasive is secondary. About 73 thousand tonnes was consumed in abrasive in 1985, out of which about two million tonnes consumed in various industries.

(e) Clays

China clay, and, sometimes, ball clay and bentonite, are used as soft abrasives for polishing soft metals, uniform buttons, belts etc. For buffing ceramic goods, ball clay is used. The consumption of clays in abrasives is insignificant compared to its use in other industries like ceramic, cement, refractory rubber etc.

(f) Lime and Limestone

Some amount of ground limestone and other calcareous materials are used in abrasives. They are used widely in putting finishing surface on metals.

(g) Chalk

It is mostly utilised for silver ware, gold, nickel, chromium plate, brass, buttons etc.

(h) Fuller's Earth

It consists mainly of montmorillonite and is greenish grey, bluish or yellowish material, soft and earthy in texture with a soapy feel. It falls to powder when placed in water and adheres to tongue. It is used in removing grease and oily materials, cleaning woollen fabrics and cloth. It is also utilised in polishing silver and chromium wares.

(i) Talc

It is used as polisher for soft metals, leather and rice grains.

(j) Rogue

It is a hydrated oxide of iron, and is used as soft metal polisher and for rubbing optical lenses.

Besides these, there are many metallic oxides which are utilised as polishing agents for metals, mineral surfaces and optical glasses in their finishing stages.

5.4 BUILDING MATERIALS AND DIMENSION STONES

India is endowed with large resources of different kinds of building and monumental stones. Amongst them granite and marble are famous for their unique properties and aesthetic values. Many other rocks like limestone, quartzite, sandstone, slate, trap, felsite etc. are also being quarried in the country for structural and building purposes. Besides being used as dimension stones, rocks are also needed in chips (broken) or crushed forms for the highway construction, concrete aggregate, rail/road blasts etc. Harder rocks, like trap rock including dolerite, basalt, andesite etc. are preferred in stone chips. Among building materials, cement, clay, sand and lime play important role. They are also being described, in brief, in the chapter.

1. GRANITE

Granite in the form of building, structural and ornamental stones has gained a vital position in modern architecture. The term 'granite' was first used by Caesalpinus in 1596, and is derived from the latin word 'granum' means grain. It is a rock capable of taking good polish and giving a pleasing appearance. In trade parlance, it includes not only true granite but many types of igneous and metamorphic rocks, namely dolerite, basalt, porphyry, syenite, diorite, gabbro, diabase, charnockite, leptynite, khondalite, schist and gneiss. Black granite is actually gabbro, diabase or dolerite rocks and is highly priced. The granite suitable for ornamental and structural purposes should be hard and compact, should have pleasing appearance and should not crumple under heavy pressure. It must be possible to quarry and dress it into blocks of desired sizes and shapes. Since it is cut to different dimensions in shape and size as per need, it is also known as dimension stone.

The dimension stone is produced in most of the countries. Italy is the world's largest producer. In the year 1989, Italy produced 7.5 million tonnes of this stone which works out to be 25% of the world's total production. Spain produced 2.6 million tonnes, U.S.A. 1.85 million tonnes, Greece 1.8 million tonnes, India 1.5 million tonnes, France 1.1 million tonnes and Brazil and China 1 million tonne each. The major importing countries were the United States, Japan and for rough stone Italy.

Physical and Chemical Characteristics

The quality and use of granite depend upon its granularity, colour, presence of structural elements, inclusions, brittleness, chemical and mineralogical compositions, physico-mechanical properties and shape and size of blocks, which are discussed below in brief.

(i) *Granularity* : The best quality granite is generally fine-grained and of uniform grain size. The orientation of grains displaying geometric fabric pattern gives a pleasant look. The composition and nature of cementing material should be such as to provide the rock hard, compact and highly cohesive character. This type of granite is utilized for ornamental and monumental works and also for inscription purposes. Fine-grained good black granite can be cut to the thickness of a post-card and letters can be engraved on it.

(ii) *Colour* : Generally fast colour with uniformity is preferred. Variation in colour is a drawback. The colour of black granite should be jet black and this fetches high price. A few buyers, however, like variegated colour.

(iii) *Colour base* : The brownish tinge base fetches a better price. The blue, the grey and the green bases come next in succession. Polished block of black granite with brownish base is most preferred.

(iv) *Structural elements* : The structural elements like joints, faults, folds etc. influence the size and shape of mineable blocks. Flow structures are mostly indicative of the presence of thin bands of different minerals of varying compositions, which do not permit good polish, and, hence, are not preferred. Hair cracks in the rock effect blasting and quarrying and decrease the value of stone.

(v) *Inclusions* : Inclusions of all types are unwanted. The presence of thin silica or calcite veins in coloured granite gives rise to white lines and is required to be avoided. White patches formed due to presence of light coloured minerals in black granite are called flowers and are undesirable. Likewise, segregations of darker minerals form patches of black or other dark colours, called moles, are also deleterious.

(vi) *Mineralogical and chemical composition* : The mineralogical and chemical compositions of rocks control the quality of rocks and, hence, should be as per specifications. Presence of metallic minerals like pyrite, marcasite etc. which rust on exposure and cause stains on the surface is avoided. The dolerite with excessive silica (say more than 55%) and lime (more than 12%) is likely to have inclusions or veins of quartz and calcite respectively.

(vii) *Physico-mechanical properties* : The physico-mechanical properties vary with the rock types. These of grey granite and the black granite (dolerite) are given below in the Table 5.14.

Table 5.14—Physico-Mechanical Properties of Grey and Black Granites

Properties	Grey Granite	Black Granite (Dolerite)
1. Hardness (Moh's Scale)	5.8-6.6	
2. Specific Gravity	2.6-2.7	2.6-3.13
3. Compressive strength (kg/cm^2)	1000-3000	2000-3500
4. Tensile strength (kg/cm^2)	70-250	150-350
5. Shear strength (kg/cm^2)	140-500	250-600
6. Modules of elasticity (kg/cm^2)	2.5	8-11
7. Co-efficient of linear expansion x 10^6	4-6	2-4
8. Porosity %	0.5-1.5	0.1-0.5

(viii) *Size and shape of blocks* : The size, in general, is classed into :

(a) random size, which depends upon customer's requirement, and

(b) monumental size, which is specific.

The general size of block is 1.5 to 2.0 m^3. Tamil Nadu Mineral Industries Ltd. produces 100 x 90 x 70 cm monumental size stones, weighing about two tonnes each. These slabs being very heavy and cumbersome to handle, tiles of smaller sizes are, hence, preferred.

The dressed blocks in rectangular and square shapes are common, though the blocks are shaped to various designs as per demand. Protrusions or depressions on dressed face are undesirable.

Uses

Granites are excavated all over the country on a large scale for use as construction material and road metal. The utilisation of processed or polished granite is, however, restricted and is employed in decorative purpose like exterior surfacing, interior wall panelling, flooring, wall cladding, kitchen platforms, sinks, table tops, monuments, tombstone, foundation stones, inaugural plaques, name plates, flower vases, ash trays, metrological aids etc. In monumental sets only visible faces are polished and for this black granite is preferred. Grey and pink granites come next in importance. The metrological aids for which granite is utilised, include surface plates, straight edges, parallels, cubes, v-blocks and work mounting tables for co-ordinate measuring machines.

Distribution

Granite deposits exist in all the states of India, but only those of Andhra Pradesh, Karnataka, Rajasthan and Tamil Nadu are in great demand in foreign countries. Grey, pink, red and black varieties are found widely disu... ted, but no systematic reserve estimation nor a study of their amenability to take good polish has so far been made. The state-wise distribution of decorative stones, so far known as important, is given below in the Table 5.15.

Table 5.15—Distribution of Decorative Stones in India

States	Districts	Remarks
Andhra Pradesh	Anantpur, Chittoor, Guntur, Hyderabad, Karimnagar, Khammam, Kurnool, Prakasham and Warangal	Good qualities including black granite are available. Export quality is being obtained at Polepalli and Edulapuram, Khammam district. There are at present over 20 producers.
Bihar	Banka, Deogarh, Dumka, Godda, Gumla, Hazaribagh, Palamau, Ranchi and Singhbhum	Black granites are available in Banka, Godda, Dumka and Singhbhum districts. Elsewhere grey and pink granites, and granite-gneiss which are porphyritic, at places, are generally found.
Gujarat	Banaskantha and Saharkantha	There are at present over 18 granite producers in the state.
Karnataka	Bangalore, Bellary, Chikmagalur, Hassan, Kolar, Mysore, Mandhya, Raichur, Tumkur & Uttar Kannad	Grey, pink and black varieties are available. There are at present over a 100 producers in the state.
Rajasthan	Ajmer, Alwar, Barmer, Bhilwara, Jalor, Jhunjhunu, Pali and Sirohi	Pink, red and black granites are available. Jalore is the most important mining centre.
Tamil Nadu	Coimbatore, Dharampuri, North Arcot, Periyar, Salem and South Arcot	Grey granite, charnockite, leptynite, gabbro and dolerite are mined. Black granite of South Arcot is considered to be similar to Ebony granite of Sweden and is the best because of its fine-grained texture, jet black colour and availability in large blocks. There are at present 26 important granite cutting and polishing units with a total capacity of 770 thousand square metre.

Method of Mining and Processing

The granite to be mined is first examined for its suitability for forming blocks or slabs. For this textural uniformity, colour, strength, durability, presence of cracks, joint planes and their frequency are noted before marking or channeling the blocks. Well spaced joints at right angles make the extraction easier. The over-burden (weathered rock portion) is removed by drilling 30 to 60 cm holes of 5 cm diameter and blasting it with gun powder. The quarry is opened as trenches taking advantage of joint systems. As far as possible, rectangular blocks of standard size are marked either by hand channeling in manual mining or by channeling machines in mechanised mining. Efforts are made to develop vertical face and granite is quarried in rectangular blocks. The blocks are separated from the parent ledge by putting closely spaced linear drill holes and charging gun powder. A series of close spaced holes are drilled underneath the blocks which are broken free with the use of feathers and wedges. Feathers are pieces of semi-circular steel wedges that are inserted into the holes. The steel wedges are driven between the features to produce a break to free a block of stone. The blocks are lifted manually with levers or by cranes and loaded into trucks.

In mechanised mining help of compressor, drilling machine, various diamond saws, wire saws, channeling machines, wedges and broaching tools, cranes, dumpers etc. is taken. Endless braided steel wires and diamond saws are employed for cutting blocks. Jet chanenling or jet piercing is quite common. In some countries flame cutting is done to cut the rocks. In India only a few mines are mechanised or partially mechanised e.g. certain mines in Dharampuri and South Arcot districts of Tamil Nadu and Chamarajnagar in Mysore district of Karnataka.

The processing of granite involves :

    (i) block squaring and dressing,
    (ii) cutting and sawing,
    (iii) surface grinding and polishing, and
    (iv) edge cutting/trimming.

The block squaring and dressing is done most commonly by means of wire saws. The sawing of granite blocks is done for cutting them into slabs of required sizes by means of large diameter vertical block saw or gang saw or horizontal slicing machine. The rough cut slabs are subjected to light milling and then polishing. The polishing machines used are single spindle machine, vertical multi-spindle or diamond surfacing disc. Finally, edge cutting of the polished slab is done by bridge type edge cutting machines.

2. MARBLE

Marble is one of the oldest building material used in monument, decoration and building construction. In modern days' culture and for better living standards in urban environment there is intensive and widespread use of this natural stone which is shaped and finished as per requirement. Italy is the world's largest producer of marble. Other important producers are Brazil, China, Finland, Norway, Portugal, Spain, Sweden, Turkey and the United States.

Physical and Chemical Properties

The term marble is named after the Latin word 'Maarmor' which itself comes from Greek root 'Marmaros' meaning thereby a shining stone. In early days any stone capable of taking polish without any regard to its chemical composition was designated as marble. Petrologically marble is a metamorphosed limestone. It is a crystalline rock, composed mainly of calcite or more rarely dolomite. It has a broad range of physical and chemical properties. At times the individual grains are so small that they cannot be distinguished with the naked eye, and again they may be quite coarse showing characteristic rhombohedral cleavage of calcite. Like limestone, it is characterised by its softness and its effervesence with acids when pure. It is white in colour, but it may show wide range of colours due to various contained impurities.

The chemical composition determines many physical properties which are essential for its marketing. The distinctive features of chemical compositions of the famous Makrana marble of India and of Carrara marble from Italy are given below in the Table 5.16.

Table 5.16—Broad Chemical Composition of Marbles.

Constituents	Makrana (India)	Carrara (Italy)
Insoluble	0.89	Traces
Fe_2O_3	0.28	0.11
$CaCO_3$	97.74	99.35
$MgCO_3$	1.22	0.87
Phosphoric Acid	0.04	Traces

Distribution and Reserves

Marble occurrences are widely distributed in India. But occurrences of economic importance are restricted to a very few states, like Rajasthan, Gujarat, Haryana and Andhra Pradesh. Marble deposits of inferior grade occur in Madhya Pradesh, Bihar, Uttar Pradesh and West Bengal. A brief description of important deposits is given below in the Table 5.17

Fig. 5.2 *Marble hillocks,*
Loc : Darkala (Kankrauli), Rajasthan

Table 5.17—Important Deposits of Marble in India

State	District	Area	Other details including reserves
Rajasthan	Nagaur	Makrana,	The estimated reserves of marble in
	Udaipur	Amet Làva, Sardargarh	Rajasthan are about 1000 million tonnes,
		Babermal, Rajnagar,	of which Makrana area accounts for 56
		Kalwa and Darkola	million tonnes. Makrana marble is known
		(Kankroli) (Fig. 5.2)	for its excellent white colour in the world.
	Pali	Bar	Of late Rajnagar area has gained impor-
	Banswara	Tripura-Sundari	tance in production. This area is estima-
	Alwar	Jhiri-Rajgarh	ted to contain more than 300 million
	Ajmer	Kishengarh	tonnes of marble.
	Bundi	Umar	
	Sikar	Moonda	
Gujarat	Banaskantha,	Bharuch, Vadodara	Ambaji area, Banskantha district is the
	Kachch	and Panchmahal	leading producer. Gujarat is estimated to
			contain 47 million tonnes of marble.
Haryana	Mahendra-garh	Antri-Beharipur and Dhanota-Dhancholi	Reserves of 8.8 million tonnes have been estimated.
Andhra Pradesh	Guntur & Khammam	-	There are several localities where good quality marble is found.
Uttar Pradesh	Mirzapur, Dehradun, Tehri, Garhwal and Nainital	-	A total recoverable reserves of about 3 million tonnes have been estimated.

Production

The total all-India production of marble slabs and tiles during 1987-88, 1988-89 and 1989-90 were 1110 thousand tonnes, 1273 thousand tonnes and 1285 thousand tonnes, respectively. About 80% of production of marble slabs and tiles comes from Rajasthan alone.

Mining and Processing

It is mined mostly by open cast method except at Makrana (Rajasthan) where underground mining to depth of about 100 m is resorted to exploit excellent white marble. Workings are mostly manual. Recently a few large mines like Rajnagar (Rajasthan) and Ambaji (Gujarat) are adopting mechanised mining. The processing of marble involves three main activities, sawing, polishing and edge cutting. Sawing is commonly carried out by gang saws, frame saws and wire saws. Diamond impregnated circular saws are used for sub-dividing the slabs and in trimming and shaping. Rubbing, gritting, buffing and polishing are generally done with the help of various machines and abrasives. Edge cutting is accomplished with the help of emery wheels, carborundum wheels or diamond impregnated wheels. There are about 250 marble processing centres in India and most of them are located in Rajasthan and Gujarat.

3. LIMESTONE

India possesses extensive deposits of limestone suitable for building purposes. Limestone is a term used generally for carbonate of lime. Marble, marl and chalk are also calcium carbonate. Limestone may be crystallines, pisolitic, oolitic or earthy. The impure limestone may be argillaceous, siliceous, bituminous, ferruginous or dolomitic. The limestone after crystallisation (metamorphism) by heat and pressure form marble. Different names are given to limestone depending upon their colour, structure, locality where found, formation in which they occur, genesis etc. e.g. fawn limestone, nodular limestone, Narji limestone, carboniferous limestone, coral limestone, crinoidal limestone etc.

Uses

Limestone including other lime material is extensively mined in India next only to coal and iron-ore. It has a wide range of uses such as in cement (about 84%), iron-steel (10%), chemical (3%), sugar 1%) and paper (1%) industries. The other uses of limestone are in fertilizer, ferro-alloys, glass manufacture, lime manufacture, foundry, refractories, textile, electrode, ceramic, sponge iron etc. In Iron and Steel Industries it has two basic functions :

(i) to lower the temperature of melting (S.M.S. - Steel Melting Shop), and

(ii) to form calcium silicate by combining with silica of iron-ore which comes out as slag from the blast furnace.

The limestone to be used as dimension stones is usually very hard, and is siliceous, dolomitic or argillaceous in nature. Flaggy limestone in varying colours and fine-grained texture is suitable for paving and flooring work.

Specifications

The specifications of limestone in respect of some important industries are given below.

(i) Cement Industry

Cement is, in general, a mixture of about 4 parts of limestone and 1 part of shale or clay, calcined to near fusion and ground to powder. The calcining releases CO_2 and the remaining constituents combined to form complex silicates, aluminates and ferrates of calcium. The addition of water gives rise to a gel of hydrous compound which later crystallises and interlocks, giving the hard set.

The specification of limestone for manufacture of cement is given in the Table 5.18 overleaf.

Table 5.18—Specification of Limestone for Manufacture of Cement

CaO	45% and above (preferred), 42% minimum
Al_2O_3	2-4%
•Fe_2O_3	1-2%
SiO_2	12-16%
MgO	4% maximum

In place of limestone some other calcareous materials like BF slag, EPIF slag, fly ash, blue dust, boiler and cyclone dust, $CaCO_3$ sludge etc., are being used. Besides, limeshell, calcareous sea sand and lime kankar are also used.

The presence P_2O_5 more than 1% considerably slows down the setting time of portland cement. The portland cement is produced by burning to a clinker a finally ground mixture containing 75% $CaCO_3$ and 25% clayey material, the later consisting of 20% SiO_2, Al_2O_3 and Fe_2O_3 and 5% MgO, Alkalies etc. 3% gypsum is added before final grinding to prevent too rapid setting.

(ii) Iron and Steel Industry

A. Blast Furnace (B.F.) Grade

The specifications of limestone required in different steel plants for use in blast furnace are given below in Table 5.19.

Table 5.19—Specification of Limestone for Blast Furnaces of Different Plants (in per cent)

Plant	CaO	MgO	SiO_2	Al_2O_3	Total insoluble
1. Bhilai Steel Plant	42 ± 1	8 ± 1	-	-	6.5 ± 1
2. Bokaro Steel Plant	45	6 (max.)	5 (max.)	-	-
3. Rourkela Steel Plant	44.5 (min.)	3.5	10	2 (max.)	9 ± 1
4. Durgapur Steel Plant	45	-	7.5 (max.)	-	12 (max.)
5. Indian Iron and Steel Co., Burnpur	46	-	-	-	-
6. Tata Iron and Steel Co.	47.5	4	5.3	1.3	12 (max.)

In general, for blast furnace use the calcium carbonate content in limestone should not be usually less than 90%, $SiO_2 + Al_2O_3$ not more than 6% though upto 11.5% is allowed. MgO should be within 4% and sulphur and phosphorous as low as possible. The required size of limestone, in general, is + 10 mm.

B. *Steel Melting Shop (SMS) Grade*

The specifications of limestone required by different steel plants for SMS purposes are given below in the Table 5.20.

Table 5.20—Specification of Limestone for SMS Purposes in Different Plants (in per cent)

Plants	CaO	MgO	SiO_2	Al_2O_3	Total insoluble
1. Bhilai Steel Plant					
(a) O.H.	49	4 (max.)	-	-	5 (max.)
(b) L.D.	53	2 (max.)	1.8 (max.)	1.7 (max.)	-
2. Bokaro Steel Plant	50	3.5 (max.)	--	--	-
3. Rourkela Steel Plant	51	-	-	-	5 ± 1.5
4. Durgapur Steel Plant	48	3.5	-	-	5 (max.)
5. Indian Iron & Steel Company, Burnpur	50	-	4 (max.)	-	-
6. Tata Iron & Steel Company					
(a) O.H.	-	2 (max.)	-	-	4 (max.)
(b) L.D.	53	2 (max.)	-	-	1 (max.)

The required size of limestone is, in general, 50 mm.

(iii) Chemical Industry

The specifications of limestone in (%) for some chemicals are given in the Table 5.21 below.

Table 5.21—Specifications of Limestone for Some Chemicals (in per cent)

Chemicals	Cao	MgO	SiO_2	Al_2O_3	Remarks
(a) Bleaching powder	54 (min.)	2 (max.)	0.75 (max.)	-	L.O.I. = 46, Fe_2O_3 = 0.15 (max), Mn_2O_3 = 0.06 (max.)
(b) Caustic soda	53 (min.)	1 (max.)	-	-	$SiO_2 + Al_2O_3 + Fe_2O_3$ = 3 (max.), CO_2 = 42 (min.)
(c) Calcium carbide	54 (min.)	0.8 (max.)	1.00	-	Fe_2O_3 = 0.25 (max.), P = 0.50 (max.), S = 0.01 (max.)

For chemicals coral limestone and limeshell, being comparatively purer, are preferred.

(iv) Sugar Industry (in %)

CaO = 50 (min.), MgO = 1.00 (max.)

SiO_2 = 2 (max.), Al_2O_3 + Fe_2O_3 = 1.5 (max.)

L.O.I. = 44, CO_2 = 41

(v) Paper Industry (in %)

$CaCO_3$ = 95, MgO = 3.00 (max.)

Limeshell is also used.

(vi) Glass Industry (in %)

For colourless glass high purity of limestone is needed.

$CaCO_3$ = 94.50 (min.), FeO = 0.5 (max.)

$CaCO_3$ + $MgCO_3$ = 97.5, P = 1.00 (max.)

SiO_2 = 1.5, Organic matter = 0.3 (max.)

Al_2O_3 = 1.5 (max.)

As per Indian Standard 997-1937, FeO content should not be more than 0.50% in marble or calcite, 0.15% in dolomitic limestone or dolomite and 0.01% in limestone. In the manufacture of glass, BF slag is also being used.

(vii) Lime Manufacture

For chalk and plaster, SiO_2 may be 10% or more but MgO should be negligible. For hydrated lime, the limestone should be solid and compact to prevent its loss in the furnace.

(viii) Fertilizer Industry (in %)

$CaCO_3$ = 84% (min.), SiO_2 = 5.00, Humidity = 0.5 (max.)

Organic matter and sulphur should be negligible. The above specification is prescribed by the Nangal Fertilizer Plant.

(ix) Textile Industry (in %)

$CaCO_3$ = 94 (min.), MgO = 3 (max.)

Fe_2O_3 + Al_2O_3 = 2 (max.) and SiO_2 = 2.5 (max.)

(x) Ceramic Industry (in %)

$CaCO_3$ + $MgCO_3$ = 97 (min.), Fe_2O_3 = 0.3 (max.)

SiO_2 = 2, SO_2 = 0.1 (max.)

B.F. slag is also used in this industry.

Mode of Occurrence and Origin

Limestone occurs as extensive beds, bands and pockets, and is a typical sedimentary rock. It may be of both marine and fresh water origins.

When magnesium replaces the calcium, in part, it forms dolomitic limestone. Silica, iron, aluminium, phosphorus, sulphur, manganese and carbonaceous matter are present as impurities.

The calcium is derived from the weathering of rocks and is transported to the sedimentary basins as bicarbonate, carbonate and sulphate. The calcium carbonate is deposited by inorganic, organic and mechanical means. Carbon dioxide plays a prominent role in inorganic process. The amount of air and water temperature controls the presence of carbon dioxide in water. Cold water holds more carbon dioxide and with rise in temperature calcium carbonate is deposited. Algae, bacteria, corals, foraminifera and larger shells bring about organic deposit of limestone. The entire limestone bed may consist of coral, foraminifera or nummulite shells. The photosynthesis of plants also gives rise to calcium carbonate. The limestone may have originated by mechanical means through deposition of coral sand or shell matters and these are cemented to form compact limestone.

Distribution

The distribution of limestone, state and district-wise, is given below in the Table 5.22.

Table 5.22—Distribution of Limestone

State	District/Area	Formation/Age	Remarks
Andhra Pradesh	Adilabad, Cuddapah, Guntur, Hyderabad, Karimnagar, Khammam, Krishna, Kurnool, Nalgonda, Nellore, Visakhapatnam, Warangal and West Godavari	Cuddapah, Kurnool, Pakhal, Sullavai and Bhima Formation/Group; Proterozoic to Lower Cambrian deposits; Deposits found in Dharwars, Archaean, Gondwana and Inter-trappeans are not of economic importance.	All grades. Crystalline limestone are noted in Khammam, Nellore and Visakhapatnam districts.
Arunachal Pradesh	Lohit (Tidding and Tezu areas)	Miri Formation, Lower to middle Palaeozoic	Cement grade
	Siang (Kabbu and Dali)	-do-	Limekiln
Assam	Karbi-Anglong, Lakhimpur, Nagaon, Sibsagar, United Mikir and N. Kachar. Between Jadukala river in the west to Lubha river in the east	Shella Formation, Jayantia Group (Sylhet) Eocene	All grades

(Contd.)

State	District/Area	Formation/Age	Remarks
Bihar	Hazaribagh (Religara, Bachara, Bicha etc.), Ranchi (Benti Bagda, Khelari etc.), Palamau (within a radius of 16 to 42 kms from Daltonganj and Latehar.), Dhanbad (Hansa, Kulbana etc.), Munger Karmatanr, Dhaktari etc.),	Archaean	All grades, crystalline, cement grade
	Singhbhum (Near Chaibasa and to its north-west-Putada spring, Ghatkuri, Kendrugutu, Putung, Rajanka etc.)	Iron Ore Formation	Dolomitic Limestone
	Singhbhum (Purna Chaibasa and to its south-west upto Jagan-nathpur and thence to Kotgarh)	Kolhan Formation	Cement grade, also B.F. grade
	Palamau (Bhavanath-pur, Panda Valley, Dhanmendra, Nawadih etc.)	Kajarahat Formation, Lower Vindhyan	Different grades (BF, SMS and Cement) available
	Rohtas (Between Ram-dhira and Jagdag for about 75 km length)	Rohtas Formation, Semri, Lower Vindhyan	-do-
Goa Daman & Diu	North Eastern part of Goa	Archaean	Mg-rich crystalline limestone
	Diu town, Malala and Nagao in Diu	Foraminiferal Limestone, Pleistocene to Sub-recent	Cement grade
Gujarat	Amreli, Banaskantha, Bhavnagar, Bharauch, Jamnagar, Junagarh, Keda, Kachch, Panchmahal etc.	Jurrassic to Tertiary (Lower Miocene)	All grades (Chemical, B.F., S.M.S., Cement and lower grade) available
Haryana	Ambala, Bhiwani and Mahendragarh	Precambrian	All grades (S.M.S., B.F. Cement) and Lower
Himachal Pradesh	Bilaspur, Chamba, Kangra, Mandi, Simla, Sirmur and Solon	Different formations, Precambrian to Eocene, Subathu limestone, Krol limestone etc.	All grades (Chemical, Cement, SMS, BF and Lower)
Jammu & Kashmir	Anantnag, Baramula, Doda, Kathua, Ladakh, Punch, Srinagar and Udhampur	Different ages, Archaean to Eocene, mostly Upper Trias	All grades (Chemical BF, SMS, Cement and Lower)

(Contd.)

State	District/Area	Formation/Age	Remarks
Karnataka	Belgaum, Chitradurga, North Kanara, Shimoga and Tumkur	Precambrian (Dharwar and Algonkian)	All grades, mostly crystalline
	Bijapur, Gulbarga	Kaladagi and Bhima Formation	All grades, Bedded non-crystalline
	South Kanara	Shell Limestone	
Kerala	Kottayam, Palaghat	Archaean (mostly) crystalline)	All grades
	Quilon, Allepey, Cannore, Koshikole and Trichur	Tertiary and Shell lime-stone (Quilon bed)	All grades
Madhya Pradesh	Bastar, Bilaspur, Chhatar-pur, Chindwara, Damoh, Dhar, Durg, Hoshanga-bad, Jabalpur, Jhabua, Mandsaur, Morena, Raigarh, Rewa, Sagar, Satna, Sehore, Sidhi and West Nimar	Precambrian (Mainly Cuddapah and Vindhyan equivalents)	All grades
Maharashtra	Ahmednagar, Chanda, Chandrapura, Dhule, Nagpur, Nanded, Sangli, Ratnagiri and Yeotmal	Different ages (mainly Precambrian, also Cretaceous) e.g. Upper Gondwana in Chandrapura, Upper Bagh bed in Dhulia, Lameta in Nagpur, and intertrappean in Nanded	All grades
Manipur	Manipur east (Ukhrul area)	Cretaceous	Cement and lower grade
Meghalaya	West Garo, Khasi and Jayantia hills	Eocene, Shella Formation of Jayantia Group	All grades
Orissa	Koraput, Sambalpur and Sundargarh	Precambrian, Gangpur, Formation, Lower dolo-mitic and Upper calcitic member, crystalline at a few places	All grades
Pondichery	Sadarapattu, Tuttipattu, Irumbai etc.	Upper Cretacious, Nerine limestone	Cement and lower grade
Rajasthan	Ajmer, Alwar, Banswara, Bundi, Chittaurgarh, Dungarpur, Jaipur, Jaisalmer, Jhunjhunu, Jodhpur, Kota, Nagaur, Pali, Sawai Madhopur, Sikar, Sirohi and Udaipur	Precambrian	All grades (Chemical, SMS, B.F., Cement and lower grades)

(Contd.)

State	District/Area	Formation/Age	Remarks
Tamil Nadu	Coimbatore, Dharmpuri Madurai, North Arcot, Ramnathpuram and Salem	Archaean, Khondalite Group	All grades
	Tiruchirapalli	Cretaceous, Uttatur basal limestone	All grades
	Chingleput, Kanyakumari, Ramnathpuram, South Arcot and Thanjavur	Sub-Recent	Limeshell, All grades
Tripura	Shakan ranges	Tertiary	Sporadic, Siliceous limestone, lower grade
Uttar Pradesh	Almora, Banda, Dehra Dun, Garhwal (Pauri), Mirzapur, Pithoragarh, Tehri Garhwal	Precambrean	All grades
West Bengal	Bankura, Darjeeling, Jalpaiguri and Purulia	Archaean	Lower grade, Crystalline, dolomitic, siliceous at places. Calctuffa in Darjeeling and Jalpaiguri

Limeshell occurs in the states of Andhra Pradesh (Nellore and Visakhapatnam districts), Karnataka (South and North Canara districts), Kerala (Alleppey, Cannore, Kottayam, Kozhikode and Malapuram districts) and Tamil Nadu, while lime 'kankar' is produced in Andhra Pradesh (Karimnagar district) and Haryana (Bhiwani and Rohtak districts). Calcareous sand occurs in Jamnagar district of Gujarat and is being used in cement plant).

Reserves

The total recoverable reserves of limestone of all grades as on 1.4.1990 are placed at 76,446 million tonnes, out of which about 11,562 million tonnes are of proved category, 16,464 million tonnes of probable category and 48,420 million tonnes of possible category. The reserves of cement grade constitute about 68% of the total reserves. The reserves of SMS, BF and chemical grades are placed at about 4,710, 5,544 and 2,685 million tonnes, respectively.

The total conditional reserves have been estimated at 360 million tonnes. Of this, reserves in proved, probable and possible categories are placed at 37 million tonnes, 32 million tonnes and 291 million tonnes respectively.

The major reserves of limestone are in the state of Karnataka, Andhra Pradesh, Gujarat and Madhya Pradesh. Other states like Meghalaya, Rajasthan, Maharashtra, Uttar Pradesh, Himachal Pradesh, Orissa, Tamil Nadu, Bihar and Assam, also have appreciable reserves.

Production

In consonance with the higher demand, particularly from the cement plants, the production of limestone continued to increase progressively and reached at 63.9 million tonnes in 1988. During 1990-91, 1991-92 and 1992-

93 its productions were about 70 million, 77 million and 76 million tonnes, respectively. Madhya Pradesh continued to remain the leading producers during 1992-93, accounting for about 28% production, followed by Andhra Pradesh 18%, Rajasthan 10.5%, Gujarat 9.5% and Karnataka 9%. The remaining 25% was shared by other states like Tamil Nadu, Maharashtra, Orissa, Himachal Pradesh, Bihar, Uttar Pradesh and others. In Rajasthan, Karnataka and Andhra Pradesh, productions of limestone to be used as dimension stone, flagstone and paving-stone have also been reported, in addition.

4. QUARTZITE AND SANDSTONE

Quartzites and Sandstones are chiefly used in building and road metals. Fine-grained compact quartzites, as found in Gurgaon district of Haryana, are suitable for building purposes.

Distribution

They are the most common rocks, widely distributed in the country. A few important occurrences are in the Rajasthan, Madhya Pradesh, Bihar and Uttar Pradesh, and they are given below in the Table 5.23.

Table 5.23—Important Occurrences of Quartzite and Sandstone in India

State	District/Area	Other Details
Rajasthan	Kota, Bundi, Sawai Madhopur, Jodhpur, Chittaurgarh, Bhilwara, Bikaner, Jaipur, Tonk, Ajmer, Udaipur, and Bharatpur	They belong mainly to Vindhyan formations. The colour varies from dark red to light pink and, sometimes, it is white. They are used as masonry stone, for making roofing and flooring tiles. Historic buildings have been constructed out of these sandstones
Madhya Pradesh	Hoshangabad	Red sandstone of Rewa formation (Vindhyan group) is used as flagstone in making roofings.
Bihar	Rohtas, Munger and Gaya	Sandstone of Kaimur formation (Rohtas) finds use in building purposes. The quartzites of Munger and Gaya are being used as building stone.
Uttar Pradesh	Mirzapur (Chunar)	Sandstones of Vindhyan Formation have been used in historic building and are still in use.

Production

Rajasthan is the leading producing state. Sandstone slabs and tiles of sizes 2 to 3 m long and 30 to 60 cm wide, are being produced. The all-India productions of quartzite and sandstone for building purposes during 1987-88, 1988-89 and 1989-90 were 4.3 million tonnes, 3.2 million tonnes and 5.3 million tonnes, respectively.

5. SLATE

Slate is a metamorphic rock of argillaceous origin. It is fine-grained in texture and splits into layers of thicknesses varying from 8 to 14 mm. It occurs in various shades of grey to steel grey, brownish ash colour, and sometimes with other pleasing colours. It is used for roofing and flooring in the local rural areas around the deposit. It is also employed as slate for the school children.

Systematic assessment of resources of slate in the country has not been made, although reserves are believed to be quite large. Andhra Pradesh, Haryana, Himachal Pradesh, Madhya Pradesh and Rajasthan are the leading producers. Mandsaur district of Madhya Pradesh contributes 60 to 70 per cent of the country's total production. In Bihar it is mined in Kharagpur hills of Munger and in eastern Singhbhum. The total slate productions in India during 1989-90, 1990-91 and 1991-92 were 30.6, 25.1 and 21.3 million tonnes, respectively.

6. CEMENT

Cement is one of the most vital materials to be used by mankind in building and structural works. It is a manufactured material which when mixed with water sets and becomes hard. It is of diversified types like ordinary portland cement, portland pozzolana cement, white cement, portland blast furnace slag cement, oil well cement, hydrophobic cement etc. The word 'Pozzolana' is after the place name Puzzolona in Italy, wherefrom the volcanic ash was first obtained for use as cement by mixing with quicklime. The name portland cement was derived from the resemblance of the substance to limestone quarried at the Isle of Portland, England. The varieties of cement such as high early strength, masonry, low heat, oilwells, high alumina etc. have been made for specialized uses. Mixed with sand and rock chips, the cement forms a mass of great strength, called concrete. Depending upon the hardness required, the cement content in the concrete varies from a tenth to a fourth of the total bulk, higher the cement content greater the strength of concrete. In reinforced concrete the ratio of cement, sand and rock chips is 1 : 2 : 4.

There remained increasing demand of cement throughout the world because of many notable activities, like plant modernisation and expansion, and ever growing building and construction works with population. The total world production of cement in 1989 was about 1.1 billion tonnes. China continued to produce the maximum amongst all cement producing countries, contributing about 18% to the world production, followed by Russia (12%), Japan (7%), USA (6%) and India (4%).

Composition and Manufacture

The cement, composed mainly of lime (CaO), silica, alumina and iron oxide, is complex silicate, aluminate and ferrate of calcium; which ultimately break down to form other compounds like tricalcium and dicalcium silicate, tricalcium aluminate etc.

Limestone is the main ingredient in the manufacture of cement. Its specifications for utilisation in cement, as mentioned earlier under topic 'Limestone', are CaO = 42% and above, Al_2O_3 = 2 to 4%, Fe_2O_3 = 1-2%, SiO_2 = 12-16%, MgO = 4% (max.) and P_2O_5 = less than 1%. It is first crushed to sizes of small stones and fed into a grinding mill together with clayey material in the ratio of 4 : 1. The resulting powder is burnt in huge revolving kilns. The clinker (fused material) produced by this process is sent to final grinding mills when it is mixed with three per cent gypsum to prevent too rapid setting. Limestone is replaced fully or partly as raw-

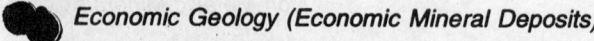

material by the supply of oyster shells, marls and blast furnace slag. The deficiencies in Al_2O_3, SiO_2 or Fe_2O_3 are met respectively by adding clay or shale, sandstone or sand and bauxite or iron-ore.

Production

The production of cement in the country continues to show increasing trend. The production in 1980-81 was about 20 million tonnes, in 1985-86 about 33 million tonnes and in 1989-90 about 46 million tonnes. In 1989-90 a total of 88 major cement plants excluding over a hundred mini-cement plants were in operation with a total installed capacity of 57.03 million tonnes.

Industry

The cement industry continued to take steps for modernization and upgradation of technology. Waste materials of other industries like blast furnace slag and fly ash are being utilised as raw materials and solving the problems of industrial waste and conservation of cement grade limestone. A number of cement plants are engaged in switching over from wet process kilns to dry process/semi-dry process kiln with the introduction of precalcining technology. They are also trying to switch over from coal to lignite as fuel. Many cement plants have set up diesel generator sets to cope up with the frequent power cuts. Efforts are also being taken to promote export of cement to foreign countries.

7. CLAY

Clay is employed in making bricks, which are invariably used in the plain country where rock exposures are not found. Sandy clays are not so suitable for bricks. Clay also finds use in cement manufacture. It is utilized in place of cement and sand, where cheaper houses, called 'Kachcha' building is needed. It occurs extensively in the country except in the rocky terrain.

8. SAND

Sand forms an important ingredient in construction of buildings. It should be clean and of specific sizes for different uses. Much of mica, felspar and iron-oxide in sand is unwanted. Coarse sand with grain sizes of 0.6 mm to 2 mm is preferred for concrete and fine sand with 0.06 mm to 0.2 mm in grain diameter for plastering work. The supply of sand is generally made from river beds, and the resource is vast.

9. LIME

Lime is no less important as building material. It is prepared from limestone, dolomite or other calcareous material which when heated in kilns to about 900°C, quicklime (CaO) is formed after removal of CO_2. The quicklime slakes with addition of water. Mixed with sand it makes a good plaster or mortar. Generally, hydrated lime ($Ca(OH)_2$) is made by pouring necessary water into it. With addition of water it sets in air and, hence, is consumed as mortar and plaster. For distribution of its raw materials, limestones/dolomites, the respective chapters may be referred to.

5.5 INDUSTRIAL MINERALS

A number of non-metallic minerals are utilised for industrial purposes. Chief among them are mica, asbestos, barytes, talc/soapstone/steatite, bentonite and vermiculite. This group does not include the minerals used in the chemical industries.

1. MICA

Introduction

India dominated the world market in mica ever since the demand of mica in electrical industry has become indispensible. The world depends on India for quality mica sheets and splittings. The USA, Japan, Germany, Russia, Belgium, Poland, France and Czechoslovakia are the principal countries where India is exporting mica. India is the leading producer of sheet mica, while USA is the principal producer of scrap mica.

Mineralogy

The chief minerals of mica are given in the Table 5.24 below.

Table 5.24—Chief Mica Minerals and their Chemical Compositions

Mineral Name	Commercial Name	Chem. Composition
Muscovite	Potassium mica (White mica, ruby mica)	$H_2K\ Al_3\ (SiO_4)_3$
Paragonite	Sodium mica	$H_2\ Na\ Al_3\ (SiO_4)_3$
Phlogopite	Magnesium mica (Amber mica)	$H_2K\ Mg_3\ Al\ (SiO_4)_3$ with fluorine
Biotite	Magnesium-Iron mica (Black mica)	$H_2K\ (Mg,\ Fe)_3\ (Al,\ Fe)\ (SiO_4)_3$ in part
Lepidolite	Lithium mica	$(OH,F)_2K\ Li\ Al_2\ (Si_3O_{10})$ in part
Zinwaldite	Lithium-Iron mica (Lithium biotite)	Li, Fe, K, F in addition to Al and SiO_2
Roscoelite	Vanadium mica	V, Mg, Fe in addition to Al and SiO_2

Paragonite and phlogopite are included under the trade name 'muscovite'.

Uses

The splitting property of mica combined with its flexibility, elasticity, resilience, toughness, low heat conductivity and high dielectric strength make the micas excellent electrical insulators. Under mica are included :

    (i) fabricated/manufactured mica, including cut condensor films, discs and washers, bridges and spacers, cut mica plates, electrical heating elements, mica capacitors, mica powders, mica paper, micanite, mica paper products, glass bonded mica, mica bricks, mica tapes etc.,

(ii) processed mica, including mica blocks, mica films and splittings of all grades and varieties, excluding manufactured or fabricated mica, and

(iii) mica waste and scrap which are obtained by processing mica and which because of size and colour are considered below the specification of processed mica.

Micanite or built-up mica is made of overlapped irregular shaped and arranged splittings, cemented together with a binder. Mica paper or reconstituted mica is paper-like, made by depositing fine flakes of scrap-mica as a continuous mat which is then dried. Some types of built-up mica have the bended splittings, reinforced with special paper, silk, linen, muslin, glass cloth or plastic. Fabricated products of different sizes and shapes are obtained from sheets and films using die-punches and press machines. The uses of different micas are given in the Table 5.25 below.

Table 5.25—Uses of Different Types of Mica

Mica types	Uses	Remarks
1. Muscovite		
(a) Natural sheet mica (High quality)	Helium neon laser as retardation plates	
(b) Sheet mica in the form of micanite, books and film mica	Electrical and electronic industries as insulating materials, such as capacitors, communicator segments, high pressure steam boilers, gauge glass, diaphragm of oxygen breathing equipments, marker dials of navigation compass, quarter wave plates of optical instruments, pyrometers, optical filters, thermal regulators, stove windows, furnace doors, lamp chimnies, window microwave, hair drier elements, washers etc.	Since mica sheets of sufficient thickness is not always available, micanite is used.
(c) Mica Powder	In manufacture of patent roofing, wall paper, automobile tyres, moulded insulators, as a filler in rubber goods, drill muds etc., and for fancy paints and lubricants	
2. Phlogopite	As for muscovite, good for manufacture of spark plugs of aeroplane, and washers for electrodes	
3. Biotite	Little used in sheets. Most of it goes into ground mica in 'Ayurveda' medicine as 'Abhrakha Vashma'	
4. Lepidolite	Source of Lithium element	
5. Zinwaldite	-do-	
6. Roscoelite	Source of Vanadium element	

Grading and Classification

The crude mica from the mine is cut to remove defects like cracks, stains, inclusions, warpings etc., and then sorted according to size and quality. Based on the size the mica is classified into thirteen grades which are given in the Table 5.26 below.

Table 5.26—Grading of Mica based on Size

Grade	Size (in sq. cm.)
Over over extra special	645.2 (Average)
Over extra extra special	516.1
Extra extra special	387.1
Extra special	309.7
Special	232.3
No. 1	154.8
No. 2	96.8
No. 3	64.5
No. 4	38.7
No. 5	19.4
No. 5½	14.5
No. 6	6.4
No. 7	4.8

Source : Industrial Minerals of Bihar and Specifications required by various industries, Mines and Geology, Bihar Govt., February 1983, p. 9.

Mica mined are of various colours such as ruby, white, silver, green, brown etc. The ruby mica is the most valuable. Based on the quality, the mica is classed into nine types, shown in the Table 5.27 below.

Table 5.27—Types of Mica Based on Quality

Types	Characteristics
1. Superfine	Completely free from stains and other defects, totally flat
2. Clear	Free from inclusions, stains, cracks and wraps
3. Slightly stained	A few bubbles of stain free gas, free from mineral inclusions, cracks and warps
4. Fairly stained	Presence of minute stains, spots and gas bubbles, free from mineral inclusions and cracks
5. Good stained	Slightly more stained, more gas bubbles, free from mineral inclusions and cracks, slight warping permissible
6. Stained	Possibility of deep stained, free from mineral inclusions and cracks, normal warping permissible
7. Heavy stained	Highly spotted and warping
8. Densely stained	Mineral inclusions, more of warping quality
9. Black spotted	Totally bad quality

The first four types are normally considered as of high quality.

Mode of Occurrence and Origin

Muscovite occurs in silicic pegmatites, mostly in association with granitic intrusives. Phlogopite comes from quartz-free pegmatites. Biotite is obtained from metamorphic rocks-biotite-schists. Lepidolite and zinwaldite are restricted to granitic pegmatites in some areas. The micas, as a whole, occur as a constituent of pegmatites and veins which penetrate the mica-schists. Commercial micas occur as zonally distributed 'books' in the pegmatites. The mica books vary in size considerably even within the short distances within the pegmatite bodies. Single books of muscovite weighing half a tonne have been mined in the Bihar mica-belt.

The pegmatites are composed normally of plagioclase felspar and quartz, but also, in places, contain orthoclase felspar and such minerals as tourmaline, garnet, apatite and very rarely beryl. They have been emplaced along favourable structures such as noses and limbs of folds, bedding, foliation, slip planes and joints, and are commonly long, narrow, lensoidal and, in places, arcuate and sinuous. They vary in size from insignificant stringers to big bodies, 500 m long and 30 m wide. Petrographically, distinct mappable units exist in a zoned or composite pegmatite with a central quartz core, followed by a zone of pegmatite with quartz, felspar and mica, and a mica zone at the contact of the pegmatite-vein and the country rock.

The mica has been formed from magmatic solutions in the pegmatite. It may be an igneo-aqueous deposit or hydrothermal replacement of earlier minerals. It has been observed that pegmatites occurring within mica-schists are richer in mica books compared to those occurring within other rock types. It is, hence, reasonable to assume that the country rock-mica-schist has contributed, in part, towards the formation of mica. The fluids responsible for the formation of mica-pegmatite, were rich in volatiles (H_2O, F, B etc.) and (K, Na, Li, Al and Si). The intrusive origin of mica- pegmatites is agreed upon by most Indian workers. These mica-pegmatites may have formed at a high temperature below a great thickness of the crust which has since been removed. It has been suggested that the formation of pegmatites is favoured in a geological setting where the mica-schist, country rock, has metamorphosed under Amphibo-lite-facies conditions. The origin of phlogopite mica is different from that of muscovite where the former forms under the action of gases, vapours and aqueous solutions of granitic magma on magnesium rich host rocks, dolomitised limestones and dolomites. Under the action of magma pure limestone re-crystallised into to marble, and dolomitised limestones turn into diopside rocks.

Distribution

World's best quality of mica comes from India and the workable deposits have been recorded mainly from Bihar, Andhra Pradesh and Rajasthan. The occurrences of less significance are noted in the states of Tamil Nadu, Karnataka, Kerala, West Bengal, Madhya Pradesh and Orissa. The geological and geographical distributions of mica are given in the Tables 5.28 and 5.29, respectively.

Table 5.28—Geological Distribution of Mica in India

Formation	Area	Geological Details
Precambrian, predominantly mica-schist besides calc-granulite and gneisses, hornblende-schist, micaceous quartzite and conglomerate (955 ± 40 million years)	Bihar Mica-belt	The mica-pegmatites are localised mainly in mica-schist and micacious quartzite, and rarely in hornblende-schist.
Biotite-schist and hornblende-schist interbanded with minor quartzite, felspathic-schist etc. (1570 million years)	Nellore mica-belt	Mica-bearing pegmatites are most frequent (70%) in biotite-schist and other rocks.
Pre-Vindhyan metasediments intruded by granites of different ages	Rajasthan mica-belt	The pegmatites carrying mica are commonly intrusive into steeply deeping muscovite-biotite-schist, micaceous quartzite and less commonly into amphibolite, dolerite and calc-silicate rocks.
Archaean gneisses	Nilgiri, Tamil Nadu	The pegmatites carrying muscovite are intrusive into gneisses.
Charnockite and leptynite	Madugula, Tamil Nadu	Irregular pockets of phlogopite are formed at the contact of pegmatites with charnockite and leptynite. The deposits are not of much significance.
Peninsular gneissic complex	Mysore and Hassan distt., Karnataka	The distribution of mica in pegmatites is very erratic.
Calc-gneiss, dunite and norite	Neyoor and Punaloor, Kerala	Phlogopite mica is concentrated at the contact of pegmatite with calc-gneiss and dunite.
Gneisses and Schists	Bankura and Purulia distts, West Bengal	Occurrences of muscovite of not much economic significance have been noted.

Table 5.29—Geographic Distribution of Mica in India

State	Location/Area	Remarks
Bihar	(a) Bihar mica-belt. The potential areas are around Koderma, Domchanch, Giridih and Chakai	The belt with an average width of 16 km extends for about 160 km from Gaya on the west through Hazaribagh and Munger districts to Bhagalpur district in the east, covering an area of about 3900 sq. km. Lepidolite, columbite, tantalite and beryl are common associates of muscovite. Clear ruby muscovite is most common.
	(b) Isolated occurrences in Palamau, Ranchi, Dhanbad and Singhbhum districts	The deposits are not of much economic value.
Andhra Pradesh	Nellore mica-belt. The main deposits are at Kalichedu, Thalpur and Gudur in Nellore district.	The belt with an average width of 16 km extends for about 96 km trending NNW-SSE. Muscovite occurs in shades of green.
Rajasthan	Rajasthan mica-belt	The belt extends for about 320 kms from Jaipur to Udaipur district through Ajmer, Bhilwara, Tonk and Pali districts with an average width of 96 kms, covering about 3000 sq. km. The belt extends NNE-SSW. Muscovite are in shades of pink, but black and green spotted varieties are also common.
Tamil Nadu	Nilgiri	Old workings are seen. The muscovite micas of 15 to 20 cm across have been recorded.
	Madugula	Phlogopite mica of not much value has been obtained.
Karnataka	Mysore and Hassan districts	The deposits are very small and poor in quality.
Kerala	Neyoor and Punaloor, Quilon district	Phlogopite mica of medium size is found.
West Bengal	Purulia and Bankura districts	Occurrences of small books of ruby mica or ordinary mica, generally, about 2.5 sq. km. or more and occasionally assuming sizes of even 205 sq. cm., have been reported.

Production and Consumption

In recent years the production of mica in India has continuously shown a downward trend. In 1975 India produced 11501 tonnes of crude mica which gradually reduced to 7934 tonnes in 1980, 4062 tonnes in 1990-91 and 2507 tonnes in 1992-93. This is possibly due to discovery of substitutes for mica by other materials and technological developments. The total world production of mica (crude and scrap) in 1990 was 245 thousand tonnes, to which India's contribution was about 4,000 tonnes

(1.7%). India enjoys the first position in production of mica blocks and splittings. However, it would descend to the ninth position if the productions of scrap, flakes and all other forms are considered together. Mica mining is mostly in private sector. In public sector, only four mines operated in 1985 producing 53 tonnes, Bihar continued to be the leading producer accounting for 50% production in 1988, followed by Andhra Pradesh with 38% and Rajasthan with 12%.

Complete picture regarding consumption of mica is not available, as there is no adequate coverage of mica user industries in small scale sectors. However, reported consumption of mica in organised sectors in 1985 was 3285 tonnes. Important consumers of fabricated mica are Indian Telephone Industries, Bangalore, Pieco Electronic and Electric Limited, Bombay and several manufacturers of elements required for electric irons and heaters. Important consumers of silvered mica for manufacturing capacitors are Peak Electrical Ltd. at Giridih and its sister concern at Bombay, and MITCO's silvered mica plant at Giridih. About 100 tonnes of splittings on an average are being consumed annually by the micanite industries. Micanite is being manufactured by half a dozen factories in India, the important of them are Indian Mica and Micanite Industries Limited, Jhumaritelaiya, Mysore Micanite Industries, Mysore, and Rao Insulating Co. Pvt. Ltd., Gudur. The BHEL, Bhopal produces micanite for its own use.

Prospecting Guides

The following are the prospecting guides for mica :

1. Since mica is formed in deep seated parts of the earth's crust, it has to be sought in the denuded mountain ranges where pegmatite-veins are exposed.

2. Granitic rocks or country rocks, whose constituent minerals have been altered by the magmatic actions, are best suited for the presence of mica. The mica-schist is the suitable host rock for development of mica. But phlogopite (magnesium-mica) is developed in magnesium-rich host rocks such as metamorphosed magnesium-rich limestone and dolomite with green diopside.

3. Mica books are mostly found at the contact of pegmatites with the country rock.

4. Presence of disintegrated pegmatite blocks and mica leaves on hill tops and slopes as well as along rivers and 'nalas' demand search for mica in the area.

5. The pegmatites bearing mica, are comparatively soft and are mostly zoned with quartz core. The presence of smoky quartz and plagioclase in pegmatite is best suited for development of mica.

6. The presence of apatite, garnet, beryl, tantalite and columbite in pegmatites is one of the guiding factors to find promising mica deposits.

Mining

Mica mining is done mostly by open cast, known as 'Uparchalla Working'. For micas occurring at depths, underground mining by means of inclines, shafts, drives and winzes is carried out. Mica is removed by stopping. Explosives and compressed air drills are being used extensively in mica mining. Bigger mines are highly mechanised. Shivshankar, Bendro and Pokharia, falling in Hazaribagh district, are worth mentioning mines.

2. ASBESTOS

Asbestos is fibrous amphibole, consisting of long, fine, flexible, soft and silky fibres that can be spun readily into threads and woven into cloth. It varies in colour from white to greenish and brownish. It is an excellent non-conductor of heat and electricity, and, thereby, resists fire and electricity. Commercial asbestos includes fibrous varieties of a number of minerals which are listed below :

    (i) Chrysotile, fibrous serpentine,

    (ii) Amosite, fibrous anthophyllite,

    (iii) Crocidolite - fibrous soda-amphibole (Riebeckite),

    (iv) Asbestos proper - fibrous actinolite-tremolite.

Chrysotile is in the most common use, but other forms are preferred because of their acid resisting qualities. Chrysotile is decomposed by HCl, while other asbestos (amphibole types) are resistant.

Russia was the largest producer of asbestos. Canada and Russia accounted for more than three-fourth of world production.

Uses

The utility of asbestos depends upon its property of being spun into yarn and its resistance to heat. The better grade (spinning type), those with long fibres are woven into fire-proof fabrics and rope. They are also used in break-lining and clutch facings, gaskets, boiler covering etc. The shorter fibres (non-spinning type) are used in the manufacture of asbestos cement, sheets, boards, paper, pipe, roofing tiles, fire proof paints, various binders, refractory (insulation) etc.

Substitutes like copper coated steel fibres, alkali resistant glass fibres, polyacrylo-nitride (PAN), polyvinyl alcohol (PVA) and cellulose have come to some use but these fibres are costlier.

Mode of Occurrence and Origin

Asbestos occurs mainly in three ways :

    (i) as cross fibres, when fibres are at right angles to the vein-walls,

    (ii) as slip fibres, when fibres are parallel with the walls and have formed along planes of movement, and

    (iii) as mass fibre when the material occurs in confused groupings as in anthophyllite type.

Chrysolite occurs in serpentine which is an alteration product of ultrabasic rocks like periodotite and dunite or magnesian limestone or dolomite. In this alteration, olivine is converted to serpentine, only water being added. This process, known as serpentinization, is an autometamorphic process which takes place, generally, along the fractures, fissures and shear planes. The serpentine so formed changes to asbestos by molecular rearrangement or recrystallization in tight fractures into fibrous form.

The amphibole varieties occur mostly in schists and gneisses. The anthophyllite type is, however, found occurring as lenses and pockets in peridotite and pyroxenite. The rocks may have undergone re-crystallisation and re-combination due to metamorphic processes, resulting in the formation of asbestos. Deep burial is thought to have generated heat and pressure, resulting in metamorphism of rock constituents into asbestos.

Distribution and Reserves

The chrysotile asbestos is found in the Cuddapah district of Andhra Pradesh and Singhbhum district of Bihar. Small deposits of crysotile asbestos are located in Mysore district of Karnataka and Udaipur district of Rajasthan. Amphibole asbestos occurs mostly in Rajasthan (Udaipur, Ajmer and Pali districts) though occurrences are also known in other places.

The recoverable reserves of all categories and varieties of asbestos (as on 1.4.1990) are estimated as 2.29 million tonnes, out of which 0.27, 0.71 and 1.31 million tonnes are in proved, probable and possible categories, respectively. In addition, about 70,000 tonnes of asbestos are placed under the category of conditional resources.

Production

The production of chrystotile asbestos was reported from Andhra Pradesh, while that of amphibole asbestos was from Rajasthan and Karnataka. The all-India productions of asbestos of all categories and grades during 1989-90, 1990-91, 1991-92 and 1992-93 were about 41, 38, 39 and 44 thousand tonnes, respectively. Rajasthan remains the chief producer which accounts for over 95 per cent output, while the rest was contributed by Andhra Pradesh and Karnataka.

A large number of mines are small, opencast. In Pulivendla taluk, Andhra Pradesh, an incline is opened along the dip (20° to 25°) keeping trap rock as floor and limestone in roof. Thus, underground mines have been developed by levels and winzes, adopting board and pillar system, and chrysotile asbestos is being mined here.

The asbestos concentrate is fed manually into hopper of a hammer mill where asbestos and other minerals are separated. The asbestos is then fed to double deck screen with 10-40 mesh sieves. The screening gives three fractions :

(i) oversize,

(ii) middling, and

(iii) tailings.

The tailing is the waste, while over size is recycled in the hammer mill.

Grading

Small size fibres, recovered through milling process, account for nearly two-third production. Present grading is as follows :

   (i) Special Grade   45 mm and above
  (ii) Grade-A   Between 25 mm and 45 mm   Hand sorted
 (iii) Grade-B   Between 12 mm and 25 mm   "
  (iv) Grade-C   Above 16 mesh
   (v) Grade-D_3 Above 24 mesh
  (vi) Grade-D_4 Above 40 mesh   ⎤
 (vii) Grade-D_6 Above 60 mesh   ⎦ Mill processed

Producers of amphibole asbestos sell their output as crude or fluff and powder.

3. BARYTES

Barytes ($BaSO_4$) is one of the important industrial minerals. Its name was derived from the Greek word 'barys' meaning heavy, and so the mineral is also named as heavyspar which also includes witherite ($BaCO_3$) to some extent in trade. It is usually colourless or white, often tinged with yellow, red and brown. It has white streak, and vitreous to resinous and, sometimes, pearly lustre. Its hardness is 3-3.5 and specific gravity 4.5.

China is the world's largest producer of barytes with output around 0.9 million tonnes a year. The present world productions of barytes is 5.6 million tonnes (1990), to which India's contribution is about 0.51 million tonnes. India ranked second in the world in order of quantum of production. The other important producing countries are USA, Russia, Mexico and Morocco. The world's largest reserves of barytes are in China, exceeding 540 million tonnes. The total world reserves are estimated at about 900 million tonnes.

Uses

Barytes, being heavy, inert and stable, finds uses in various industries like oil and gas well drilling, paint, chemical, asbestos and asbestos products, glass, rubber, ceramic etc.

In drilling, rubber, glass, linoleum, oil cloth, paper, plastics and resins, it is used in powder form as an inert volume and weight filler, and in paint as a paint extender. Lithopone, a brilliant white paint, is 70% barium sulphate and 30% zinc sulphide. Mixed with TiO_2 it makes a white paint of outstanding opacity. It is employed in tanning, making highly surfaced papers like playing cards, for buttons, printer's ink, face powder, glass, glazes and enamels. Barium sulphate is used in medicine. It is opaque to X-rays. Mixed with flour, sugar, cocoa and water, it is drunk by patients whose intestines are to be X-rayed.

Barytes is an important raw material for extraction of barium metal which does not occur in native state. Barium is useful in removing oxygen as a 'getter' from electron tubes and promoting vacuum. It is alloyed with nickel, magnesium and aluminium.

Mode of Occurrence and Origin

Barytes occurs mostly as veins, often associated with galena, blende, fluorpar and quartz. The fissure and replacement veins of barytes in limestone and dolomite are known in many places in Andhra Pradesh, where they occur in Vempalle limestone (Cuddapah age) near their contact with basic sills or in the sills themselves. It also occurs here as residual nodules, resulting from the decay of limestone. Near Alwar, Rajasthan, fissure veins of barytes are in Alwar quartzite of Delhi Formation (Cuddapah age), while at Arsomang, Kinnaur district, Himachal Pradesh, they occur in quartzite-phyllite sequence of Tiwri formation (Silurian-Ordovician).

Barytes veins appear to have formed in various ways, some being of hydrothermal origin and others arising by leaching of barium compounds from rocks containing these. Bedded deposits of barytes (sedimentary origin) are also known e.g. Mangampeta, Cuddapah district, Andhra Pradesh, where it is associated with volcanic and volcanoclasts.

Distribution

The deposits of barium mineral, barytes, are known in many states of India, e.g. Andhra Pradesh, Bihar, Himachal Pradesh, Karnataka, Madhya Pradesh, Maharashtra, Rajasthan, Tamil Nadu, Uttar Pradesh and West Bengal. State-wise distribution of the deposits are given below in the Table 5.30.

Table 5.30—Distribution of Barytes in India

State	District/Area	Geological Details
Andhra Pradesh	Anantpur, Cuddapah, Khammam, Krishna, Kurnool, Nellore and Prakasam districts, the important deposits being in Pulivenda taluk of Cuddapah district. Some important deposits are also in Anantpur (Tadpatri taluk) and Kurnool (Dhone taluk) districts.	Vempalle limestone formation of Cuddapah age contains barytes veins near contact with basic sills or in sills themselves. The basic sills are of the post Cheyair age. The Mangampeta barytes deposits of Cuddapah district is a bedded deposit with a total reserves of over 20 million tonnes of all grades.
Bihar	Singhbhum (Kolpatka, Dhanopal etc.), Ranchi (Tantisilwai, Bahea, etc.), Palamau (Singhitali, North of Bhawnathpur)	Barytes occurs in discontinuous lenticular patches within sericite-schist with probable reserves of 4100 tonnes at Kolpatka (Singhbhum). In Tanti-Silwai area (Ranchi) barytes veins, a few cm in thickness, are found in a zone, 2.5 km x 200 m. In Singhitali area Palamau, thin veinlets of barytes associated with quartz-veins in silicified shale and dolomitic limestone (lower Vindhyan) and with probable reserves of 33 thousand tonnes up to 1.5 m depth have been noted.

(Contd.)

State	District/Area	Geological Details
Himachal Pradesh	Sirmur, Kinnaur, Presently it is being mined in Sirmur district (Kanti and Tatyana).	It occurs as concordant linear and disconnected patches within Krol limestone at Kanti and Tantya where a reserve of 15,000 tonnes have been estimated. In Kinnaur district (Arsomang) barytes veins occur in quartzite-phyllite sequence of Tiwri formation (Silurian-Ordovician).
Karnataka	Chitradurga	Not of much economic significance.
Madhya Pradesh	Dhar, Dewas, Jabalpur and Sidhi	Barytes is produced only in Dewas and Dhar districts. The other occurrences do not seem to be of much economic significance.
Maharashtra	Chandrapura (Phutana and Mahadwari)	Veins and lenses of barytes, varying in width from a few cm to 4 m and extending upto 140 m in length, occur within granite-gneiss of Archaean age.
Rajasthan	Alwar, Udaipur, Jaipur, Pali, Sikar, Chittaurgarh, Bhilwara and Bundi	The production comes mostly from Alwar and Udaipur districts. It occurs as veins in Alwar quartzite of Delhi formation (Cuddapah). In Umar area (Chittaurgarh) barytes lenses, upto 100 m x 0.5-4 m, occur in association with vein-quartz in dolomitic marble of Pre-Aravalli age, while in Rawat Bhatta area it occurs as thin veins, stingers, vug fillings and laminations in the shales of upper Vindhyan age.
Tamil Nadu	Coimbatore (Kurichi), North Arcot (Alangayam), Ramanathpuram, S. Arcot, Tiruchirapalli	Barytes occurs as veins and lenses, associated with vein-quartz, in pegmatites and gneisses.
Uttar Pradesh	Dehradun, Garhwal and Kumaon	Barytes is reported from the Precambrian-lower Palaeozoic sequence in Tethyan zone of Garhwal and Kumaon.
West Bengal	Purulia/Malthol-Belma-Raghudih and Chas-road areas	Barytes occurs as narrow veins within mica-schists, enclosed by the granitic rocks.

Reserves

All India recoverable reserves of barytes as on 1.4.1990 are placed over 70 million tonnes. Reserves are located mostly in Cuddapah district of Andhra Pradesh. The Mangampeta barytes deposits (Andhra Pradesh) alone accounts for a total recoverable reserves of 20 million tonnes of all grades.

Production

Andhra Pradesh continued to be the principal producer of barytes (over 95%) of total production. The all-India production figures for 1989-90, 1990-91, 1991-92 and 1992-93 were about 645, 509, 684 and 381 thousand tonnes, respectively.

4. TALC, SOAPSTONE AND STEATITE

The term talc includes steatite, a massive compact variety, and soapstone, massive granular type, with impurities like chlorite, serpentine, magnesite, antigorite, enstatite etc. It is characterised by extreme softness, soapy feel, common foliated structure and pearly lustre. It is flexible, but inelastic. It is classified into soft and hard, flaky and fibrous, and is marketed as crude, ground and sawed. The impure variety is comparatively harder, and talc with tremolite is fibrous. Potstone is impure soapstone, greyish-green or brownish black in colour.

The total world production of talc, soapstone and steatite is about 7.7 million tonnes (1989). The main producers were USA, China, USSR, Brazil, India, Finland and Australia. USA was the largest producer, contributing 15% to the world production, while India was the fifth largest producer, accounting for 5.3%.

Uses

Talc is largely used in powdered form as filler in paper, textile and rubber, as an extender in paints, and in the manufacture of ceramics. In the form of soapstone it is used for wash tubes, sinks, table tops, switch boards, hearthstones and furnace linings. It is also employed in the tips of gas burners, tailor's chalk, slate pencils, carved ornaments etc. Other uses are as a carrier for fertilizers and in the manufacture of insecticides and detergents. Steatite cut into bricks is utilised for lining alkali tanks in paper industry. Body and face powders are prepared from the finest quality of talc after adding deodorants and perfumes. Above 40% of talc consumption is in paper industry, followed by soap and detergent (38%), insecticide (20%) and the remaining 2% is shared by ceramic, cosmetic, fertilizer, paint, rubber, vanaspati and other industries.

Grading

Talc is graded into grades A, B, C and D, based mainly on its whiteness. Grade-A is the first quality materials, pure white to slightly greenish with whiteness in the range of 90-95%. It is used in pharmaceuticals and cosmetics. In grade-B the whiteness is in the range of 85 to 90%, and is used in superior quality paper, textile and ceramics, while in grade-C whiteness ranges from 78-85% and is used in inferior paper, paint, rubber, plastic and detergents. Grade-D is of very poor quality, having whiteness less than 78%, and is used in DDT. The colour of this grade is dark greenish-grey to reddish-green.

Mode of Occurrence and Origin

Talc occurs commonly as lenses and pockets in metamorphosed dolomite, crystalline schists or gneisses and as large bodies in ultrabasic rocks: It is often associated with serpentine, chlorite, asbestos, actinolite, tourmaline and magnetite.

It is formed as secondary mineral resulting from hydro-thermal alteration of magnesium-bearing rocks, particularly non-aluminous magnesian silicates. The change may have arisen by the contact action of granitic magma, by the action of stress during regional metamorphism or by action of magmatic waters.

Distribution and Reserves

The state-wise distribution of talc is shown below in the Table 5.31.

Table 5.31—Distribution of Talc, Soapsone and Steatite

State	Districts	Other details, if any
Andhra Pradesh	Kurnool, Chittoor, Cuddapah, Anantpur, Karimnagar, Nalgonda, Nellore, Warangal and Medak	It is being mined in the state.
Bihar	Dhanbad, Hazaribagh, Ranchi and Singhbhum	Deposit of Khejurdari and Digha (Singhbhum) are being exploited.
Karnataka	North Canara, Shimoga, Hassan, Chitradurga and Tumkur	Small quantity is being produced.
Kerala	Cannanore and Calicut	Talcose rock occurs as detached outcrops in linear fashion and is reporting production.
Madhya Pradesh	Jabalpur (Bheraghat)	It is being worked.
Maharashtra	Ratnagiri	Steatite and talc-schists occur as small bands in Dharwar schists.
Orissa	Sundargarh, Sambalpur and Korapur	Production is being reported.
Rajasthan	Udaipur, Bhilwara, Dungarpur, Sawai Madhopur, Jaipur, Ajmer, Alwar and Tonk	The state is the chief producer of steatite.
Tamil Nadu	Salem and North Arcot	It is being worked for making utensils and toys.
Uttar Pradesh	Pithoragarh and Almora	The production is reported.

The known recoverable reserves of talc/soapstone/steatite are placed at 83.665 million tonnes (as on 1.4.1990) of which proved, probable and possible reserves are 6.96 million tonnes, 15.32 million tonnes and 61.36 million tonnes, respectively. Substantial quantities of reserves are located in Rajasthan (44.33 million tonnes), Uttar Pradesh (16.05 million tonnes), Kerala (8.13 million tonnes), Maharashtra (5.88 million tonnes) and Madhya Pradesh (4.97 million tonnes). The all-India conditional reserves are 755 thousand tonnes.

Production

Rajasthan is the chief producer of steatite. In 1992-93 as much as 84% output was from this state, while the remaining 16% was contributed jointly by eight states. The individual share of Uttar Pradesh and Andhra Pradesh in the total output during the year was 7.4% and 6.1% respectively. The total all-India production during 1990-91, 1991-92 and 1992-93 were about 431, 427 and 382 thousand tonnes, respectively.

The deposits are being worked both by opencast and underground methods of mining. Except a few mines in Rajasthan and Andhra Pradesh which are worked by underground methods, almost all mines are opencast.

5. BENTONITE

Bentonite is a naturally occurring clay of great commercial significance. It is composed mainly of montmorillonite and minor beidellite with small amounts of igneous rock minerals, alkaline earth and ferric iron. It possesses inherent bleaching properties and, hence, commonly known as bleaching clay. There is another type of bentonite which swells enormously in water by absorbing about 8 times its own volume of water, and is non-bleaching. The bentonitic bleaching clays are characterised by a waxy appearance and by rapid slaking in water without swelling. It is heavier and denser than fuller's earth.

Uses

Bentonite is utilised largely in foundry facings, civil construction work for making embankments and other porous formations watertight, oil well drilling, fertilizer filler, insecticide and pharmaceuticals. It also finds use in refractory, ceramic, chemical, abrasive, alloy-steel, electrode, paint, rubber, sugar, textile and vanaspati. The norm of bentonite consumption in oil drilling, although varies from field to field and also depends upon depth of drilling, it is usually 15 to 24 tonnes per 1,000 metre drilling. The present all-India consumption of bentonite in different industries is about 125 thousand tonnes a year.

Mode of Occurrence and Origin

Bentonite is found associated with lake and marine shales and sandstone, and volcanic glass shards. In some bentonites angular pieces of accessory felspar, quartz, biotite, pyroxene, zircon and other minerals of volcanic igneous rocks are noticed. It may have formed by alteration/decomposition of volcanic ash or tuff. It is deposited mechanically in lakes or ocean waters, and may also be air-borne and air-deposited.

Distribution and Reserves

Bentonite deposits occur mainly in Barmer district of Rajasthan, Kuchch and Bhavnagar districts of Gujarat. Kuchch bentonite is regraded as one of the world's best because of its high swelling property. Other minor deposits are in Sabarkantha district of Gujarat, Nagaur, Bindi, Jaisalmer and Sawai-Madhopur districts of Rajasthan, Sahebganj district of Bihar and Udhampur district of Jammu and Kashmir.

Recoverable reserves of bentonite of all grades as on 1.4.1990 are placed at about 368 million tonnes of which Barmer district (Rajasthan) has 267 million tonnes, and Bhavnagar and Kuchch districts (Gujarat) 37 and 50 million tonnes, respectively. The all-India conditional reserves are placed at about 142 million tonnes.

Production

The all-India productions of bentonite during 1987-88, 1988-89, 1989-90 were about 119, 226 and 247 thousand tonnes, respectively. Gujarat continued to be the leading producer, followed by Rajasthan. A small amount of bentonite to the tune of about 3000 to 3500 tonnes per year was produced by Bihar. Bentonite is obtained mainly from manually operated mines. Shovels and dumpers are used for mining, haulage etc.

6. VERMICULITE

Vermiculite (hydrous magnesium-iron-aluminium silicate) is a mica which exfoliates under strong heat into very thin sheets. The degree of exfoliation and the development of golden, bronzy or silver lustre on heating determines the quality of vermiculite. Its normal expansion is 8 to 12 times its original volume. The superior quality vermiculite expands as much as 20 times. Crude vermiculite resembles mica in appearance, and occurs as glistening broad crystals or shining flakes of brown, greenish-black or greenish-yellow colour. It is soft, pliable, inelastic, and opaque with pearly or bronze like lustre. It has lower refractive index than that of biotite and higher compared to chlorite. Vermiculite books vary in size from 400 mm in length and 50 mm in thickness to thin hardly discernable flakes less than 1 mm in diameter.

The USA and the Republic of South Africa, contributing over 90% of world output, continued to be the major producers of vermiculite, followed by Brazil, Japan, India, Kenya and Zimbabwe.

Uses

Vermiculite has wide range of applications because of the unique combination of its properties, like good thermal insulator and sound absorption, high melting point (about 1400°C), non-combustibility, compressibility, extreme lightness etc. The chief utility are in the construction industry for fire resistant boards, plaster aggregate, acoustic plaster, light-weight concrete and insulation. It finds use in horticulture as composts, lawn care products, fertilizers and for soil improvements. Other uses are in paints, coatings, feeding stuffs, pesticides, moulded products, friction materials, packing, pumping aids, release agent, fire-proof safes, furnaces, fire protection, acoustic treatments, bricks, sealing material, drilling mud, sound deadening compounds, biomass support, cation exchange, water treatment, heavy metal fixation, foundry and steel works etc. Practically all uses of vermiculite are for the expanded material.

Mode of Occurrence and Origin

Vermiculite occurs in tabular, radiating and granular forms. It is found associated with biotite and phlogopite in highly silicic igneous rocks, quartz-veins and pegmatites. It also occurs associated with pyroxenite and peridotite, as found in Chatra district of Bihar. The vermiculite bearing ultrabasics is generally magnetite-rich and shows alteration to actinolite rocks.

It may have originated by hydrothermal alteration of biotite or phlogopite. The pyroxenes in ultrabasics may possibly have first altered to amphibole and then to biotite and vermiculite. The hydrothermal actions on biotite or phlogopite may have given rise to vermiculite in various rocks.

Distribution and Reserves

The major deposits of vermiculites are in Tamil Nadu, Andhra Pradesh, Gujarat and Rajasthan. The state-wise distributions are given below in the Table 5.32.

Table 5.32—Distribution of Vermiculite Deposits in India

State	District/Area	Other details
Andhra Pradesh	Nellore, Visakhapatnam, Srikakulam	
Bihar	Ranchi, Hazaribagh and Chatra	
Gujarat	Vadodara	
Karnataka	Tumkur, Hassan, Kolar	
Madhya Pradesh	Jhabua	
Rajasthan	Ajmer	
Tamil Nadu	North Arcot, Coimbatore, Tiruchirapalli	
West Bengal	Bankura	

The all-India in-situ reserves of vermiculite are estimated as 2.07 million tonnes. The recoverable reserves of all grades, however, are about 312 thousand tonnes, of which Tamil Nadu, Andhra Pradesh and Karnataka are having reserves of about 162 thousand, 83 thousand and 42 thousand tonnes, respectively.

Production

The production in Andhra Pradesh fell down considerably, and Tamil Nadu was the leading producer in 1992-93 and accounted for 61% production in 1992-93, followed by Andhra Pradesh 24%, Gujarat 12% and Rajasthan 3%. The all-India productions during 1990-91, 1991-92 and 1992-93 were about 1800 tonnes, 1650 tonnes and 1400 tonnes, respectively.

5.6 METALLURGICAL AND REFRACTORY MINERALS

The minerals, which withstand high temperature say above 1500°C, may be moulded into bricks or other forms, resist cracking and spalling under temperature change, and are non-reactive with materials being melted, are refractory minerals. The metallurgical processes, involving high temperature, need refractory minerals for furnace lining. Besides furnace lining, these minerals are required for other high temperature purposes such as retorts, ceramics, electrical purposes etc. The important refractory minerals are fireclay, graphite, dolomite, magnesite, chromite, bauxite, sillimanite-kyanite group of minerals, quartzite and quartz-schist, diaspore, pyrophyllite and zircon. Magnesite (magnesium mineral), chromite (chromium mineral) and bauxite (aluminium mineral) have been dealt separately under metallic minerals. The details of other refractory minerals are included in this chapter.

1. FIRE CLAY

Fire clays are high-alumina clays with some non-plastic refractory flint and moderately refractory plastic clays, which withstand temperature rise of 2714°F to 2984°F.

Uses

Fire clay is chiefly consumed in refractory industry. Several kinds of fireclay bricks are manufactured by admixing with calcined bauxite or kyanite in suitable proportions to meet the different insulation requirements. These bricks are utilized in most of the major industries, like iron and steel, ferro-alloys, cement, foundries, glass etc. The other uses of fireclay are in the ceramics, sanitary ware products, insecticide, manufacture of sugar, abrasive, chemical, electrical crucibles, paper, rubber, textile, vanaspati etc.

Mode of Occurrence and Origin

Fireclays occur mainly as underlying the coalseams. The beds are small, lens-shaped and exhibit little lamination. They are considered to have originated from suspended matter carried by low-gradient streams into coal swamps. In this process the coarser sediments are filtered out earlier by the marginal fringe of vegetation, while the finest materials reach and settle in the interior of the basin. The organic acids present may have acted upon the clayey matter to decompose the impurities. The heat and pressure, acted upon to form the coal, may have given rise also to the associated fireclay.

Distribution and Reserves

Fireclay is distributed mainly in the coal measures of both Gondwana and Tertiary periods. Important deposits are found in Jharia and Rainganj coal-fields of Bihar and West Bengal, Korba coalfields of Madhya Pradesh and Neyveli lignite field of Tamil Nadu. The notable occurrences of fireclay not associated with coal measures are known in Gujarat, Jabalpur region of Madhya Pradesh and Belpahar-Sundergarh area of Orissa. The distribution of fireclay, state-wise, is given in the Table 5.33 below.

Table 5.33—Distribution of Important Fireclay Deposits in India

State	District	Other details
Andhra Pradesh	East Godavari, West Godavari and Adilabad	
Bihar	Bhagalpur, Dhanbad, Giridih, Hazaribagh, Deogarh, Palamau, Ranchi, Santhal Parganas	
Gujarat	Kuchch, Rajkot, Sabarkantha, Surat, Surendranagar, Valsad	The important deposits are in Surendranagar, Rajkot, Sabarkantha and Valsad
Karnataka	Chitradurga, Hassan	
Madhya Pradesh	Jabalpur, Narsingpur, Satna, Shahdol, Sidhi	
Maharashtra	Amravati, Kolhapur, Sindhudurg	
Meghalaya	West Garo Hills	
Orissa	Cuttack, Dhenkanal, Puri, Sambalpur, Sundargarh	
Rajasthan	Alwar, Bikaner, Jhunjhunu and Sawai Madhopur	
Tamil Nadu	Chengalpattu, South Arcot and Tiruchirapalli	
West Bengal	Birbhum, Burdwan and Purulia	

The reserves of fireclays are substantial, but those of high grade (non-plastic) fireclay containing more than 37 per cent alumina are rather limited. The recoverable reserves of fireclay as on 1.4.1990 are placed at about 697 million tonnes, of which proved, probable and possible categories are about 57, 41 and 599 million tonnes, respectively.

Production

Orissa was the leading producer, accounting for about 18% of output in 1992-93, followed by Gujarat 17.5%, Rajasthan 13%, West Bengal 12%, Bihar 12%, Madhya Pradesh 10%, Tamil Nadu 9% and Andhra Pradesh 5%. Remaining 3.5% was shared by Maharashtra and Karnataka. The total production during 1992-93 was about 439 thousand tonnes, while during 1990-91 and 1991-92, 539 thousand and 420 thousand tonnes, respectively, were produced.

2. GRAPHITE

Introduction

Graphite is allotropic form of pure carbon, and is chemically the same as diamond and charcoal. It received its name from "Graphein" means 'write'. It was earlier mistaken for lead and so called "black lead" (also plumbago) and the pencils made from it are still called lead pencils. The 'lead' used in the pencils is really a mixture of graphite and clay. In nature it is in two forms :

(i) crystalline, consisting of thin, nearly pure black flakes or lumps of small crystals, and

(ii) amorphous, a non crystalline impure variety.

Graphite as mined contains between 10 and 60% fixed carbon (FC). The world resources of graphite are believed to be quite extensive. Rough estimates indicate possible reserves of 840 million tonnes, of which China alone holds over 685 million tonnes. China, Republic of Korea and Russia continued to be leading producers. Despite of self-sufficiency in this mineral, India imports some quantities of flaky and amorphous graphite of high fixed carbon to blend with locally available material and for manufacturing special qualities of mineral based products.

Properties and Uses

Graphite usually occurs in scales, laminae or columnar masses, sometimes granular, and rarely earthy. It is soft (hardness 1 to 2) and iron-grey to dark steel-grey in colour with black shining streak and sp. gr. of 2 to 2.3, depending upon its purity. It has a metallic lustre and feels cold like metal, when handled, owing to its being a good conductor of heat.

The oldest use of graphite is in marking pencils which still persists. Primitive man used it as a black pigment to paint his body and later to colour pottery before firing. Its high melting temperature (3000°C) and insolubility in acid create many uses of it. It is used as a material for crucibles, heat resistance paints and in foundry facings. It conducts electricity almost as well as metals and so it is extensively used in electrical engineering in dry and storage batteries, electrodes etc. Owing to its greasy feel and plasticity, it is used for lubricants for heavy machine and as antifriction material for polishing gun powder grains. It is an ideal material for making heavy duty moulds. Powdered graphite is used in boilers to prevent incrustation. It also finds application in atomic reactors to slow down neutrons and as an absorber of radiation in atomic energy research.

The major consumers of graphite are the crucible industries and foundries. Refractory industry has begun to consume more and more graphite in recent years for manufacturing magnesium-carbon bricks for use in steel making in basic oxygen and electric arc furnaces. Synthetic graphite has replaced natural graphite in making electrodes, carbon brushes and graphite bricks required by blast furnaces.

Specifications

The specifications of graphite for different types of industries are tabulated overleaf in Table 5.34.

Table 5.34—Specifications of Graphite for Different Industries

Industries	Specifications
(a) Crucible industry	Flaky graphite with F.C. 85% (min.), size = –20 to + 80 mesh. Bigger the size of flakes, better is the quality of crucibles.
(b) Foundry industry	Preferably flaky with F.C. 50% (min.).
(c) Refractory industry	Flaky graphite with F.C. 85% (min.) - lower grade, flaky graphite with F.C. upto 95% (min.) - higher grade; fusion point > 1400°C - lower grade, > 1550°C - higher grade.
(d) Pencil industry	Amorphous graphite with F.C. 95% (min.), free from gritty particles. Micronised amorphous graphite preferred
(e) Paints/Lubricants	Amorphous graphite, synthetic graphite made from petroleum, coke has virtually replaced amorphous graphite in manufacture of lubricants

Mode of Occurrence and Origin

Graphite occurs in three forms - flakes, dusts and lumps. It is mainly found in metamorphic rocks like gneisses, schists, marble, quartzite and altered coal beds. It also occurs in igneous rocks, veins and pegmatites. The crystalline variety is mostly in minute flakes, disseminated through metamorphic rocks. The scales are usually 2 cm to 3 cm in size. The amorphous type is in dust like form and graphite content of these ores is usually high, ranging from 30-40 to 90 per cent. The graphite content of the former type varies from 2.5 to 17 per cent, but is usually around 4 to 8 per cent. The dense crystalline graphite, consisting of small crystals, occurs in pockets within magmatic rocks, granites, syenites etc. These graphite lumps contain as much as 60 to 85% carbon.

The origin of graphite has been assigned to the following processes :

1. **Regional metamorphism :** Associated with schists, gneisses, kondalites etc., as found in Palamau district, Bihar and Kalahandi district, Orissa. This is the most common type and are, generally, flaky in nature.
2. **Crystallization from magma :** Found in granite, syenite, basalt, etc. In Siberia it occurs in nepheline-syenite.
3. **Contact metamorphism :** Associated with contact metamorphic silicates in limestone adjacent to an igneous intrusion as found in Ontario, Canada.
4. **Hydrothermal solutions :** This accounts for vein-deposits and deposits in pegmatites as found in Ceylone.

The deposits of scaly graphite possibly originate from organic matter or inorganic carbonaceous matter disseminated in highly altered sedimentary rocks, which have been converted by high temperatures and pressures into gneisses and crystalline schists. The break down of calcium carbonate also yields carbon (graphite) in the rock during regional metamorphism. The amorphous graphite may have formed from coal under the action of high temperatures of molten igneous masses that rose from depths of earth and came into contact with coal beds. These graphite beds are in sandstones, clayey-schists and other sedimentary rocks. The carbon of the graphite found in igneous rocks, dykes and veins was possibly picked up from the underlying carbonate rocks or has resulted from gaseous carbon compounds given off by the magma. The occurrences of graphite in Precambrian rocks suggest an inorganic rather than an organic origin for carbon.

Distribution

The graphite deposits of India are found associated with Precambrian rocks mainly with schists and gneisses. The geographic distribution of graphite, state-wise, along with their chief geological characteristics are given in the Table 5.35 below.

Table 5.35—Distribution of Graphite in India

State	District	Chief Geological characteristics
Andhra Pradesh	East Godavari, Khammam, Srikakulam, Visakhapatnam, and West Godavari	Graphite occurs as a constituent of schists, belonging to Khondalite group of rocks. It is also found with pegmatites and quartz-veins traversing khondalites, where it is flaky in nature.
Bihar	Palamau	In most occurrences graphite occurs as disseminations in schists. The graphite content varies from 15 to 30%.
Gujarat	Panch Mahal	Graphite is associated with schists and gneisses of Dharwar Super-Group.
Kerala	Ernakulam, Iduki, Quilon, Kottayam, and Trivandrum	Graphite is in association with Khondalite group of rocks. It also occurs in the form of lenses, thin veins, streaks and disseminations (mostly flaky) in charnockite and gneisses along the contacts of quartzite band with other formations.
Orissa	Bolangir, Dhenkanal, Kalahandi, Phulbani and Sambalpur	The graphite deposits are found along the contact zones of khondalite rocks with granite gneiss, associated with pegmatite bodies. The mineral is flaky and crystalline, and its carbon content varies from 30 to as much as 80 per cent.
Rajasthan	Ajmer and Banswara	The graphite deposits are associated with schists and gneisses.
Tamil Nadu	Madurai and Ramanathapuram	Graphite disseminations in garnetiferous gneisses have been noted.
Arunachal Pradesh	Subansiri, Siang and Lohit	Graphite is associated with schists and gneisses of Bombdila group. Both flaky and amorphous graphites are found.
J & K	Baramula and Doda	Graphite is found in amorphous forms impregnating schists of Salakhala group.

Reserves

The reserves of graphite in India have been reassessed by the I.B.M. in 1990 and the total recoverable reserves have been estimated as 3.11 million tonnes. Almost entire reserves under the proved category are from Ramanathpuram district of Tamil Nadu. The fixed carbon content of the occurrences ranges between 10 and 40 per cent, occasionally rising to 60 per cent.

Production

Production statistics of graphite relate to the run of mines, containing fixed carbon range from 10 to 60 per cent. The run-of-mine is later beneficiated by crushing and froth flotation to obtain useable concentrates containing upto 92% fixed carbon. The production of run-of-mine in 1985 was at about 33,850 tonnes, in 1987 at 42,590 tonnes and 1988 at 58360 tonnes. The productions during 1991-92 and 1992-93 were about 77 thousand and 68 thousand tonnes, respectively. Orissa continued to be the leading producer, followed by Bihar and Tamil Nadu.

Graphite mines in India are mostly opencast workings, few to them are worked underground. Almost all the mines are small mines. The acting mining centres are Bolangir, Kalahandi, Phulbani and Sambalpur in Orissa, Palamau in Bihar, East Godavari and Khammam in Andhra Pradesh and Banswara in Rajasthan. Graphite of Orissa, Bihar and Andhra Pradesh is flaky, while that of Rajasthan is amorphous. The run-of mine which sometimes contains as low as 10% F.C. has to be invariably beneficiated before marketing. The beneficiation process adopted consists of crushing the run-of-mine in ball mills and floating the crushed material in flotation cells with the help of kerosene, using pine oil as a collector. After collecting the higher size fractions, the beneficiated material is ground, the finest fraction being 300 mesh size.

3. DOLOMITE

Dolomite is a double carbonate of calcium and magnesium ($CaCO_3$ = 54.35% and $MgCO_3$ = 45.65%), and is of great industrial importance. Some of the magnesium in the dolomite is replaced by iron or manganese, and with lesser proportion of magnesium carbonate, it is dolomitic limestone. Dolomite has also been described, in brief, under magnesium minerals (see Metallic Minerals).

Mineralogy

It is generally yellowish-white or brownish-white in colour and some times tinged with red, green or black. It crystallises in hexagonal trirhombohedral forms with curved faces and also found as massive and granular with saccharoidal texture. The cleavage is perfect and lustre is vitreous inclining to pearly in crystals, but massive types are dull and opaque. It shows conchoidal or uneven fracture, and has hardness of 3.5 to 4 and sp. gr. 2.8-2.9. It is distinguished from cacite in that it does not effervesce readily in cold acid, and has slightly higher refractive index and birefringence. It is also distinguished from calcite by Lemberg's test which involves boiling with a solution of aluminium chloride (or ferric chloride) and logwood. Calcite is stained pink, while dolomite undergoes no such staining.

Uses

Dolomite is used as a flux and as refractory material in iron and steel industry, accounting for over 94% of the total consumption. It is an important building material. It is also consumed in glass (2%), ferro-alloys (2%), alloy-steel (1%) and other industries (1%). Its use in chemicals and soft abrasives are detailed below.

Chemicals

It is utilised in the preparation of magnesium and calcium bisulphites, used for paper making. It is also used in the preparation of certain magnesium salts. The roasted dolomite is employed to neutralise acid water. Dolomite is also used in production of carbon-dioxide.

Soft Abrasive

It is calcined to unhydrated calcium and magnesium oxides, called 'Vienna lime', which is a common buffer for various metals, pearl and celluloid. It is used to give the deep under-surface blue colour in highly polished nickel articles.

Specifications

With the introduction of LD process of steel making, the requirement for low silica dolomite is increasing. Generally, insolubles like SiO_2, Fe_2O_3 and Al_2O_3 are considered to be the deleterious constituents for any industrial use. High purity dolomite with less than 1% insolubles is preferred in refractory brick making for use in the lining of LD furnaces. Similarly, high grade dolomite with less than 0.15% iron content is needed in glass making. The ISI specifications of dolomite for use in iron and steel and glass industries are given in the Tables 5.36 and 5.38 below. Besides, specifications of dolomite consumed in different steel plants both in BF and SMS are given below in the Table 5.37.

Table 5.36—Specifications of Flux Grade Dolomites for use in Steel Plants (IS : 10346 - 1982)

Grade	Constituents			Size
	CaO	MgO	Acid insolubles	
Grade-I				
For making tar bonded dolomite bricks for LD converters	30% (min.)	20% (min.)	4.5% (max.) out of which SiO_2 should not exceed 2%	50-80 mm
Grade-II				
For use in blast furnace/sintering plant and for production of sintered dolomite for fritting purpose in open hearth furnace	28% (min.)	18% (min.)	8% max. out of which Al_2O_3 should not exceed 2%	1. Ballast furnace 25-75 mm, tolerance ± 10% 2. Sintering plant 0-60 mm, tolerance ± 5% 3. Refractory 70-80 mm, tolerance ±10%

Table 5.37—General Specifications of Dolomite consumed in different Steel Plants (in per cent)

Plant	Constituents	Sintering Plant (SP) Blast furnace (BF)	SMS (Steel) Melting Shop)	Refractory
1. Bhilai Steel Plant	MgO	19 (min.)	20 (min.)	20 (min.)
	CaO	29 (min.)	30 (min.)	30 (min.)
	SiO_2	-	-	1.7 (max.)
	Al_2O_3	-	-	2.5 (max.)
	Fe_2O_3	-	-	
	Acid insoluble (A.I.)	6 (max.)	5 (max.)	
	Size	0-60 mm	60-100 mm	50-80 mm
2. Bokaro Steel Plant	MgO	20 (min.)	-	20 (min.)
	CaO	30 (min.)	-	30 (min.)
	SiO_2	5 (max.)	-	1.5 (max.)
	Al_2O_3	-	-	1.0 (max.)
	Fe_2O_3	-	-	
	A.I.	-	-	-
	Size	25-80 mm	-	5-22 mm

(Contd.)

Plant	Constituents	Sintering Plant (SP) Blast furnace (BF)	SMS (Steel Melting Shop)	Refractory
3. Rourkela Steel Plant	MgO	19 (min.)	20 (min.)	21 (min.)
	CaO	-	-	-
		-	2.5 (max.)	1.5 (max.)
	Al_2O_3	-	1.5 (max.)	0.75 (max.)
	Fe_2O_3	-	1.0 (max.)	1.0 (max.)
	A.I.	8 (max.)	-	-
	Size	0-6 mm	40-80 mm	-
4. Durgapur Steel Plant	MgO	18 (min.)	(20 min.)	-
	CaO	-	30-35	-
	SiO_2	6 (max.)	2.5 (max.)	-
	Al_2O_3	-	0.8 (max.)	-
	Fe_2O_3	-	1.0 (max.)	-
	A.I.	10 (max.)	-	-
	L.O.I.	-	44.0	-
	Size	-	3-16 mm	-
5. Indian Iron & Steel Co.	MgO	19.5 (min.)	19.5 (min.)	-
	CaO	-	-	-
	SiO_2	-	-	-
	Fe_2O_3	-	-	-
	A.I.	8.7 (max.)	8.7 (max.)	-
	Size	25-75 mm	50-125 mm	-
6. Tata Iron & Steel Co.	MgO	20 (min.)	20 (min.)	20 (min.)
	CaO	-	-	-
	SiO_2	-	3.5 max.	1.7 (max.)
	Al_2O_3	-		
	Fe_2O_3	-	-	-
	A.I.	6 (max.)	6 (max.)	1.5 (max.)
	Size	20-75 mm	25-50 mm	5-25 mm

Table 5.38—Specification of Limestone & Dolomite for Glass Industries (IS 997-1973)

Constituents	Requirement on dry basis (per cent)
1. SiO_2	2.5 (max.)
2. Total iron as Fe_2O_3	
(a) Calcite or marble	0.05 (max.)
(b) Limestone	0.10 (max.)
(c) Dolomitic limestone and dolomite	0.15 (max.)
3. CaO	53 (min.) may be fixed by mutual agreement
4. Total CaO + MgO	54.50 (min.)

Mode of Occurrence and Origin

Dolomite occurs in extensive beds at many geological horizons. It may also occur as bands and pockets. It may be deposited directly from sea-water, but most dolomite beds have been formed by the alteration of limestone, the calcite of which replaced by dolomite. Thus, many dolomites are not of sedimentary origin, but are epigenetic replacement of limestone. Dolomitisation is often related to joints and fissures through which the solutions penetrated, and thick beds of limestone may be changed to dolomite. As a result of this change, a shrinkage takes place and useful minerals may afterwards be deposited in the cracks so caused. The solutions giving rise to dolomitisation are mainly drived from the sea, and an example of this change is seen in the conversion of the calcite and aragonite of coral reefs into dolomite by reaction with magnesium-salts contained in sea water. Sedimentary dolomites are, generally, considered to be sea-floor replacements of calcareous ooze.

Distribution and Reserves

The dolomite occurrences are widespread in almost all parts of the country. The distribution of dolomite (state/district-wise) is given in the Table 5.39 below.

Table 5.39—Distribution of Dolomite

State	District/Area	Other details
Andhra Pradesh	Anantpur, Khammam	Vempalle formation of Cuddapah Super group and the Pakhal formation (Archaean) contain dolomite. All grades are available. Flux grade is known in Khammam district.
Arunachal Pradesh	Kameng (Rupa and Dedza areas)	Tengu formation (lower Palaeozoic) shows presence of workable dolomite. All grades are available.

(Contd)

State	District/Area	Other details
Bihar	Palamau (Tulsidamar, and Kauriya)	Dolomite occurs in Archaeans. All grades.
Gujarat	Bhavnagar, Vadodara	Dolomite is found in Jurassic, Cretaceous and Tertiary formations of Gujarat. All grades are available.
Haryana	Mahendragarh	The dolomite belongs to Precambrian.
Karnataka	Belgaum, Bijapur, Chitradurga, Shimoga, Tumkur and Uttrakanad	All grades of dolomite in Pre-cambrian formations are found.
Madhya Pradesh	Balaghat, Bastar, Bilaspur, Chhindwara, Durg, Hoshangabad, Jabalpur, Jhabua, Mandla, Narsimhapur, Raigarh, Satna and Seoni	The dolomite of all grades, belonging to Precambrian age, is available. This is the leading state so far reserves are concerned.
Maharashtra	Chandrapur, Nagpur and Yeotmal	All grades of dolomites, belonging to Precambrian formations are available.
Orissa	Koraput, Keonjhar, Sambalpur, Sundargarh	Sizeable reserves of all grades of dolomite of Precambrian age are available.
Rajasthan	Ajmer, Alwar, Bhilwara, Jaipur, Jaiselmer, Jhunjhunu, Jodhpur, Pali, Sikar, Udaipur	Dolomite belongs to Precambrian age.
Tamil Nadu	Salem and Tirunelveli	Dolomite is found in Khondalite group of rocks.
Uttar Pradesh	Banda, Dehradun, Garhwal, Mirzapur, Nainital, Tehri Garhwal	All grades of dolomite, belonging to Precambrian age are available.
West Bengal	Jalpaiguri	The dolomite belongs to Buxa formation (Algonkian). It is of all grades.

The total recoverable reserves of all grades of dolomite as on 1.4.1990 are placed at 4967 million tonnes of which the proved reserves are about 11% (523 million tonnes), probable reserves about 20% (1010 million tonnes) and possible reserves about 69% (3434 million tonnes). The reserves of BF and SMS grades dolomite are large, but refractory grade reserves at 116 million tonnes comprise only 2.3% of the all-India reserves. Reserves of low silica (less than 1% SiO_2) used as refractory in L.D. converters, are limited. Of the total all-India recoverable reserves, about 92% are distributed in seven states - Madhya Pradesh, Orissa, Arunachal Pradesh, Karnataka, Maharashtra, West Bengal and Gujarat.

Production

The all-India productions of dolomite during 1990-91, 1991-92 and 192-93 were about 2.6 million tonnes, 2.7 million tonnes and 3.1 million tonnes, respectively. Orissa remained the leading producer during 1992-93, contributing about 55 per cent to the total production, followed by Madhya

Pradesh 19 per cent, Gujarat 10.5 per cent, Bihar 6 per cent, Andhra Pradesh 4 per cent and West Bengal 3 per cent. The remaining 2.5 per cent was shared by Uttar Pradesh, Maharashtra, Rajasthan and Karnataka.

4. SILLIMANITE GROUP OF MINERALS

The sillimanite group of minerals, comprising andalusite, sillimanite, and kyanite with identical chemical composition (Al_2O_3 SiO_2) and the fourth mineral dumortierite, a basic aluminium borosilicate, are important high grade refractory minerals. Andalusite, sillmanite and dumortierite crystallise in orthohombic form, while kyanite is triclinic and expands as much as 16% of its original volume. These minerals convert at high temperature, 1100°C to 1650°C, to mullite (cristobalite), which is also an excellent high temperature insulator. The world resources of these minerals are large, but very little of these are in the form of high grade, coarse-grained material, which is in great demand. The major producers of sillimanite group of minerals in the world are South Africa, the USA, Russia, Czechoslovakia, Romania, India and France.

Uses

The major utility of these minerals is in refractory and to a limited extent in electricals and ceramics. The main consumers of these refractories are iron and steel, glass, ceramic and cement industries. Because of the change in refractory technology, andalusite has greater demand and better market.

Mode of Occurrence and Origin

The sillimanite group of minerals occurs in metamorphic rocks, mostly of argillaceous composition. Andalusite is formed under conditions of high temperature and low stress, as in the case of andalusite - hornfelses in thermal aureoles and in regional metamorphic rocks such as the andalusite-schists. For example, it occurs as porphyroblasts in schists and phyllites, belonging to Pre-Vindyan formation of Mirzapur district, U.P. and its continuation in Palamau district in Bihar, and is also available as placers in this area. It is also found in pegmatites. The usual associates are tourmaline, garnet, corundum, topaz, quartz and mica.

Sillimanite occurs as slender prisms in argillaceous crystalline rocks. It is produced at high temperature and moderate stress, and is found in the rocks of innermost zone of thermal metamorphosed or in regional meta-morphism of high grade. In Sonapahar area of Meghalaya it occurs associated with corundum in lenses of varying size and shape within folded quartz-sillimanite-schist. Kyanite is characteristic of argillaceous rocks, metamorphosed under high stress and moderate temperature, as in many gneisses and schists and also in eclogites. Occurrences of kyanite in lenses are also known in pegmatites and as bunches in quartz-veins. Dumortierite occurs in pegmatites or quartz-veins, traversing the aluminous rocks. Many of the sericite-schists contain dumortierite where andalusite has altered partly or fully to dumortierite. It may have possibly resulted due to pneumatolytic actions combined with hydrothermal metamorphism.

Distribution and Reserves

The aluminous refractory minerals occur in several states in India.

Fig. 5.3 *Kyanite boulders.*
Loc : Lapsaburu, Singhbhum district.

The distribution of significant deposits, mineral-wise and state-wise, is shown in the Table 5.40 below.

Table 5.40—Distribution of Important Alumina Refractory Minerals in India

Minerals	State	District/Area	Other details
Audalusite	**Bihar**	Palamau/Nagar Untari	This continues to Wyndhamganj area in U.P.
	Uttar Pradesh	Mirzapur/Wyndhamganj	Andalusite content in phyllite and schist varies from 5 to 15%.
	Rajasthan	Jhunjhunu/Usri	The basal part of Ajabgarh formation, north of Khetri copper belt, is andalusite bearing.
Sillimanite	**Andhra Pradesh**	Khammam and Nellore	It occurs associated with kyanite and khondalite along the east coast.
	Bihar	Ranchi, Monghyr and Singhbhum	It is found associated with kyanite.
	Karnataka	Mysore	It is found associated with kyanite.
	Kerala	Quilon	It contains 5 to 6% sillimanite in beach sand.

(Contd.)

Minerals	State	District/Area	Other details
	Madhya Pradesh	Sidhi/Pipra	The deposit is associated with corundum and is about 800 m long and 64 m wide.
	Maharashtra	Bhandara/Pohra	It is associated with kyanite.
	Meghalaya	West Khasi Hills, Sonapahar	Best quality sillimanite in world is found. Altogether 27 deposits have been located in the area.
	Orissa	Ganjam, Sambalpur	Low grade occurrences are known in Khondalites in the Easter Ghat, and in schistose rocks in Sambalpur. Beach sand of Chattarpur, Ganjam district contains recoverable reserves of sillimanite.
	Rajasthan	Udaipur/Madar	No economic deposit has been located.
	Tamil Nadu	Tiruchirapalli	Beach sand is sillmanite-bearing.
	West Bengal	Purulia	Minor occurrences associated with kyanite are reported in schistose rocks.
Kyanite	**Andhra Pradesh**	Khammam/Garibpet, Nellore/Malakonda	The deposits are associated with Archaean schists and gneisses.
	Bihar	Singhbhum/Lapsaburu, Kanyaluka, Sirboi-Jyoti Pahari. Ranchi and Monghyr	In Lapsaburu it occurs in a belt, 160 km x 16 km, on the northern flank of copper belt (Fig. 5.3).
	Karnataka	Chikmagalur, Hassan, Mysore and Croog	It is found associated with sillimanite at places.
	Maharashtra	Bhandara/Pipalġaon, Dahegaon, Girola, Pohra and Nawgaon	Kyanite-sillimanite rock occurs as segregation along the axial portion of a syncline. Quartz, topaz and dumortierite occur as associates.
	Orissa	Dhenkanal, Mayurbhanj, and Sundergarh	It is found associated with dumortierite at Panijia (Mayurbhanj). The occurrences are with quartz-veins, quartz-mica-schist and talc-tremolite-schist. The quality is generally poor.
	Rajasthan	Udaipur, Ajmer	Minor occurrences of kyanite, associated with sillimanite, are reported.
Dumortierite	**Bihar**	Singhbhum copper belt, Rakha area	It is found associated with quartz-schist in minor amount.
	Meghalaya	West Khasi hills, Mawshinrum,	Two bands of quartz-dumortierite-tourmaline-schist with a little sillimanite, upto 800 m x 60 m in dimension, are located.

The total recoverable reserves of kyanite and sillimanite in the country as on 1.4.1990 are placed at 2.7 million tonnes and 507 million tonnes, respectively. Besides, conditional resources of 175.84 million tonnes for kyanite and 5.9 million tonnes for sillimanite have been estimated. For sillimanite prospective resources of 0.8 million tonnes are also estimated. The possible reserves of andalusite in Mirzapur district of UP have been put at 14 million tonnes. There is no economically viable deposit of dumortirite in the country

Production

Bihar continued to be the leading producer of kyanite, accounting for nearly 54 per cent production in 1992-93, followed by Maharashtra (36 per cent). The remaining 10 per cent production was shared by Karnataka and Rajasthan. In respect of sillimanite, Kerala was the leading producer during 1992-93, contributing about 51 per cent output, followed by Maharashtra 46 per cent and Meghalaya 3 per cent. No production of andalusite in the country has been reported since 1987.

The total all-India productions of kyanite and sillimanite during the years 1990-91, 1991-92 and 1992-93 were about 37 thousand and 13 thousand, 19 thousand and 14 thousand, and 10 thousand and 20 thousand tonnes, respectively.

5. DIASPORE

Diaspore ($Al_2O_3.H_2O$) is a non-silicate aluminium mineral and is grouped under high alumina refractory. It is distinguished by its hardness (6.5 to 7) and pearly lustre on cleavage faces. It is mostly used as refractory material in making high alumina refractory bricks. The indigenous refractory manufactures use diaspore, analysing Al_2O_3 = 56 to 62%, Fe_2O_3 = 1 to 4%, TiO_2 = 0.8 to 1.5, PCE (orton) 36 min., and size between 75 and 100 mm. It may also be used by bonding with flint or plastic clay as per content of alumina needed in the finished products. It is, thus, consumed in minor amounts in ceramic industry.

Mode of Occurrence and Origin

It commonly occurs associated with corundum or emery, probably as an alteration product of the oxide. It occurs similarly in bauxite deposits, and some of the bauxite may be diaspore. It has been noted as an accessory mineral in metamorphic limestone and dolomite, and in some manganese mines as mangan-diaspore. It has been found in the form of geodes and lenses associated with pyrophyllite and quart-reef in Hamirpur, Jhansi and Lalitpur districts of Uttar Pradesh and Chhatarpur, Shivpuri and Tikamgarh districts of Madhya Pradesh.

Distribution, Reserves and Production

Systematic and planned exploration of diapsore has not yet been carried out to prove its potentiality in the country. Madhya Pradesh (Chattarpur and Tikamgarh districts) and Uttar Pradesh (Jhansi, Lalitpur and Hamirpur districts) are the only two states where it is mined. Manual opencast method is being adopted in the dispore mining. The average Al_2O_3 percentage varies from 42 to 72 in the mined ore, where as Fe_2O_3 ranges from 0.7 to 4.45 per cent. The productions made during 1990-91, 1991-92 and 1992-93 were about 8 thousand, 10 thousand, 13 thousand tonnes, respectively. There were altogether 23 mines reporting production in 1992-93. Madhya Pradesh produced about 6 thousand tonnes, while the rest about 7 thousand tonnes were produced by Uttar Pradesh.

6. PYROPHYLLITE

Pyrophyllite $(Al_2Si_4(O_{10}(OH)_2)$, hydrous aluminium silicate, resembles talc in physical properties, though differs in chemical composition. It occurs, generally, in foliated, radiated, lamellar or somewhat fibrous forms. The laminae is flexible, but not elastic. It has greasy feel with hardness 1 to 2 and sp. gr. 2.8 to 2.9.

The uses of pyrophyllite are similar to talc except that it is not utilized for making body and face powders. It is harder than talc, and unlike talc, does not flux when fired. It is, hence, employed as a refractory material. Besides refractory, it is also used in ceramic industry.

Mode of Occurrence and Origin

It occurs chiefly as foliated masses in crystalline schists. It is found as gangue for kyanite mineral. Compact massive variety with white, greyish or greenish colour resembles compact steatite, and is found in schistose rocks as base material. It also occurs in slates and tuffs with interbedded volcanic breccias and flows, all of which have been metamorphosed. It is a product of hydrothermal metamorphism of aluminous sediments.

Distribution, Reserves and Production

The distribution of important deposits of pyrophyllite, state-wise, is given below in the Table 5.41.

Table 5.41—Distribution of Important Pyrophyllite Deposits of India

State	District/Area	Other details
Madhya Pradesh	Chhatarpur, Gwalior, Shiv puri, Surguja and Tikamgarh	Reserves in Khari and Khera areas of Tikamgarh district are placed at 1.1 million tonnes, while these of Shivpuri and adjoining areas of Bundelkhand are estimated as 0.18 million tonnes.
Maharashtra	Bhandara	
Orissa	Keonjhar	Reserve estimates are not available.
Rajasthan	Udaipur	
Uttar Pradesh	Jhansi and Lalitpur, Hamirpur and Garhwal	3.05 million tonnes reserves have been estimated. Large deposits are found in Jhansi district.

The production of pyrophyllite during 1990-91, 1991-92 and 1992-93 were about 82 thousand, 90 thousand and 81 thousand tonnes, respectively. Madhya Pradesh continued to be the leading producer, accounting for about 42% production during 1992-93, followed by Orissa 22%, Uttar Pradesh 17%, Maharashtra 10.5% and Rajasthan 8.5%. There were altogether 41 mines reporting production during the year.

7. QUARTZITE AND QUARTZ-SCHIST

Quartzite and quartz-schist are mined largely for use in refractory industry. They have high refractoriness and, hence, are utilised in the manufacture of acid-silica bricks. Refractoriness is lowered by presence of fluxes such as lime, iron-oxide, magnesia and alkalies. High silica content is essential with least possible Al_2O_3. Silica-rock of metamorphic origin is better than that of sedimentary origin. Silica grains cemented with cristobalite and tridymite are stable phases of silica. Physically, quartzite should be fine-grained, compact and crypto-crystalline.

The Raw Materials Sub-Committee of the Director General of Technical Development on Refractories in its report in September, 1985 has stipulated the following specifications for quartzite.

(i) Chemical Composition

	Grade I	Grade II
SiO_2	Above 98%	Above 96%
Al_2O_3	Less than 0.75%	Less than 1%
Fe_2O_3	Less than 0.75%	Less than 1.5% in distributed form

(ii) Physical Characteristics

Grade I	Grade II
Medium to fine-grained, compact, granular texture, homogeneous, free from iron-bands, patches, pyrite spots, pyrophyllite and mica-coatings	Occasional iron patches may be allowed. Free from iron bands

The thermal characteristics of quartzite should be such that :

(a) there is uniformity during thermal conversion,

(b) specific gravity of fired quartzite lumps in conventional kilns at 1430°C with proper firing schedule should be less than 2.45, and

(c) fired quartzite lumps should be clean, white and spot free.

Besides refractories, quartzite and quartz-schist are consumed in iron and steel for correcting silica-alumina ratio in the blast furnace charge, in ferro-alloy for ferro-silicon production and in foundries in the form of silica flour. They are also used in glass and ceramic industries where natural silica sand and quartz of required specifications are not available. For use in different industries, there are prescribed specifications which are required to be followed.

Mode of Occurrence and Origin

Quartzite and quartz-schist occur in the form of beds, pocket, lenses etc. They usually form hillocks or mounds. The quartzose-rocks are stubborn and resistant to alteration and may reach only the cataclastic stage of metamorphism under a given intensity. But with the increase in the degree of metamorphism (dynamothermal metamorphism), these rocks are recrystallised into quartzite, schistose quartzite and quartz-schist depending upon the presence of felspathic impurities. The thermal metamorphism of pure quartz-sandstones also give rise to quartzite with granoblastic texture and vitreous lustre. Sedimentary quartzite also occur in nature. When the cementation of sand grains in sandstone is such that interspaces are completely filled, a very solid compact rock known as quartzite is formed.

Distribution, Reserves and Production

The occurrences of quartzite and quartz-schist are found in every state of India.

The deposits of significance, state-wise, are included below in the Table 5.42.

Table 5.42—State-wise Distribution and Reserves of Quartzite and Quartz-schist Deposits

State	District/Area	Recoverable Reserves (in thousand tonnes)	Other details
Bihar	Monghyr, Singhbhum	317,206.6	
Madhya Pradesh	Durg, Raigarh,	15,744.6	
Orissa	Bolangir, Koraput, Keonjhar, Kalahandi, Mayurbhanj, Sundergarh, Sambalpur, Phulbani	6,773.5	
Punjab	Hoshiarpur	10,605.0	
Rajasthan	Sawai Madhopur	76.2	
Sikkim		15,000.0	
Uttar Pradesh	Uttar Kashi		Possible reserves around 100 million tonnes have been estimated. It is in the grade of good quality.

The all-India recoverable reserves of quartzite and quartz-schist as on 1.4.1990 have been placed at 365.5 million tonnes. The deposit of Uttar Pradesh requires further detailed work to determine its recoverable reserves.

There were altogether 23 mines producing quartzite in 1992-93. Orissa was the leading producer accounting for 49% production, followed by Madhya Pradesh 41%, and then Bihar and Rajasthan 5% each. The total productions of quartzite during 1990-91, 1991-92 and 1992-93 were about 85 thousand, 91 thousand and 81 thousand tonnes, respectively. The decline in production during 1992-93 compared to the previous years was due to less production reported from Bihar and closure of some mines in Orissa and Madhya Pradesh.

8. ZIRCON

Zirconium metal is of great industrial importance. It was discovered in 1789 by Klaprath in its principal ore mineral - zircon. The world resources of zircon are large relative to its present and future demands. Australia led the world production and reserves of zirconium minerals. The total world production of zirconium minerals is approaching one million tonnes a year to which Australia's contribution is about 50 per cent, followed by Republic of South Africa, USA and Russia. China, Brazil, India, Malaysia, Srilanka and Thailand are the other countries which also produce zircon to certain extent. During 1989, the total world production of zirconium minerals was about 993 thousand tonnes to which Australia's contribution was 546 thousand tonnes and India's about 16 thousand tonnes. The world zirconium reserves base is estimated to be about 44 million tonnes of ZrO_2 content, to which the shares of Republic of S. Africa, Australia, Russia, USA, India, Brazil and China are 15.3, 12.0, 6.0, 3.4, 2.4, 1.1 and 1.0 million tonnes, respectively.

Mineralogy

Zirconium (Zr) metal is obtained in crystals or in the powder form. In powder form, it burns readily in air. It has a specific gravity of 4.08 with melting point of 1300°C. The chief minerals of zirconium with their diagnostic characters are given below in the Table 5.43.

Table 5.43—Chief Minerals of Zirconium and their Characters

Minerals	Chem. Comp.	Characters
Zircon	$Zr.SiO_4$ (contains hafnium-oxide upto 4 per cent in some cases)	Reddish brown, light yellow, greenish grey or colourless, lustre admantine, H = 7.5, sp. gr. 4.7
Baddeleyite	ZrO_2 (contains small amount of hafnium-oxide	Colourless to yellow, brown or black, H = 6.5, sp. gr. 5.5-6

The pure zirconium metal may be produced by reduction of zirconium-oxide with magnesia, while the oxide is prepared by fusing zircon with acid potassium fluoride, treating with hydrochloric acid and precipitating the oxide with ammonia.

Test

A dilute hydrochloric acid solution, containing zirconium, imparts an orange-yellow colour to turmeric paper, moistened by the solution.

Use

The metal is used in alloys with iron, silicon, tungsten, nickel, copper, aluminium, vanadium etc. Zirconium alloys with aluminium and vanadium have been used in steel manufacture for removing oxides and nitrides. Zirconium steels make good armous plate and projectiles. With nickel, it is used in the manufacture of high speed or sharp cutting tools. The metal is also used in flashlight, bulbs, electronic tubes, electrical condensers, X-ray filters, lamp filaments, welding electrodes, rayon, spinnerets etc.

Its oxide, zirconia, is used as refractory, in abrasive, in enamels, as an insulator for heat and electricity, for gas mantles and certain incandescent lamps. It is also employed in the manufacture of laboratory wares, crucibles, furnace bricks and high temperature cements. Its use is rapidly expanding in production of artificial diamonds.

The mineral zircon is used in refractory, foundry, abrasive, ceramic, electrode and gemstone industry. Refractory and foundry industries account for over 97 per cent of the total consumption of zircon. It is the principal ore for the production of metal and alloys. Owing to its low thermal neutron absorption cross-section and low radio-activity after radiation exposure, it is used as structural material in nuclear reactors. For nuclear application the hafnium must be separated from zirconium. The present principal application of hafnium is mainly as control rods in nuclear submarines.

Mode of Occurrence and Origin

In nature zirconium does not occur in free state. It is found in the form of silicates and oxides. The most common and wide spread source of zirconium is zircon ($ZrSiO_4$). It is also found in baddeleyite (ZrO_2) which is mined in Brazil. It enters into the composition of some other complex silicates and oxides which are rare in occurrence.

Zircon occurs as accessory constituent of igneous rocks of acidic composition, which is derived from magmas containing much soda, such as syenite, diorite etc. It is one of earliest minerals to crystallise from the cooling magma. It occurs as larger crystals in pegmatitic bodies. It is also found in metamorphic rocks, such as gneisses, crystalline limestone etc. Zircon, being heavy and resistant to weathering, is a common constituent of heavy residues of some sedimentary rocks like sandstone. It is readily concentrated in placers and beach sands, and is associated with ilmenite, rutile, monazite, sillimanite etc.

Distribution and Reserves

It is known that the beach sands of Indian coasts contain zircon to certain extent. But there are four main belts where its concentration along with other heavy minerals is prominently seen. They are :

(i) a 22-km long stretch between Neendakara and Kayamkulam, Quilon district, Kerala with 5 to 8% zircon;

    (ii) a zone of about 6-km length extending from the mouth of Valliyar river to Colachel in Manavalakurichi and adjoining beaches in Kakulam and Vilavankode Talukas in Kanyakumari district, Tamil Nadu with 4 to 6% zircon;

    (iii) a 12-km long belt along Cauvery delta between Sirkuli and Keveripattanam in Thanjavur district, Tamil Nadu with higher content of ilmenite, zircon, monazite and garnet compared with those at Munavalakurichi; and

    (iv) a 20-km long stretch on Chattarpur coast, Ganjam district, Orissa with an average of 20% heavies.

Other new areas with extensive deposits of heavy minerals have been delineated in Andhra Pradesh and Maharashtra by the Atomic Energy Department. The Andhra coast was estimated to contain 1,185,000 tonnes (1.185 million tonnes) of zircon, while the Ratnagiri coast in Maharashtra is reported to have about 1000 tonnes of zircon. Few small inland deposits have also been reported from Ranchi in Bihar and Purulia in West Bengal. The recoverable proved reserves of zircon in the coastal beach sands of Kerala, Tamil Nadu and Orissa are placed at 1.2 million tonnes.

Production

Indian Rare Earths Ltd. (IRE) and Kerala Minerals and Metals Ltd. (KMML) are the two mining concerns engaged in processing of beach sands and production of zircon. There are four plants in the country producing zircon from the beach sand. The details of the plants are given below in Table 5.44.

Table 5.44—Plants Producing Zircon from Beach Sand

Producer	Deposit/Location of Plant	Installed capacity	Specification of products
IRE Ltd.	Manavalakurichi, Kanyakumari district, Tamil Nadu	7000	65% ZrO_2, 32% SiO_2 and 0.6% TiO_2
IRE Ltd.	Chavara, Quilon district, Kerala	9000	65.4% ZrO_2 and 33% SiO_2
IRE Ltd.	Orissa Sand Complex, Ganjam district, Orissa	2000	-do-
KMML	Chavara, Quilon district, Kerala	1200	N.A.

They altogether produced 12,613 and 11,587 tonnes of zircon, 290 and 721 tonnes of zirconium oxychloride and 5844 and 6535 tonnes of zirconium oxide in 1984 and 1985, respectively. The production of zircon during 1989-90 was 15,095 tonnes as compared to 23,899 tonnes in 1988-89. According to Department of Atomic Energy, consumption of.zircon is estimated at around 12000 tpy in various uses.

5.7 CERAMIC MINERALS

Clay is the vital raw material to be used in ceramic industry. Felspar is employed both in the body and glaze for chinaware. Wollastonite prevents shrinkage and other dimensional and glaze defects. Other minerals, like bauxite, andalusite-sillimanite group of minerals, borax, magnesite, lithium minerals, fluorspar, pyrophyllite, talc etc., are utilised to supply certain desired ingredients to clay or to manufacture special ceramic materials. Clay, felspar and wollastonite are being described in brief in the chapter, under reference.

1. CLAY

Clay is an aggregate of minerals and colloidal substances which commonly become plastic when wet, and stone like hard under fire. The constituent minerals are so fine (0.002 mm or so) that they can be recognised only by the use of electron microscope, X-ray and thermal analyses curves. Kaolinite, montmorillonite, saponite, beidellite, illite and allophane are some of the important clay minerals. Besides clay minerals and colloidal substances, clay also contains rock fragments and hydrous oxides. Plasticity, shrinkage, fusibility and transverse strength are the important physical properties upon which depend its various uses. Low grade clays start fusing at 1000°C or so, while the refractory clays are fusible at 1300° to 1400°C.

Uses

Clay has diversified uses, e.g. in ceramics, cement, refractory, paper and textile, rubber, cosmetic, pharmaceutical, insecticide, electrical, building and other industries. Based on its utility, clay is classified into following main types as given below in the Table 5.45.

Table 5.45—Types of Clay

Types	Characteristics	Chief Uses
1. Kaolin (China clay and paper clay)	Residual deposits, fine grained, white - burning, high-grade refractory	Crockeries, insulators, spark plugs, white-wares, sanitary wares, glazed tiles, pipes, jars etc. As filler and sizing material in paper and textile, as reinforcing and filler in rubber and as suspending agent in paints. Also in cosmetic, pharmaceutical and insecticide industries
2. Ball clay	Highly plastic variety of kaolin, having high binding power, tensile strength and considerable shrinkage, white burning	White ware, mixing with non-plastic clay to impart desired plasticity
3. Fireclay (Flint clay and Diaspore clay)	High alumina clay, endures high temperature	Refractories

(Contd.)

Types	Characteristics	Chief Uses
4. Bentonite	Mainly montmorillonite and minor beidelite, swelling in water, and is non-bleaching	Iron steel works, reconditioning, revivifying, moulding sands, oil drilling, medicines, cosmetics, leakage prevention, filtering. Activated bentonite having same use as fuller's earth.
5. Fuller's Earth (Bleaching clay)	Absorptive quality, decolouring oils, fats and greases	Petroleum refining, also in refining vegetable oils and animal fats, water purification, printing, abrasive, filler and cosmetics
6. Brick and tile clay	Common clay, low value types, aluminous and ferruginous, burns to creamy or red colour, mined in bulk and undergoes no processing except for removal of stones	Bricks and tiles, also sometimes in stonewares, cement plants

Mode of Occurrence and Origin

China clay occurs associated with Precambrian granites, gneisses, pegmatites, phyllites and schists, Gondwana shales and sandstones as bedded deposits, and within Tertiary sandstones and underlying laterites in many parts of the country.

It results from chemical weathering of aluminous rocks. Hydrothermal actions also play an important part in formation of clays. These may be formed in-situ to give rise to residual clay deposits or be transported and deposited as sediments. The tropical climate favours the formation of residual clays. The sedimentary clays may be divided into marine, lacustrine, flood-plain, estuarine, deltaic, loess and glacial. Loess clays are wind formed, while glacial clays are those resulting from glacial erosion and deposition from melt water.

The residual clays assume roughly the form of the source rock, e.g. dyke-like or veins, if derived from pegmatitic dykes, and as blanket or mantle if from crystalline rocks. Some residual clays occur under the cover of later formation. The sedimentary clays are in bedded forms, and show bandings, laminations and composition variations, both vertically and laterally. They are also pockety in nature, if deposited in protected places.

Distribution

The clay occurrences are found in almost all states of India. The state-wise distributions of important clay deposits are shown below in the Table 5.46.

Table 5.46—Important Clay Deposits in India

State	District	Other details
Andhra Pradesh	Cuddapah, East Godavari, Kurnool, West Godavari, Visakhapatnam, Srikakulam, Nellore, Nalgonda and Hyderabad	Kaolin (China clay) is mined. Ball clay is produced in West Godawari district.
Assam	Lakhimpur, Sibsagar, Kamrup, Karbi-Anglong	Normally, Kaolin (China clay) is found. Fuller's earth of inferior quality occurs in northern part of Kamrup district, while oil well-drilling clay exists in Sibasagar district.
Bihar	Singhbhum (Dumaria, Sararia, Karanjia), Sahebganj (Patharghatta), Bhagalpur, Santhal, Parganas (Mangalhat, Manjhi-tola, Pondongri), Dhanbad, Ranchi (Haridih, Sarsu), Gaya, Hazaribagh and Palamu	China clay is mined. Clay associated with sandstone of Barakar formation occurs in Santhal Parganas district, and is under lease.
Goa	Penda and Gupim Taluk	China clay and refractory clay are known.
Gujarat	Ahmedabad, Baroda, Sabarkantha, Mehsana, Kuchch, Bharauch	Deposits of China clay exist, and this is used by pottery and refractory units. Ball clay is mined in Bharauch.
Haryana	Faridabad	China clay is mined.
J & K	Udhampur, Srinagar, Anantnag, and Laddakh	The clays of Ladakh are of bentonic character. The clays of Udhampur, Srinagar and Anantnag districts are suitable for use in manufacture of cement.
Karnataka	Bangalore, Belgaum, Chickmagalur, Chitradurga, Hassan, North Kanara, Mandya and Shimoga	Kaolin (China clay) deposits are found.
Kerala	Trivandrum, Quilon, Ernakulam, Kozhikode, Palghat, Cannore, Kasaragod	Mostly China clay is mined. Ball clay occurs in Cannore, Quilon, and Trivandrum districts.
Madhya Pradesh	Betul	China clay deposits are found.
Maharashtra	Ratnagiri, Amaravati, Chandrapura, Nagpur	China clay deposits are in Ratnagiri, elsewhere it is of inferior grade.
Meghalaya	Garo and Khasi hills	Normally the basal portions of Tertiary rocks contain clay beds.

(Contd.)

State	District	Other details
Orissa	Koraput, Cuttack, Dhenkanal, Ganjam, Phulbani, Keonjhar, Balasore, Mayurbhanj, Bolangir and Sambalpur	Occurrences of China clay exist, though in some parts fire clay is also found.
Tamil Nadu	Chingleput, Ramanathapuram, South Arcot and Terunevelli	It is suitable for porcelain, sanitary wares and refractories.
Rajasthan	Ajmer, Udaipur, Jaipur, Alwar, Barmer, Bikaner, Chittaurgarh, Nagaur, Sawai Madhopur, Bhilwara, Bundi etc.	Mostly China clay is found. Ball clay occurs in Barmer, Bikaner, Pali, Jaisalmer and Nagaur districts. Refractory grade clay is found in Bikaner, Barmer and Bhilwara districts.
West Bengal	Bankura, Birbhum, Burdwan, Midnapur, Purulia, Darjeeling and Jalpaiguri	Terra Cotta of Birbhum is well known. Mostly China clay is found. It is suitable for porcelain and white wares, refractories, rubber, paper, textile, paint and chemicals.

The common clays are ubiquitous in nature and occur in numerous places. The higher grades are somewhat restricted. The clays, mined in Chikmagalur district of Karnataka, Amreli district of Gujarat, Phulwana district of Jammu and Kashmir and many other places, are used in the neighbouring cement factories. Clay, obtained from Theranipayam, Tiruchirapalli district, Tamil Nadu, is utilised in stoneware pipe factory.

Reserves

The all-India recoverable reserves of kaolin including white clay of all categories (as on 1.4.1990) have been placed at about 986 million tonnes of which only about 20% are under proved category. Rajasthan, West Bengal, Tamil Nadu, Kerala, Orissa, Andhra Pradesh, Meghalaya and Bihar together posses over 90 per cent of the reserves.

Production

The principal producers of kaolin were Rajasthan, West Bengal, Kerala, Delhi, Gujarat, Andhra Pradesh and Bihar which together accounted for about 90 per cent production. The total all-India productions of different types of clay for the years 1990-91, 1991-92 and 1992-93 are given below in the Table 5.47.

Table 5.47—Production of Clays During 1990-91 to 1992-93

Clay types	Productions in thousand tonnes		
	1990-91	*1991-92*	*1992-93*
Kaolin (China clay)	724	694	712
Ball clay	291	297	252
Common clays (Tile and brick making clay)	42	50	50

2. FELSPAR

Felspar is the most abundant rock forming mineral, and comprises a group of minerals which is composed of aluminous silicates with potassium, sodium and calcium or barium, as isomorphous mixtures. The important minerals of the group are potash felspar (orthoclase and microcline), sodium felspar (albite) and calcium felspar (anorthite). They are found in various colours, whitish, greyish or pale shades of red. Both potash and soda felspars are the commercial varieties, the former being more important. Lime varieties are undesirable. The world production of felspar is around 5.6 million tonnes, the major producers being Italy, the USA, Japan and Thailand. India's contribution to world production is about 1%.

Uses

Felspar (alkali felspar) is mainly utilised in the manufacture of ceramics, glass, pottery, vitrified enamels, special electrical porcelain, wind plates, opalescent glass as well as glassware. In ceramic industry felspar is used both in the body and as glaze for China ware. It is generally potash felspar which is used for fluxing, and soda felspar for glazing purpose. It is also used as refractories, in electrode manufacture, as binding agent in the manufacture of abrasive and in coal washeries. Ceramic and glass industries are the major consumers, accounting for 60% and 30%, respectively. In 1989-90 about 72 thousand tonnes of felspar was consumed.

Mode of Occurrence and Origin

The alkali felspars are the essential constituents of acid igneous rocks such as granite, syenite and diorite and their hypabyssal and volcanic equivalents. It is a common mineral in metamorphic rocks, in felspathic sandstone and arkoses, and occurs also in vein-stone. It occurs in large crystals as a constituent of pegmatites, and occurrences of this type are exploitable and are being mined for felspar. Quartz is the usual associate, and is required to be discarded.

Distribution and Reserves

Mica pegmatites of Rajasthan, Andhra Pradesh and Bihar are the important sources of felspar. Besides, numerous pegmat̃ ̃-veins with alkali felspar exist in other states. The state-wise distribution of important occurrences is incorporated in the Table 5.48 below.

Table 5.48—Important Occurrences of Felspar in India

State	District/Area	Other details
Andhra Pradesh	Nellore and Nalgonda	There are numerous pegmatites, rich in good quality felspar.
Bihar	Hazaribagh, Giridih, Chatra, Munger, Santhal Paragnas, Singhbhum and Dhanbad	Good quality pink felspar is available around Kodarma and Tisri areas.
Maharashtra	Ratnagiri	A few occurrences of good quality potash felspar in pegmatites are known.
Rajasthan	Ajmer, Bhilwara, Banswara, Tonk, Jodhpur, Jaipur, Pali, Udaipur and Alwar	Felspar is quarried at numerous localities. Important ones fall in Ajmer, Bhilwara and Tonk districts.
Tamil Nadu	Coimbatore, Salem and Tiruchirapalli	Numerous pegmatite-veins with alkali felspar are known.

The known recoverable reserves of felspar of all grades are of the order of about 16 million tonnes to which Rajasthan accounts for over 60 per cent.

Production

Rajasthan is the leading producer of felspar, followed by Andhra Pradesh, Tamil Nadu and Bihar. The total all-India productions for the years 1990-91, 1991-92 and 1992-93 were about 74, 71 and 75 thousand tonnes, respectively.

3. WOLLASTONITE

Wollastonite (CaO, SiO_2) is an important ceramic mineral. It varies in colour from milky-white to brownish-grey and occurs as radiating, prismatic and bladed crystals. It is transluscent, highly cleavable and brittle. The length of crystal varies from a few cm to 50 cm. Its utility depends upon acicularity or the aspect ratio i.e. ratio between length and width of a crystal. Wollastonite with aspect ratio in the range of 1 : 15 to 1 : 20 is considered as a semifibrous replacement for asbestos and finds high potential growth as a performance filler for strengthening various plastic and resin systems of daily use. Somewhat fibrous type is replacing asbestos in the manufacture of asbestos cement products. The whiteness of the mineral and low loss on ignition are important for its use in filler, coating and ceramic application.

The USA, India, Finland and Mexico are the major producers of wollastonite. The USA, besides being the largest producer, is also the largest consumer.

Uses

Wollastonite, because of its low thermal expansion, finds major use in ceramics in which it reduces shrinkage and other dimensional instability and glaze defects. As a filler it reduces the volume of the expensive plastic or resin media and contributes to physical and chemical properties of the finished products. It works as a natural low temperature fluxing material and, hence, is utilised in powder form in casting of steel for better surface finish. The major domestic demand for wollastonite in the country is in ceramic industry for the manufacture of floor and wall tiles. Other uses are in asbestos cement products, as filler in insecticide industry and as a coating material for welding rods. Researches are in progress for the new uses of wollastonite.

Mode of Occurrence and Origin

Wollastonite occurs in certain igneous rocks which have been contaminated with limestone, and in some nepheline-bearing basic igneous rocks. It is found often associated with diopside, grossularite, calcite etc. In Pali district, Rajasthan, it occurs in the form of several elongated lenticular bodies within Erinpura Granite massif and is traversed by thin calcite and quartz-veins. It may have originated as a product of contact metamorphism of impure limestone or other lime-rich rocks. It may also form by the action of silicic acid on limestone. Its occurrence in eruptic rocks is due to the inclusions of blocks of limestones.

Distribution and Reserves

The major deposits of wollastonite are in Sirohi, Dungarpur, and Pali districts of Rajasthan, and Banaskantha (Goddha area) district of Gujarat. Minor deposits are found in Dharampuri and Tirunelveli districts of Tamil Nadu. The total recoverable reserves of wollastonite are placed at 4.29 million tonnes. About 50 million tonnes of reserves have been estimated in the Pali district, Rajasthan.

Production

India is self-sufficient in this mineral. It is replacing asbestos to a considerable extent in the manufacture of asbestos-cement products. The production figures for 1990-91, 1991-92 and 1992-93 are 60 thousand tonnes, 62 thousand tonnes and 55 thousand tonnes, respectively. Two mines in Sirohi district of Rajasthan are producing wollastonite presently for domestic market and export. Both the mines are being worked by opencast method.

5.8 GLASS MANUFACTURING MATERIALS

Quartz and silica-sand are the chief glass manufacturing materials. Minor amounts of soda and lime are also needed. Quartz or silica sand is mixed with sodium carbonate (soda ash) or sodium sulphate in the prescribed proportion for the easy manufacture of glass. Calcium in the form of lime or limestone is added to give strength to the glass. Borax is employed for transparency, while manganese dioxide, nickel oxide, celenium, cobalt and chromium are meant for providing different colours to the glass. Coal is utilised for firing purposes.

Quartz and silica sand, being the main raw materials in glass manufacture, are dealt with in this chapter.

1. QUARTZ AND SILICA SAND

Quartz and silica sand are used largely in glass, foundry, ferro-silicon alloy and cement industries besides being used in many other industries like ceramic, fertilizer, alloy steel, abrasive, chemical, coal washery, electrode, insecticide, paint, rubber, textile, water filteration, brick, mortar etc. Specifications for their use in different industries are prescribed.

The natural silica sand is preferred in glass industry, but in cases when silica deposits are located far away, crushed quartz is used. The silica sand should be fairly free from contaminations like clay material, pebbles and other extraneous matter, and should not contain more than 4% moisture. Depending upon the types of glass to be manufactured, silica sands have been classed into four grades, namely :

(i) Special grade for high grade colourless glass, such as crystal glass, tableware, and decorated ware,

(ii) Grade-I for decolourised glassware, such as contain-ware, lampware etc.,

(iii) Grade-II for glassware where slight tint is permissible, and

(iv) Grade-III for coloured glasses.

The prescribed specifications of silica sand and quartz for glass foundary and ferro-silicon alloy are mentioned in the Table 5.49.

Mode of Occurrence and Origin

Quartz occurs in veins and pegmatites and has wide distribution. Vein-quartz is quite common, forming small hillocks in the Precambrian terrain. It is also prevalent in trap rocks. Hydrothermal processes may be responsible for the origin of such quartz deposits. Silica-sand occurs in abundance in river beds and beach deposits. Glass sands are obtained chiefly from sandstone and also from unconsolidated deposits. Sandstone contains more than 80% silica-sand. The sandstone must be friable and break readily around the grains. Vindhyan and Gondwana sandstones are source of silica sand. These are of sedimentary origin. The process of mechanical concentration gives rise to sand placers.

Distribution, Reserves and Production

Quartz and silica sand occur in most of the states of India. The distribution of important deposits of quartz and silica-sand, where produc-

Table 5.49—Specifications of Silica Sand/Quartz in different Industries Specifications (I.S.)

Industry	Physical	Grades	SiO₂ (min.)	Al₂O₃ (max.)	Fe₂O₃ (max.)	CaO (max.)	TiO₂ (max.)	As	LOI (max.)	Remarks
Glass	Size : + 0.6 mm (1% max.) 0.3-0.6 mm (50%) 0.125 mm (5% max.) Sand should not be coarser than 20 or 30 mesh nor finer than 100 to 120 mesh. Uniform grain size promotes even melting.	Sp. Gr.	99	—	0.02	—	0.10	—	0.5	For optical glass Fe₂O₃ should not be more than
		Gr-I	98.5	—	0.04	—	0.10	—	0.5	0.08%
		Gr-II	98	—	0.06	—	0.10	—	0.5	
		Gr-III	97	1.5	0.10	—	—	—	0.5	
Foundry	(i) Sintering temp. 1685-1710°C (ii) Grain-shape subangular to rounded (iii) Grain fineness well defined, grading with 70% and above retained by three adjacent sieves	Gr-I	98	1.0	1.0	1.0 (max.)	0.5	1.0	—	—
		Gr-II	95-98	1.5	1.0	1.0	0.5	2.0	—	—
		Gr-III	90-98	5.0	1.5	2.0	0.5	2.0	—	—
Ferro-silicon alloy	Quartz should have good thermal stability at 1200°C or more.	—	98	0.4	0.2	0.2 (max.)	0.2	—	—	Quartz with Al₂O₃ upto 0.8% and Fe₂O₃ upto 1.2% is also used

tions are being made, is given below in the Table 5.50.

Table 5.50—Distribution of Important Deposits of Quartz and Silica-Sand in India

Minerals	State	Districts
Quartz	Andhra Pradesh	Khammam, Nalgonda, Cuddapah, Guntur, Ranga Reddy, Kurnool, Nellore, Prakasam, Visakhapatnam, Medak, West Godavari, Vijayanagaram etc.
	Bihar	Singhbhum, Palamau and Hazaribagh
	Gujarat	Panchmahal, Baroda
	Haryana	Mohindergarh
	Karnataka	Bellary, Shimoga, Tumkur
	Madhya Pradesh	Dhar
	Maharashtra	Bhandara, Nagpur, Gadchiroli
	Rajasthan	Ajmer, Alwar, Bharatpur, Bhilwara, Jaipur, Jhunjhunu, Pali, Sikar, Tonk, Udaipur
	Tamil Nadu	Periyar, Salem, Tiruchirapalli, Anna, Dharampuri
Silica sand	Andhra Pradesh	Nellore, Prakasam
	Bihar	Sahibganj
	Gujarat	Junagarh, Kuchch, Rajkot, Sabarkantha, Surendranagar, Bharauch, Bhavnagar, Surat
	Haryana	Faridabad, Gurgaon
	Karnataka	Belgaum, Chitradurga, Gulbarga, Dakshin Kannad, Dharwar, Uttar Kannad
	Kerala	Alleppey
	Madhya Pradesh	Balaghat, Dewas, Khargone, Murena
	Maharashtra	Sindhuburg, Ratnagiri
	Rajasthan	Bundi, Jaipur, Alwar, Barmer, Bharatpur, Bhilwara, Bikaner, Jaisalmer, Jhunjhunu, Sawai Madhopur, Sikar, Tonk
	Uttar Pradesh	Allahabad, Banda
	Tamil Nadu	Chengalpattu, Tiruchirapalli

The resources are quite extensive. The total recoverable reserves as on 1.4.1990 are placed at about 983.5 million tonnes, of which proved, probable and possible reserves are about 73.7 million, 244.6 million and 665.2 million tonnes, respectively. The major reserves of these silica minerals are in Haryana (185 million tonnes), Bihar (164 million tonnes), Rajasthan (137 million tonnes), Kerala (110 million tonnes), Uttar Pradesh (73 million tonnes), Tamil Nadu (72 million tonnes), Maharashtra (61 million tonnes), Madhya Pradesh (53 million tonnes), Gujarat (35 million tonnes), Andhra Pradesh (30 million tonnes) and Karnataka (26 million tonnes).

The productions of quartz and silica-sand during 1990-91, 1991-92, 1992-93 were 216 thousand and 1457 thousand, 177 thousand and 2117 thousand, and 218 thousand and 1046 thousand tonnes, respectively. During 1992-93, Andhra Pradesh was the leading producer for quartz, followed by Karnataka and Rajasthan. Tamil Nadu, Maharashtra, Bihar, Madhya Pradesh and Gujarat also made some contributions to the production. In respect of silica sand, Haryana was the leading producer, followed by Rajasthan, Maharashtra, Karnataka, Kerala, Andhra Pradesh and Uttar Pradesh. The rest was shared by the other states.

5.9 FERTILIZER MINERALS

Fertilizer minerals play a very important role in enhancing food production. They may be utilised directly in crude state or may form basic raw materials for the manufacture of fertilizers. The different types of fertilizers, normally being used, are phosphates, potash, nitrates, sulphur, lime and gypsum. They are described in this chapter.

1. PHOSPHATES

The natural phosphates of importance are apatite which is phosphate of calcium with some chlorine and fluorine, and rock phosphates like phosphorites, phosphatic limestone, guano, basic slag, bone beds etc. having no definite chemical composition. They are the most important fertilizer minerals. The total world production of these minerals in 1990 was about 155 million tonnes to which India's contribution was 0.4%.

Uses

The largest consumer of phosphates is the fertilizer industry (about 95%). Treatment with sulphuric acid produces super phosphate, triple super phosphate and dicalcium phosphate which are more valuable for plant life. Phosphate is also utilised in manufacture of elemental phosphorus, chemicals, glass, sugar, and iron and steel industries. It finds use in safety matches, fire works, shells, grenades, tracer bullets, smoke screens, distress signals, medicines, soft drinks, baking powder, photography, cement, ceramics etc.

Mode of Occurrence and Origin

The phosphatic rocks may be fragmental, pelletal, nodular, oolitic, pisolitic, lenticular, platy, granular and massive in form. They may occur as marine sedimentary beds, phosphatic marls and limestone beds, reworked pebles, residual concentration of phosphatic materials and apatite deposits. The marine sedimentary beds of phosphate may have originated by marine chemical deposition in large enclosed basins like rock phosphate deposit near Mussorie, Dehradum district of UP, where it is associated with chert and black shale (Lower Tal formation). Phosphatic limestone and marl are merely sedimentary beds with high phosphate content, and are mostly of low grade, as found in Palamau districts of Bihar, Pithoragarh district of UP and elsewhere. In Pithoragarh district the calcarious beds appear to have been replaced by phosphates. The reworking of phosphatic limestone and subsequent erosion and transportation give rise to pebble deposits. The apatite deposits are formed due to concentration of apatite in pegmatites, pneumatolitic veins and magmatic segregations.

Distribution

(i) Apatite

The apatite occurs as a primary constituent of igneous rocks in accessory amount, and in small amount in metamorphic rocks like marbles, kodurites etc. But workable deposits as mentioned earlier, occur in pegmatitic and pneumatolytic veins. The state-wise distribution of important deposits are given below in the Table 5.51.

Table 5.51—Apatite Deposits of India

State	District/Area	Other details
Andhra Pradesh	Nellore, Srikakulam and Visakhapatnam (Kashi-patnam area)	Pegmatite veins are the main source of apatite. Kodurite of Srikakulam also contains apatite. Apatite vermiculite-magnetite veins in Kashipatnam area have been worked for apatite.
Bihar	Singhbhum (Pathergora), Gaya, Hazaribagh and Munger	In Singhbhum it occurs as veins and lenses of apatite and magnetite within apatite-schist of the copper belt, while in other places it is found in mica-pegmatite.
Rajasthan	Ajmer, Bhilwara, Tonk, Pali, Udaipur	Associated with pegmatite and quartz-veins, light-green and bluish-green apatites occur as accessory mineral.
Tamil Nadu	Coimbatore, Salem, Tiruchirapalli, Nilgiri, Kanyakumari	Mica-pegmatites are the source of apatite.
West Bengal	Purulia, (Beldih and Panrkidih)	Quartz-apatite veins and apatite magnetite-quartz breccia are the source of apatite. They are the important deposits in the country.

(ii) Phosphorite (Phosphate Rock)

The state-wise distribution of important deposits of phosphorite is given in the Table 5.52.

Table 5.52

State	District/Area	Other details
Himachal Pradesh	Simla, Solan, Sirmur, and Mahasu	Sporadic occurrences of phosphorite exist in Krol formation and they are not of economic value.
Jammu & Kashmir	Pir Panjab Range (Khansibal and Nunkhul section)	Phosphorite occurs as nodules in Upper Permian limestone and shale as thin bands and is not of economic significance.
Meghalaya	East Khasi, Garo and Jayantia Hills	Phosphatic shales and marl occur in the basal part of Kopili formation. The occurrences are un-economical.
Madhya Pradesh	Jhabua, Chattarpur, and Sagar	Dolomite-limestone-chert formation of Aravali group is phosphorite-rich.
Orissa	Ib river coal-field	Barren measures show the presence of thin layers of phosphatic rocks with an average P_2O_5 content of 9% to 15%.
Rajasthan	Udaipur, Banswara, Jaisalmer	Phosphorite occurs associated with limestone, chert, black shale and siltstone.
Uttar Pradesh	Dehradun, Tihri and Lalitpur	It occurs in the form of nodules and layers, associated with chert and black shale of Lower Tal Formation overlying Krol limestone in Dehradun district.

Reserves

(i) Apatite

The recoverable reserves of apatite in India (as on 1.4.1990) are placed at 13.29 million tonnes of which 0.61 million tonnes are in proved category, 11.62 million tonnes in probable category and the remaining 1.06 million tonnes in possible category. In addition about 12.89 million tonnes of apatite are placed under category of conditional resources. The major part of the reserves are of low grade, suitable for soil reclamation and beneficiation.

(ii) Phosphorite

The total recoverable reserves of phosphorites are 115.17 million tonnes, of which 71.13 million tonnes fall in proved category, 15.84 million tonnes in probable category and the remaining 28.20 million tonnes in the possible category. Besides, 152.88 million tonnes of reserves are placed under the category of conditional resources. Of the total recoverable reserves, about 9% are of chemical and fertilizer grades, (+30% P_2O_5) and the remaining will need blending or beneficiation before use.

Production

(i) Apatite

The total all-India production of apatite during 1989-90, 1990-91, 1991-92 and 1992-93 were about 14.5, 15.8, 16.6 and 16.9 thousand tonnes, respectively. The productions were mainly from Purulia, West Bengal and Visakhapatnam, Andhra Pradesh.

(ii) Phosphorite

The productions of phosphorite during 1989-90, 1990-91, 1991-92 and 1992-93 were 683.8, 667.6, 520.1 and 625.2 thousand tonnes, respectively. Rajasthan continued to be the leading producer, followed by Uttar Pradesh and then Madhya Pradesh. The mines are open-cast and semi-mechanised to mechanised. The mining in Dehradun district is mostly by under-ground method. The soft mineral mined (6 mm) is sold as such. The lumpy-ore is crushed by jaw crushers and grinding units. The low-grade ore is beneficiated at Jhamarkotra (Rajasthan).

2. POTASH

Potassium is another essential plant food. It is an important nutrient for protein synthesis and in helping plants to use water more efficiently. It also protects crops from bacterial and fungal diseases. The potassium as such does not occur in native state. Formerly, it was obtained from vegetable materials which being burnt were dissolved in water. The soluble portion yielded potash salts after evaporation. The potassium minerals of economic importance other than silicates are its chlorides such as sylvite, carnallite, kainite, sulphate like polyhalite and alunite, and nitrate namely nitre. The extraction of potash from its silicates is a complex and costly process. The mineral glauconite, a hydrous silicate of iron and potassium, also forms a potential source of potash and be utilised as a slow acting potash manure. Russia, Canada and Germany are the main potash producing countries in the world. There is no commercial production of potash in India and the requirements are met by imports.

Uses

The most important use of potash sa'. is as fertilizers. The other uses of potassium minerals and compounds are in the manufacture of glass, explosives, chemicals and metallurgical processes. Potassium nitrate and carbonate are used in glass, while nitrate is utilised in explosive manufacture. Minor uses of potash are in ceramics, dyeing, tanning, soap, meat curing, matches and photography.

During 1989-90 ten fertilizer units in India were utilizing potash muriate (KCL) for manufacture of complex fertilizer. Indian Potash Limited, Bombay is the main supplier of imported potash, and about 10 lac tonnes of potash muriate and potash fertilizer are consumed annually. The potash muriate (KCL) is the principal raw material for manufacturing potash-bearing fertilizers.

Mode of Occurrence and Origin

Marine evaporites, potash-rich brines, saline lakes, playa deposits and bitterns at salt works, are the various modes of occurrences of potash deposits. The saline water on evaporation becomes concentrated as super saturated in soluble salts which finally precipitate as solid salt in order of their solubility. The most soluble salt is precipitated last. In other words, these salts represent the residual products of evaporation of saline water.

Distribution and Reserves

The potash and potash-rich brines are known to occur in Tsokar lake, Leh district, Jammu and Kashmir, and Rann of Kuchch, Gujarat. Potash in the form of sylvite has been recently found in Sriganganagar district of Rajasthan. Glauconite, an important source of potash, is found in Banda district of Uttar Pradesh and Satna district of Madhya Pradesh, associated with Semri and Rewa groups of Vindhyan Super Group. Potash contents in Panna Shale (Rewa group) analyse 5.27% average potash (K_2O).

Total conditional resources of potash in Jammu and Kashmir and Gujarat have been placed at 739 million tonnes. Tentative reserves of about 78 million tonnes with 1.7 to 19.2% potash have so far been estimated in Ganganagar district of Rajasthan. The work is still in progress to prove the total reserves of potash in the basin.

Production

The production of potash in the country is negligible. Tata Chemicals is presently recovering minor amounts of crude potash from bitterns at its salt works in Gujarat.

3. NITRATES

The nitrogen compounds are essential for plant life. About 78% of nitrogen gas by volume is present in the atmosphere. The organic refuse absorbs nitrogen from the air and supplies to the soil, but the amount is negligible. It is obtained mainly from sodium and potassium nitrates, ammonium sulphate and calcium cynamide. Sodium and potassium nitrates, the description of which included in the chapter, occur in the nature in the form of sodanitre (caliche) and saltpetre respectively. Ammonium sulphate is produced as by-product in the manufacture of coke and gas from coal. It is also chemically produced from ammonia which is a synthetic product (combination of nitrogen and hydrogen). Some ammonium salts are also found in nature e.g. Sal Ammonioc (Ammonium Chloride), Mascognite (Ammonium Sulphate $(NH_4)_2SO_4$) etc. Calcium cyanamide is produced from atmospheric nitrogen and calcarious materials. The nitrates are salts of nitric acid and are restricted in occurrence because of their soluble properties in water.

Uses

The principal use of nitrate is in fertilizer. In fertilizer, it is used in the form of sodium nitrate, potassium nitrate, ammonium sulphate, calcium cyanamide, calcium nitrate, urea etc. A small amount of nitrate is being used in explosive, nitric acid, chemical salts and refrigeration.

(i) Saltpetre (KNO₃)

Saltpetre is white in colour with saline taste and is having sp. gr. of 2.1. It has vitreous, substransparent lustre. It occurs as a mixed efflorescence of potassium nitrate, sodium chloride, sodium sulphate, sodium carbonate and magnesium nitrate in silky tufts and thin crusts.

The nitrates of the saltpetre may have accumulated in the course of century in soil of highly populated areas. Villages built on the site of the previous abondoned villages of mud huts have generally their floor made of mud and cowdung, ashes, animal refuse etc., which form various nitrates. These nitrates drain from the raised village sites into the surrounding fields of low level, and in dry season these salts are brought to the surface as a mixed efflorescence.

The distribution of saltpetre is mostly in the Indo-genetic plains. Muzaffarpur, Siwan, Chapra, Gopalganj, Champaran and Samastipur districts of North Bihar were once famous for the production of saltpetre. The saltpetre contents in the soil of North Bihar is about 1 to 29%, but normally it is less than 5%. Some of these nitrates are directly used as manure, but most are used for extraction of saltpetre. The collected nitrate bearing earths are put in sufficient water so as to dissolve the nitrate salts. On evaporation of the liquor, sodium chloride separate first, followed by nitrate salt. The industry is almost dead now due to availability of alkali salts at cheaper rates. The production is still being made in remote villages on a very small scale for which there is no record.

(ii) Soda Nitre (NaNO₃)

Soda nitre ($NaNO_3$) is another nitrate mineral which is white, grey, yellow, greenish, purple and reddish brown in colour and vitreous, transparent in lustre, having specific gravity of 2.29. It is soluble in water, and occurs generally as efflorescence in crust or massive granular form.

Chile deposit of sodium nitrate, known locally as 'Caliche', is the only large deposit. It occurs associated with potassium nitrate, sodium chloride, sulphates of sodium, calcium and magnesium and minor amounts of iodates, borates, bromides and phosphates, and form source of nitrates. It is also used for iodine extraction besides the source of nitrate. It is a highly soluble mineral and as such the workable deposits occur only in regions of low rains. The nitrates may have originated by the evaporation of the ground water which may have carried those salts underground and through the capillary action these salts are brought up during summer months.

The arid and semi-arid regions such as plains of Gujarat, Rajasthan, Punjab, Haryana, UP and Bihar where evaporation is higher than precipitation are the ideal sites for the formation of nitrates.

No production of this mineral is reported in India. However, this is chemically manufactured to meet the internal demand.

4. LIME

Calcium is also an important element required for plant life. It may be supplied by lime, limestone, marl, oyster shells or gypsum. Calcium being soluble in surface water, is leached, and the soil becomes deficient in lime and acidic. It, therefore, becomes imperative to neutralise soil by adding lime to it. Besides correcting the soil acidity, the lime granulates heavy clay soil, provides plant food, promotes digestion of other fertilizers and counteracts some soil poisons. For fertilizer industry the limestone should contain $CaCO_3$ 84% (min.), SiO_2 (5%) and humidity 0.5% (max.), as prescribed by the Nangal Fertilizer Plant.

The consumption of limestone and other calcarious material for fertilizer is meagre (roughly about 0.2%) compared to its total consumption. Only about 163, 169 and 161 thousand tonnes of the materials were utilised in fertiliser industry during 1987-88, 1988-89 and 1989-90, respectively.

For mode of occurrence, origin, distribution, reserves and production the heading 'Limestone' dealt under chapter 'Building Materials' may be referred to.

5. GYPSUM

Gypsum ($CaSO_4.2H_2O$) is an important mineral, required for cement and fertilizer industries. It is of five types :

(i) rock gypsum,

(ii) gypsite, a mixed porous type with sand and clay,

(iii) alabaster, a massive, fine grained, snow white or light coloured,

(iv) satin spar, a fibrous silky form, and

(v) selenite, a transparent crystal form.

It has a specific gravity of 2.3 and can be scratched easily with the finger nail. Indian gypsum is mostly of gypsite type.

The world reserves of gypsum are placed at 2600 million tonnes which are adequate to meet the total world demand for nearly 30 years. The United States of America remained the world's largest producer of crude gypsum, accounting for about 17% of the world production. About 91.8 million tonnes was the total world production during 1989.

Uses

Gypsum is used as a retardant in cement manufacture to control the setting time. Gypsum of less purity may be employed for this purpose. High purity type is utilised in manufacturing ammonium sulphate fertilizer. For conditioning alkaline soil, low grade is utilised.

Gypsum makes an excellent plaster of paris for which purpose the mineral is calcined, ground and water added. Calcined gypsum also finds use for making partition blocks, sheets and tiles, insulation boards, and for stucco and lattice works. In pottery, gypsum is used for moulding purposes. Selenite, a crystalline variety, is utilised for producing white cement and for gypsum plate in microscope. Besides, gypsum is used in many other industries like pharmaceuticals, textiles, asbestos products, paint etc. The Indian Standard Specifications of mineral gypsum for different purposes are given below :

(i) Surgical Plaster

(a) $CaSO_4.2H_2O$ = 96% (min.),

(b) Free water = 1% (max.)

(c) CO_2 = 1% (max.)

(d) SiO_2 and other insolubles = 0.7% (max.)

(e) Iron and Al-oxide = 0.1% (max.)

(f) MgO = 0.5% (max.)

(g) NaCl = 0.01% (max.)

(ii) Ammonium Sulphate Fertilizer

(a) $CaSO_4$ $2H_2O$ = 85-90% (min.)

(b) SiO_2 and other insoluble = 6% (max.)

(c) Iron and Al-oxide = 1.5% (max.)

(d) MgO = 1% (max.)

(e) NaCl = 0.003% (max.)

(iii) Pottery

(a) $CaSO_4$ $2H_2O$ = 85% (min.)

(b) Free water = 1% (max.)

(c) CO_2 = 3% (max.)

(d) Insolubles = 6% (max.)

(e) Iron and Al-oxide = 1% (max.)

(f) MgO = 1.5% (max.)

(g) NaCl = 0.1% (max.)

(iv) Cement

(a) $CaSO_4$ $2H_2O$ = 70-75% (80-85% for export quality)

(b) MgO = 3% (max.)

(c) NaCl = 0.5% (max.)

(v) Reclamation of Soil

(a) $CaSO_4$ $2H_2O$ = 70% (min.)

(b) Na_2O = 0.75% (max.)

(c) Fineness = Residue on 2 mm sieve : nil, on 0.25 mm sieve : 50% (max.)

For extender paints $CaSO_4.2H_2O$ should be 75% (min.) and free water 0.5% (max.) when heated for 2 hours at 45°C. It should be very fine (–240 mesh B.S.T. sieve) in the form of dry powder.

Mode of Occurrence and Origin

Gypsum occurs as an evaporite in regular beds or lenses. It may be associated with anhydrite depending upon the temperature and salinity of the solution. It may originate by the evaporation of sea water in the enclosed basin. It may also form by dolomitisation of limestone in the sea. The origin of gypsum has also been assigned to the action of sulphuric acid, generated by decomposition of pyrite, on calcium carbonate present in the shells, limestone etc.

Distribution and Reserves

Gypsum deposits are reported from several states of the country. Besides, gypsum is obtained as by-product from sea water during processing of common salt and from several chemical plants like phosphoric acid, hydrofluoric acid, borax and boric acid plants. The recovery of by-product phospho-gypsum, fluoro-gypsum and marine gypsum together is substantial, slightly higher than the production of mineral gypsum.

The state-wise distribution is given below in the Table. 5.53.

Table 5.53—Distribution of Gypsum Deposits

State	District	Other details
Andhra Pradesh	Nellore, Prakasam, Guntur	Selenite occurs as irregular lumps in marine clays at depths ranging from a few cm. to about 1.5 m below the surface.
Gujarat	Jamnagar, Kuchch, Surendranagar, Bhavnagar and Junagarh	It occurs associated with marine clay and is of sub-recent age.
Himachal Pradesh	Sirmur and Chamba	
J & K	Baramula, Doda and Ladakh	In Baramula and Doda districts, it is within phyllites and schists of Salkhala formation, while in Ladakh, it is seen within Permo-Carboniferous limestone and shale.
Karnataka	Gulbarga	
Madhya Pradesh	Shahdol	
Rajasthan	Nagaur, Bikaner, Barmer, Churu, Jaisalmer, Ganganagar and Pali	Gypsum and gypsite occur as a part of evaporite sequence of sub-recent age.
Tamil Nadu	Tiruchirapalli, Coimbatore, Ramanathapuram and Tiruneveli	The mineral occurs with clay.
Uttar Pradesh	Tehri Garhwal, Dehradun, Pauri Garhwal and Nainital	It occurs as interbanded with limestone or dolomite belonging to Krol formation.

The recoverable reserves of mineral gypsum as on 1.4.1990 are placed at about 239.3 million tonnes of which 19.7 million tonnes are proved, 40.3 million tonnes probable, and 179.3 million tonnes possible reserves. The reserves of various grades out of the total recoverable reserves, are surgical plaster grade: 70 thousand tonnes, fertilizer and pottery grade: 39.3 million tonnes, cement/paint grade: 38.6 million tonnes, soil reclamation grade: 4 million tonnes and the rest is unclassified.

Production

Rajasthan is the chief producer of gypsum, contributing about 96% to the total production. The remaining 4% is accounted for jointly by Tamil Nadu, Jammu and Kashmir, Uttar Pradesh, Gujarat and Himachal Pradesh. The total productions of gypsum during 1989-90, 1990-91, 1991-92 and 1992-93 were 1.67, 1.54, 1.56 and 1.58 million tonnes, respectively.

Gypsum is worked by open cast mining and the production is classified into four grades based on $CaSO_4. 2H_2O$ contents :

1. above 90% $CaSO_4.2H_2O$,
2. 85-95% $CaSO_4.2H_2O$,
3. 80-85% $CaSO_4.2H_2O$, and
4. less than 80%. $CaSO_4.2H_2O$,

High grade gypsum is mined in Bikaner and Jaisalmer districts of Rajasthan. A part of gypsum mined in Bikaner district is of selenite type.

The productions of by-product phospho-gypsum, fluro-gypsum and boro-gypsum from various chemical plants in the country during 1988-89 and 1989-90 were 1.49 and 1.60 million tonnes, respectively. They find use both in cement and fertilizer plants. The limits generally specified for usage in cement are P_2O_5 - 0.5% max. and F - 0.15 max. Marine gypsum is recovered in Tamil Nadu .and Gujarat. Total productions of marine gypsum in 1988 and 1989 were about 111 and 96 thousand tonnes, respectively. These are quite suitable for use in different industries.

5.10 CHEMICAL MINERALS

A number of minerals are utilised chemically either in raw state or in prepared form to supply required ingredients such as rock salt (NaCl), borax ($Na_2O.2B_2O_5.10H_2O$), fluorspar (CaF_2), lithium and strontium minerals etc. The chemical minerals find use also in other industries. The descriptions in respect of rock salt, borax and fluorspar (fluorite) are included in this chapter.

1. ROCK SALT

Rock salt (NaCl) is the solid salt deposit, consisting mainly of sodium chloride with minor amounts of calcium sulphate, calcium chloride, magnesium chloride and, sometimes, magnesium sulphate. It is colourless or white when pure, but turns to yellow, red, pink and blue with impurities.

Uses

It is the most important raw material in chemical industry in the production of basic chemicals like caustic soda, soda ash, chlorine, hydrochloric acid and sodium metal. These basic chemicals, in turn, find numerous applications like production of soap, detergents, DDT, carbon tetrachloride etc. Rock salt is also used in dyes, emulsions, tanning, food and wood preservative, cement, glass making, water purification, cotton and paper bleaching, refrigeration, medicines etc. besides being used as food salt. It is employed in metallurgical industries for treating, smelting and refining of ores and metals, in ceramics for glazes, shrinkage prevention, vitrifiers etc., and in agriculture for cattle food, fertilizer, hay preservation, insecticide etc.

Mode of Occurrence and Origin

The commercial salt occurs in sedimentary bedded deposits, brines, sea water, surface playa deposits, and salt domes. The bedded deposits of salt are found associated mostly with gypsum and anhydrite or potash minerals. Brines include ocean water, salt lake water, sub-surface natural and artificial brines. They supply a considerable part of salt and also yield bromide and calcium chloride. They may be of ocean and salt lake origin or buried sea water in porous rocks. The playa deposits of rock salt are from the desicated salt lakes and occur associated with many other chemical minerals like borax, potash etc. The salt domes also contain great thicknesses of salt, associated with gypsum and anhydrite. The evaporation of saline water gives rise to salt beds. Gypsum or anhydrite is deposited first followed by salt, potash and other minerals.

Production

The occurrence of rock salt in the country is restricted to Mandi district of Himachal Pradesh. It is obtained in Mandi area by board and pillar method of mining, while solution method of mining is under active consideration. The productions of rock salt so made during 1989-90, 1990-91, 1991-92 and 1992-93 were about 2800, 3000, 3100 and 2900 tonnes, respectively.

The main source of common salt is, however, the sea and the lake waters. A number of by-products like gypsum, magnesium chloride etc. are obtained during recovery of salt from marine sources. Rajasthan saline lake yields a variety of chemical minerals like sodium sulphate, magnesium sulphate, magnesium carbonate, calcium chloride, liquid bromine, sodium bromide etc. as by-products besides the common salt. A portion of salt produced is ionised to meet the requirements of goitre epidemic areas in the country.

2. BORAX

Borax, a hydrated sodium borate (Na_2O $2B_2O_3$ $10H_2O$), is one of the important chemical mineral. Economically workable deposit of borax has not yet been established in the country, and the domestic need is met by imports of crude borates which are refined to produce borax and boric acid. The USA, Turkey, Russia, Argentina, Chile and China are the main borate producing countries.

Uses

Borax finds domestic use as a good cleanser, either directly or in soaps. It is an important component in baking powder, food preservatives, flavouring extracts, syrups and pickles, and insecticides. It is consumed in various industries like glass, ceramic, chemical, graphite, crucible, abrasive, refractory, paint, rubber, sugar, pharmaceutical, foundry, cosmetics, paper, textile, vanaspati etc.

Mode of Occurrence and Origin

It occurs as deposits from volcanic emanations (fumaroles), hot springs and as dried up shallow basin (playa) or saline lake. It occurs associated with other borates, ulexite ($Na_2O.2CaO.5B_2O_5.16H_2O$), colemanite ($2CaO, 3B_2O_3, 5H_2O$) etc. The origin of borax mineral involves simple concentration and evaporation. This is accompanied by some chemical and mineralogical transformations to give rise to borax mineral.

Distribution

Borax occurrences are known in Leh district of Jammu and Kashmir, Surendranagar district of Gujarat, and Jaipur and Nagaur districts of Rajasthan. But these occurrences for the present are not economically exploitable. Conditional sub-marginal resources of about 74,000 tonnes of borax have been estimated in Leh district of Jammu and Kashmir.

Production

There are a few units in the country which produce borax from the imported crude borates. The main among them are Borax Morarji Ltd., Bombay and Southern Borax Ltd., Madras, having installed capacities of 17,000 tpy borax and 3,000 tpy boric acid, and 15,000 tpy borax and 3,000 tpy boric acid, respectively. The consumption of borax in the organised sectors during 1989-90 was 9200 tonnes, glass and ceramic industries being the major consumers (95%).

3. FLUORITE

Fluorite (CaF_2) or fluorspar is an important chemical mineral because of its fluorine content. Cryolite and aluminium fluoride are the most important compounds manufactured from fluorite. It is also a mineral of critical significance in the metallurgical industries because of its demand in iron and steel, ferro-alloys and alloy-steel. It is of colourless, white, amethyst, purple, yellow, blue or green colour, having hardness of 4 and sp. gr. of 3 to 3.25. Apatite and rock phosphate and industrial gases are the other useful sources for the recovery of fluorine. Human beings and animals living in the fluorine-rich areas are known to suffer from fluorosis.

The world reserves of fluorite are estimated at about 336 million tonnes, compared to which India's reserves are insignificant, about 0.6%. China is the leading producer, while the other major producers are Mexico, Mangolia, Russia and South Africa.

Uses

There are a number of fluorite-based chemical industries in India, manufacturing cryolite, aluminium flouride, silico-fluorides, calcium carbide and cynamide, fluoro-carbon gases and other products. The chemical industry alone accounts for about 85% consumption, followed by ferro-alloy and alloy-steel. It is used as a fluxing agent in iron and steel, ferro-alloys and alloy-steel industries, where it improves the fluidity of the melt and removes phosphorus and sulphur. In aluminium industry, it is used in cryolite bath as an additive to balance the fluorine content, while it is utilised as vitrifying agent in glass industry. It is also employed in foundry, electrode and ceramics. It is also finds use in extraction of potash from felspar, making portland cement, bond for emery wheels, insecticides, preservatives, dyestuffs, refrigeration fluids etc.

Mode of Occurrence and Origin

It occurs as fissure-veins, vug fillings and as metasomatic replacement beds in calcarious rocks as found in Ambadongar area, Gujarat. It is found associated with quartz, cassiterite, galena, sphalerite and barytes. It also forms the cementing material in some sandstone. It is normally of hydrothermal replacement and pneumatolytic origin.

Distribution and Reserves

The fluorite occurrences of India are distributed in three main types of formations, namely :

(i) Precambrian granites, gneisses, quartzites, pegmatites and quartz-veins, as found at Dongargaon (Chandrapur) in Tamil Nadu, Nellore in Andhra Pradesh, Chandidongri (Durg) in Madhya Pradesh, Vadodara in Gujarat, and Dungarpur, Jallore and Jhunjhunu in Rajasthan,

(ii) Carbonatite suite of rocks, as in Ambádongar, Gujarat, and

(iii) Proterozoic/Early Cambrian sedimentary diagenetic type of formation, as in Ramanwara-Rewa area in Madhya Pradesh.

Some of these fluorite occurrences are associated with base metal mineralisation like those in Chandidongri and Imalia in Madhya Pradesh, Ambadongar in Gujarat, Devarakonda (Nalgonda district) in Andhra Pradesh and other places.

The total all-India reserves of fluorite as on 1.4.1990 has been placed at 2.15 million tonnes, the major reserves being in Gujarat (1.08 million tonnes) and Rajasthan (0.97 million tonnes).

Production

Ambadongar mine (Gujarat) worked by Gujarat Mining and Development Corporation is the largest mine in the country. Rajasthan State Mineral Development Corporation is also producing fluorite from its mines in Kaliha, Gehuwara and Malwa areas of Dungarpur district and Karara area of Jallore district. The total fluorite (graded and concentrated together) produced during 1989-90, 1990-91, 1991-92 and 1992-93 were about 29200, 32500, 35200 and 22100 tonnes, respectively.

4. SULPHUR AND PYRITES

Sulphur and pyrites are mainly chemical minerals, but are also of their great importance in fertilizer. In the form of sulphuric acid they are utilised in conversing phosphate rock into super-phosphate, and ammonia gases into ammonium sulphate. Pyrites works as a substitute for sulphur in the manufacture of sulphuric acid, and as such pyrites and sulphur are being described together.

Elemental sulphur is obtained commercially either from native sulphur or pyrites (FeS_2) or recovered as by-product from smelters and industrial plants. Native sulphur is purified from the associated gangue by melting in oven or by distilling in closed vessels. It is sulphur-yellow in colour with a reddish or greenish tinge, and has resinous lustre with specific gravity of 2.07 and hardness varying from 1.5 to 2.5. It is insoluble in water, but is dissolved by carbon disulphide.

The world production of all forms of sulphur in 1989 have been placed at 60.3 million tonnes. USA was the leading producer (19.2%), followed by Canada (12.3%), USSR (12.1%), Poland (8.5%), China (8.1%) and then others.

Uses

Sulphur is an important plant food. Raw sulphur and sulphuric acid are mixed with soil to neutralise its alkalinity. The main utilization of sulphur is, however, in the manufacture of sulphuric acid which finds numerous uses in fertilizer, oil refinery, chemicals, rayon, paints, coal products, explosives, textiles, iron and steel and other metallurgical purposes. Other important consuming industries are chemical, sugar, paper and rubber. In chemical industries, sulphur is used for manufacturing carbon di-sulphide and dye stuffs. Carbon disulphide is employed in making synthetic fibres. In sugar industry, it finds use in the form of sulphur dioxide for refining cane-juice. Paper industries utilise sulphur in the form of calcium disulphide solution for digesting wood pulp. In rubber industry, it is used for vulcanizing.

Mode of Occurrence and Origin

Sulphur occurs in the crators and crevices of extinct volcanoes, around thermal springs, bedded or layer deposits, and in salt dome caprocks. The deposits associated with volcanism may have originated due to condensation of sulphur vapours, reaction between H_2S and SO_2, and oxidation of H_2S to H_2O and S. In hot springs, it is formed by oxidation of H_2S present in the water. In bedded deposits (sedimentary formation) organic matter by its decay into H_2S and changed to sulphur by action of bacteria, may have given rise to sulphur beds or layers. In salt dome, it may have originated through the reduction of gypsum or anhydrite with which it is found associated.

Pyrite occurs as an accessory mineral in igneous rocks. It is also a common constituent of ore-veins which may form commercial deposits. The commercial deposits also occur as massive replacement bodies with associated pyrrhotite, chalcopyrite and sphalerite. The pyrite deposits also form due to magmatic segregation, pyrometasomatism or contact metamorphism. The Oolitic pyrite may be of sedimentary origin. The Amjhore pyrite deposit (Rohtas district, Bihar), occurring within Bijaigarh shales, belonging to the Kaimur Group of Vindhyan Super Group, has been assigned to syn-sedimentary origin.

Distribution and Reserves

The deposits of sulphur in India is very much limited. The only known occurrence of importance is that of Tsokar lake, Leh district, Jammu and Kashmir, where total conditional (sub-marginal) reserves of native sulphur are estimated as 0.21 million tonnes, all under possible category. Presence of native sulphur layer, about 15-30 cm in thickness, has been recorded in the northern face of the younger cone of the Barren

Island Volcano which is still active. This does not have any economic importance.

The pyrite is widely distributed in the country. The deposits of importance are, however, in the states of Andhra Pradesh, Bihar, Himachal Pradesh, Karnataka, Madhya Pradesh, Rajasthan and Tamil Nadu. In most of the cases, it is found associated with pyrrhotite, chalcopyrite, galena and sphalerite. In Bastar district (Mundatikra area) of Madhya Pradesh, it is found associated with copper mineralisation, while in Chingleput (Narayanpuram and Polur areas) and Kanyakumari (Arumanallur area) districts of Tamil Nadu and in Doda district (Assar area) of Jammu and Kashmir state, it is found associated with pyrrhotite and chalcopyrite.

The potentialities of pyrite deposits in the above areas have not so far been systematically and fully assessed. The only deposit of Amjhore, Rohtas district, Bihar is being worked presently. The sulphur contents of pyrite beds in this area vary from 35.08 per cent to 44.66 per cent, with an average of 41.23 per cent. The all-India recoverable reserves of pyrites are estimated as 91.52 million tonnes, of which 19.46 million tonnes are under proved category. Bihar is having about 52 million tonnes of recoverable reserves out of the total. For soil reclamation, pyrites with 30-38% sulphur is used, and this is about 4.82 million tonnes, and the remaining is beneficiable (22 to 30% sulphur) and of unclassified grade. Besides, conditional (sub-marginal) resources of pyrites are placed at about 1,558 million tonnes of all grades, of which about 9.6 million tonnes are in the proved category.

Production

There is no mine of sulphur in India. It is, however, recovered as by-product from fertilizer plants and oil refineries. National Fertilizer Ltd. at its plants at Panipat (Haryana) and at Naya Nangal and Bhatinda (Punjab), and Indian Oil Corporation Ltd. at its plant at Mathura (Uttar Pradesh) produced together 10.775, 10.536, 14.218 and 14.581 thousand tonnes of sulphur during 1989-90, 1990-91, 1991-92 and 1992-93, respectively.

The productions of pyrites during 1989-90, 1990-91, 1991-92 and 1992-93 were about 87.39, 105.52, 130.65 and 130.33 thousand tonnes, respectively. The almost entire productions were from Bihar (Amjhore, Rohtas district) except a small quantity of about 50 tonnes from Rajasthan during exploratory operations.

5.11 MINERAL PIGMENTS

Mineral pigments may be either natural or manufactured.

The natural mineral pigments consist essentially of limonite or hematite with or without mixture of clay and manganese oxides and form ochres, umbers and siennas. Ochres are mixture of hematite, limonite and clay with 15 to 80 per cent iron oxide, and provide yellowish and reddish brown colours. Umber is a brown ochre with 11 to 25 per cent manganese oxide and more of limonite. In addition, ground red and black slate and shale, greenstone and other coloured rocks also yield natural pigments. Gypsum, barytes, talc and natural white clay produce white pigments.

Manufactured pigments include those derived by direct treatment of minerals, say by roasting ochres, umbers, seinnas, iron-ore, copper-ore etc. They are also chemically produced such as those manufactured from carbon, chromium, barium, titanium, zinc, lead etc. and from organic matters. All pigment minerals require to be finely ground and must possess hiding power (opacity) and ability to absorb oil.

Ochre which is found in various colours like yellow, red, brown, white etc., is the natural pigment mineral and is described below.

1. OCHRE

Ochre is principally hydrated iron-oxide, and its pigmentary quality is mainly due to the presence of iron-oxides. Hydrated iron-oxides yield yellow colour and anhydrous red. A mixture of ferrous and ferric oxides imparts mainly brown besides other shades. The quality and value of ochre are judged by its staining power, brilliance and fine texture.

Uses

The ochres of various shades have a reasonably good covering power, are permanent colours and have no effect on other pigments. They are extensively used in the manufacture of colour-washes, distempers, oil paints, lacquers, primers and also for imparting colours to cement and paper. They also find use in fertilizer, foundry, linoleum, ceramic, asbestos products, electrical, glass, rubber, sugar and textile industries.

Ochres dominate the market because of their cheapness, abundance in occurrence and good pigmentary quality. In India, they are largely consumed (about 98%) in cement industry. Synthetic products, like synthetic ferric oxides and other colouring pigments, having better pigmentary properties, are fast replacing the natural ones.

Mode of Occurrence and Origin

Ochres occur as an alteration product of other iron minerals or by degradation of highly ferruginous rocks in the form of weathered residual concentrations. They are largely ferric hydroxide mixed with clay and other impurities. The above rocks and minerals succumb to mechanical disintegration and chemical decomposition. The minerals that are unstable under weathering conditions suffer chemical decay, the soluble parts, like silica, may be removed and the insoluble residues, mainly iron-oxides and other impurities, may accumulate in the form of ochres.

Distribution, Reserves and Production

Occurrences of ochre are widely distributed in the country. Deposits of red ochre are chiefly found in Gujarat, Karnataka, Madhya Pradesh and Rajasthan, yellow ochre in Andhra Pradesh, Madhya Pradesh, Rajasthan, Uttar Pradesh and West Bengal. Jaitwara area in Satna district of Madhya Pradesh is the well known area for yellow ochre. Red oxide mined in Bellary-Hospet area, Karnataka is also marketed as ochre.

The state-wise distributions of important ochre deposits with reserves, where known, are given below in the Table 5.54.

Table 5.54—Distribution of Important Ochre Deposits

State	District	Reserves (in million tonnes)	Remarks
Andhra Pradesh	Kurnool, Guntur, Cuddapah and Anantpur	1.7	The reserves relate to yellow ochre in Guntur district.
Bihar	Singhbhum	N.A.	-
Gujarat	Rajpur, Banas-kantha, Jamnagar and Kuchch	3.05	Reserves pertain to red ochre in Rajpur area and is suitable for paint industry.
Karnataka	Bellary, Bidar (Ubbalgundi)	0.23	Reserves (Probable) pertain to red ochre for Bellary and Ubbalgundi mines.
Madhya Pradesh	Gwalior (Behat, Dhiroli), Jabalpur (Jonli), Satna (Jaitwara, Amrithi, Partappur, Lindra and Khoga hills, Kailashpur, Madhogarh, Birpur etc.), Mandla, Rewa, Shahdol	2.23	About 0.23 million tonnes of red ochre and 2 million tonnes of yellow ochre have been estimated. Reserves estimates for many localities are not available.
Maharashtra	Nagpur	N.A.	Production is being made.
Rajasthan	Udaipur, Alwar, Bikaner, Chittaurgarh Sawai Madhopur	0.27	Reserves pertain to yellow ochre for Udaipur (Baneti and Iswal).
Uttar Pradesh	Banda	N.A.	
West Bengal	Mednipur	0.09	Reserves of yellow ochre for Thakurani pahari area have been estimated.

The productions of ochres for the years 1990-91, 1991-92 and 1992-93 were about 167 thousand, 191 thousand and 218 thousand tonnes, respectively. Rajasthan contributing over 74% to the production in 1992-93, continued to be the leading producer, followed by Madhya Pradesh 15% and Karnataka 6%. The remaining 5% is shared by Andhra Pradesh, Madhya Pradesh and Bihar. Rajasthan was the main producer of red ochre, Madhya Pradesh of yellow ochre and Karnataka of red-oxide. The entire productions were in private sectors.

5.12 MINERAL WATER AND GROUND WATER

1. MINERAL WATER

India is endowed with a large number of mineral springs. Many of them have medicinal and radioactive properties. With the exception of a few springs, like Phillips springs (Monghyr), Bihar, exploited for aerated water, these springs have not been harnessed either for medicinal use or as table waters. They exhibit great variation in surface temperatures, 20°C (Parasnath, Bihar) to as much as 98°C (Manikaran, Kulu, Himachal Pradesh).

Study of thermal springs in India was started as early as 1848 when Capt. T.J. Neobold gave an account of thermal springs of Mahanadi and adjacent country. T. Oldham (1982) published "The Thermal Springs of India", wherein he gave a comprehensive account of the locations of hot springs in India. La Touch (1918) described their medicinal qualities. P.K. Ghosh (1948-54) was first to carry out systematic study of the geology, flow of water, temperature, radon content and chemical composition of the hot springs of India and also compared them with some world famous "Spas". Since then study on thermal springs of India has been continued by various workers. Some of them were Seitz and Tewari (1959), S.K. Guha (1960-70), Deb (1964), V.S. Krishnaswami (1965, 1974, 1976, 1982) and Ravi Shankar (1977, 1989). The above work helped in having details of about 3400 springs in India. Many more springs are likely to be discovered with further intensified geological work.

Classification

The spring waters may be of two types :

(a) radioactive waters, and

(b) medicinal water.

The radioactivity of spring waters is due to emission of radium, and is measured in terms of radon per litre of water. Based on the degree of radioactivity, the springs may be grouped into :

(i) very high radioactivity (> 5 mμc radon content),

(ii) high radioactivity (between 3 and 5 mμc radon),

(iii) moderately high radioactivity (between 1 and 3 mμc radon),

(iv) mild radioactivity (<1 mμc radon), and

(v) negligible radioactivity (radon in traces).

The Indian springs compare favourably with some of the most radioactive waters overseas. They may be used for their radioactive properties in curing rheumatism, arthrites, gouts, spondylites, skin diseases, acidity, tonsilites and anaemia. The spring water of Rajgir, Bihar is used for such diseases.

The medicinal values of water are due to various salts present in the water and water temperature. Based on the presence of salts, the various types of waters recognised are :

(i) sulphur waters, both cold and warm types,

(ii) high and low content carbonate waters,

(iii) chloride (saline) waters,

(iv) mild chloride waters, and

(v) others which don't fall in the above groups.

The total salt contents of the spring waters can be determined by measuring its electrical conductivity, expressed in micro-mhos/cm at 25°C. The elements like Si, Fe, Al, Ca, Mg, Na, K, CO_3, HCO_3, SO_4, NO_3, Cl, F, H_2S and CO_2 are principally determined to find out the medicinal values of waters. Trace element analyses for Mn, Li, Sr, Ba, Cu, Sn, Zn, Pb, Bi, B, Be, Ti, MO etc. may also be carried out.

Origin

Ground water emerges in the form of a spring or seepage in the hilly terrain where the slope surface meets or intersects the water-table. In the case of topographic depression, marshes or ponds are formed when the water-table rises above its floor and ground water issues out through myriads of openings. Water may also come out at the contact point of the permeable and the impervious layers or where a dyke (impervious) cuts permeable layer (contact spring). Water may also move along the fracture or fault before surfacing. This is fracture spring. When the pressure built up within the aquifer forces the water out, artesian spring is formed. When the surface or rainwater penetrates great depths, it gets heated and comes out in the form of heated water. This is known as thermal spring like those in Parbati, Sutlej and Alaknanda Valleys. At places the vaporization of water throws out heated water in form of forceful fountains, called geysers.

The concentrations of different chemical constituents in spring waters and also their ratios by weight indicate that they have been derived from the country rocks by circulating mateoric water. The spring waters have also possibly been heated by heat liberated from exothermic chemical reactions and radiogenic source besides being heated by deep circulation.

Distribution

The list of known thermal/mineral springs in India is quite exhaustive and is beyond the scope of the book to present an account of them. These springs occur in Andhra Pradesh (43 nos.), Arunachal Pradesh (11), Assam (4), Andaman & Nicobar (1), Bihar (60), Gujarat (13), Haryana (3), Himachal Pradesh (34), Jammu and Kashmir (37), Karnataka (6), Kerala (3), Madhya Pradesh (13), Maharashtra (40), Orissa (11), Rajasthan (8), Sikkim (2), Tamil Nadu (2), Uttar Pradesh (37), West Bengal (6) and other states. The state-wise distributions of some springs with their surface temperatures are shown in Table **5.55** below.

Table 5.55—State-wise Distribution of some Springs

State	Springs	Location	Temperature (°C) Surface
Andhra Pradesh	Agnigundala	17°38' : 80°56'	60-80
	Unapdeo	19°57' : 78°18'	43
	Tegugudam	17°30' : 80°41'	28
Arunachal Pradesh	Takshing (Subanshri)		51
	Chetu (Subanshri)		38
Assam	Garampani (N. Cachar)	25°30' : 92°41'	54
	Nambar	26°24' : 93°56'	35
Bihar	Tanloi (Sahibganj)	24°23' : 87°16'	66-70
	Surajkund (Hazaribagh)	24°09' : 85°41'	42-88
	Tapoban (Gaya)	24°55' : 85°19'	47
	Rajgir Group	25°01' : 85°29'	36-42
	19 sprs. (Nalanda)		
	Parasnath (Dhanbad)	-	20
Gujarat	Tui (Panch Mahal)	22°48' : 73°34'	57-63
	Harson (Ahmedabad)	23°22' : 73°05'	30
Haryana	Sohna (Gurgaon)	28°15' : 77°08'	47
Himachal Pradesh	Manikaran (Kulu)	32°02' : 77°21'	63-98
	Baijnath (Kangra)	32°5'50" : 76°44'	49-55
	Lansa (Kangra)	32°23' : 76°05'	22
Jammu & Kashmir	Puga (Ladakh)	33°13' : 78°22'	30-85
	Tattapani (Udhampur)	33°14' : 74°25'	70-80
	Islanabad (Anantnag)	33°44' : 75°13'	17
Karnataka	Irade (S. Kanara)	12°43' : 75°13'	39
	Wujul (Gulbarga)	16°28' : 76°36'	32
Kerala	Warkali (Quilon)	8°44' : 76°42'	30
Madhya Pradesh	Sarguja	23°41' : 83°42'	86-98
	Deori	23°33' : 80°36'	27-28
Maharashtra	Mat (Ratnagiri)	16°58' : 73°33'	69
	Khair (Yeotmal)	19°54' : 78°51'	30-31
Orissa	Tarabela (Puri)	20°16' : 85°19'	60
	Lohagudi (Ganjam)	19°28' : 84°04'	43-44
Rajasthan	Mora (Sawai Madhopur)	26°43' : 76°33'	49
	Gangra (Chittaurgarh)	25°03' : 74°40'	26-27
Uttar Pradesh	Jamnotri (Uttarkashi)	31°00' : 78°31'	90
	Tapoban (Chamoli)	30°29' : 79°40'	53-89
	Rameri (Pithoragarh)	29°58' : 80°06'	30-33
West Bengal	Bakreswar (Birbhum) Group	23°54' : 87°26'	42-72
	Sarsakund (Purulia)	23°39' : 86°35'	40

Some of the spring waters contain negligible amounts of salts and minerals and are as pure as distilled waters, while others contain appreciable quantities of salts and are highly radioactive. The spring waters of Kawa Gandhari and Surajkund, Hazaribagh district and Rajgir, Nalanda district, Bihar are high in mineral substances and are also of very high radioactivity with 6.42 to 9.51 mµc radon content.

The hot springs contribute greatly to geothermal energy resource of the country. The total heat potential of 113 hot spring areas, examined by G.S.I. is placed at 40.91×10^8 calories which is equivalent to heat energy obtainable by combustion of 5730 million tonnes of coal. 46 of these hot springs are of high temperature type with base temperature more than 150°C, and could be utilised for electric power generation based on known technology and their cumulative power generation potential is estimated at 1838 MW for a 30 year period of utilisation (Krishnaswamy, V.S. and Shankar, Ravi, 1982).

2. GROUND WATER

Nature and Occurrence of Ground Water

Ground water or subsurface water constitutes one of the important mineral resources upon which development of civilization has been dependent. When it rains, a part of the precipitation evaporates into the atmosphere, a portion of it flows on surface down slopes to stream channels as surface water or surface run-off, and the balance infiltrates into the ground percolating slowly as sub-surface or ground water. The amount of infiltration is dependent upon the texture of the soil, nature and extent of vegetal cover and degree of ground slope. The soil is able to absorb greater amounts of rain water in open textured soil (porous or fractured or sheared), greater cover of vegetation, and flat ground. Gentle and prolonged rainfall favours more absorption of water in the soil.

Water Table, Permeability, Aquifer and Porosity

The upper surface of ground water is called the water-table. Between this and the surface is the zone of the aeration, where descending waters sink to meet the water table. In dry seasons the water table sinks, and in wet seasons it rises. In a flattish area the water table parallels the surface, and in a rolling humid region the water table roughly parallels the hills and vales, although it is less accentuated. The movement of ground water is towards the places of lower pressure, and thus it continually moves down slope from the intake at a rate determined by the head and the permeability of rock or soil. The permeability is the ability of soil or rock to allow passage of water. This depends upon the interconnections of the pores and cavities and cracks within the rocks and soils. The porous or fractured and permeable zone of soil or rock saturated with interstitial water is known as aquifer. The capability and capacity of an aquifer for holding water depends on its porosity, the volume of empty space between the grains constituting soil or sediment, or the cracks, fissures, fractures, joints and faults in igneous and metamorphic rocks. The porosity governs the total volume of water, a given equifer can hold. Fracture and fault zone, subsurface barriers like dykes and sills, colluvial-alluvial deposits, weathered zones and vesicular lava are the favourable locations of ground water repositories or aquifers.

Withdrawal and Recharge of Ground - Water

The withdrawal of ground water causes depletion of reserves and lowering of water-table, which depend on the rate, quantity and duration of withdrawal. The over withdrawal or over draft of ground water causes drying up of shallow wells, springs and streams. This is happening in many parts of the country like in parts of northern Gujarat and Western Rajasthan. In coastal belts saline sea water encroaches upon the depleted reservoirs, rendering the reserves unfit for consumption as found in the coastal parts of Saurashtra, Visakhapatnam and other places.

In the flood plains the aquifers are naturally recharged by rivers and streams. The recharge can be induced or accelerated by pumping in water under pressure or through numerous wells sunk along or in the river bed. In desert terrain like that of Rajasthan, the aquifers are confined under high pressure and high temperature and are recharged in the distant hills of Aravali bordering the desert.

Ground Water Provinces of India

There are four major ground water provinces in India. They are as follows :

(i) *The alluvial belts :* These are thick accumulation of porous unconsolidated sediments in the river systems, like Indo-Gangetic plain, plains of Brahmaputra, Mahanadi, Godavari, Kaveri, Narmada, Mahi, Sabarmati etc. They cover about 905 thousand sq km and are the richest ground water province of the country. This regime is characterised by large lensoid and locally interfingering aquifers. The thickness of aquifers varies from less than 20 m to as much as 330 m in the Ganga basin.

(ii) *Continental margin sedimentary belts :* These are thin covering of porous sedimentary rocks, Mesozoic (Gondwana) to Cenozoic in age, exposed in patches in the coastal belts from Orissa through Kerala to Saurashtra. They cover about 55 thousand sq km and contain smaller discontinuous aquifers. The aquifers in this belt, like those in Cuddalore sandstone (Cretaceous) in Tamil Nadu, Umia sandstone (Jarassic) in Gujarat, Gondwana (Masozoic) sandstone in central India etc., are being tapped profusely by wells for the ground water. The yield potential of ground water, though variable, is quite satisfactory.

(iii) *Fissured and weathered hard rocks :* They occupy the vast terrain of southern, central and eastern India, covering about 1700 thousand sq km area. The thick mantle of weathered material makes good aquifers with the water-table at shallow depths of 2 to 10 m. The weathered mantle is also associated with the buried erosion surfaces (unconformity) where porosity is high and permeability appreciable. In the Deccan trap the lavas are characterised by a high degree of jointing, interconnecting visicles and pipes or tunnels, and are interbedded with porous-permeable volcanic ash and sediments. There are, thus, multiple layered aquifers upto the depth of 15 to 20 m.

(iv) *The Himalayan region :* The Siwalik ranges are fringed on the southern side by an apron of coalescing fans of colluvial and fluvial debris. This piedmont belt is known as "Bhabhar" where mountain rivers loose their water and recharge the unconfined aquifers which extend down to a depth of 90 to 150 m. The depth of water-table is extremely variable. The 'Bhabhar' is fringed to the south by a line of countless numbers of perennial springs and seepages, making the whole belt very wet. This is known as 'Tarai' (means wetness) belt. In the vertical column of 300 m, there can be as many as three to four different confined aquifers under varying artesian conditions. In Dehradun valley in the Siwalik sub-province, a big synclinial valley has been filled with gravels. The water table is at a depth of 2 to 14 m and the aquifers intersected at a depth of 75 m. The Kashmir valley in the Tethys Himalaya is filled with lake deposits - the Karewa formation. The wells reaching a depth of 150 m have been discharging water at the rate of 5000 1/minute.

Ground Water Reserves

The total ground water reserve is placed at 37,00,000 million m^3 down to the depth of 300 m. Much of this stored water is restricted to the Sindhu-Ganga-Brahmaputra plains and the coastal sedimentary belts. The total annual replenishable ground water resource of the country has been estimated as 4,22,860 million m^3, out of which presently only 1,00,040 million m^3 (23.06%) per year is being used, leaving thus a balance of 3,22,820 million m^3 per year available for further use.

Prospecting for Ground Water

Water is a special kind of mineral and requires special conditions for search. For prospecting or locating ground water aquifers, the following few points may be taken care of:

(i) A careful study and proper interpretation of geology is needed to determine the nature, distribution, porosity and structure of water bearing formations.

(ii) The maps showing water strata of the area under examination is to be prepared. These maps depict the depth to aquifers, the heights to which water will rise in wells, the expected flow and quality of waters.

(iii) The surfacial gravel and sand beds where natural charging is available, are to be looked for unconfined water.

(iv) In the alluvial terrain the catchment basin with known precipitation, where total infall is not discharged in streams, may be expected to have underground supplies.

(v) Inventory of nearby wells has to be prepared since this yields valuable informations as regards ground water supplies.

(vi) In crystalline rocks, fractures, joints and other structural elements are studied to locate the aquifers.

(vii) Geophysical examination by gravitational, magnetic and seismic methods are carried out to determine favourable structural features, while conductivity and resistivity methods are employed to find out stratigraphic details.

(viii) Concentrated tree growth in the area of sparse trees point out the presence of hidden water.

(ix) Certain types of plants, whose roots thrive only in moist condition, may serve as water indicator in semi-arid and arid regions.

(x) Ground mist, seepages, water tanks etc., where surface pools of water accumulated to form them, suggest presence of water beneath.

Appendix I

Identification of Important Economic Minerals Based on their Physical Properties

Colour, Form, Lustre, Hardness, Specific Gravity, Streak, and Fracture are the main physical properties which help in identification of minerals. When it is not possible to identify any mineral based on above characters, other studies like microscopic, chemical, X-ray etc. are made to identify them.

For the sake of convenience in identification of some important economic minerals a Table is given, hereunder, based on their physical properties. Considering Lustre, Colour and Hardness as the main physical properties, the economic minerals have been divided in the following classes in the Table :

Class A : Non-metallic minerals

1. Commonly white or colourless minerals

    Hardness :

    (i) Minerals being scratched with nails (Hardness 1 to 2.5).

    (ii) Minerals being scratched with the help of knife, but not with nails (Hardness 2.5 to 5.5).

    (iii) More harder minerals (Hardness more than 5.5).

2. Coloured minerals :

    Hardness as (i), (ii) and (iii) above.

Class B : Metallic minerals : These minerals are commonly coloured or deeply coloured.

   Hardness as (i), (ii) and (iii) above.

Class A : Non-metallic minerals - 1. Commonly white, light coloured or colourless. (i) Hardness : Minerals being scratched with nails (Hardness 1 to 2.5)

Sr. No.	Name of minerals and chem. composition	Colour	Form and nature	Hardness	Sp. gravity	Streak	Cleavage	Other significant characters
1	2	3	4	5	6	7	8	9
1.	Talc $Mg_3Si_4O_{10}(OH)_2$	White, green, grey etc.	Massive, foliaceous, granular, compact, cryptocrystalline; Monoclinic	1	2.7-2.8	White	(001) perfect	Greasy feel, pearly lustre, occurs as metamorphism of Magnesium-rich rocks
2.	Gypsum $CaSO_4$, $2H_2O$	Colourless, white, sometimes red, yellowish grey etc.	Prismatic crystals, twinned, granular, massive laminated and fibrous; Monoclinic	2	2.3	White	(010) perfect	Shining and pearly on cleavage faces, fibrous forms silky, massive variety glistening but sometimes dull and earthy
(ii) Minerals being scratched with knife but not with nails (Hardness 2.5 to 5.5)								
3.	Muscovite $K_2Al_2(AlSi_3)O_{10}$ OHF_2	Colourless, white, pink, green, yellow etc.	Tabular crystals, large plates massive or in scales. Monoclinic	2.5-3.00	2.8-3.00	White	(001) perfect	Laminae flexible and elastic, and exhibits asterism in bright light, large crystals occur in pegmatites

(Contd.)

1	2	3	4	5	6	7	8	9
4.	Gibbsite $Al_2O_3 3H_2O$	White, brown, reddish brown etc.	Concretions, pisolitic or earthy. Monoclinic	2.5-3.00	2.3-2.4	White	(001) perfect	Occurs associated with bauxite and as alteration product of aluminium silicates
5.	Calcite $CaCO_3$	Colourless or white, grey, yellow, blue, red, brown etc.	Crystals common, massive, granular, lamellar, stalactitic etc. Hexagonal	3.0	2.7	White	(10$\bar{1}$1) perfect	Effervesces with cold dil HCl, gives brick red calcium flame
6.	Barytes $BaSO_4$	Colourless, white, grey	Flat crystals, massive, granular, lamellar, columnar. Orthorhombic	3-3.5	4.5	White	(001) and (110) perfect	Colouring the flame yellow green. Very common veinstone in lead and zinc veins
7.	Cerrusite $PbCO_3$	White, grey, tinged with blue or green	Prismatic crystals, massive, granular, sometimes stalactitic, Orthorhombic	3-3.5	6.55	Colourless, white	(110) and (021), distinct	Effervesces with HCl, occurs in oxidation zone of lead-vein, associated with galena and anglesite
8.	Dolomite $CaCO_3 MgCO_3$	White, tinged with yellow brown, red	Massive, granular. Tri-rhombohedral	3.5-4	2.8-2.9	White	(10$\bar{1}$1) perfect	Effervesces with warm HCl

(Contd.)

1	2	3	4	5	6	7	8	9
9.	Magnesite $MgCO_3$	White, greyish white, yellow, brown	Massive, compact, also fibrous, granular. Hexagonal	3.5-4.5	2.8-3.00	White	(1011) perfect	Effervesces with hot HCl, occurs as irregular veins in serpentine masses or as replacing dolomite or limestone
10.	Smithsonite $ZnCO_3$	White, greyish, greenish, brownish white	Massive, reniform, botryoidal, encrusting granular or earthy. Hexagonal	5.5	4.0-4.5	White	(1011) perfect	Effervesces with hot HCl, occurs associated with zinc, lead and copper ores
11.	Scheelite $CaWO_4$	Yellowish white or brownish, sometime orange yellow	Massive or granular, reniform with a columnar structure, crystal Tetragonal pyramids	4.5-5	5.9-6.1	White	(111) good	In ultraviolet light gives bright white or yellow colour; occurs in veins of pneumatolytic origin

(iii) More harder minerals (Hardness : More than 5.5)

1	2	3	4	5	6	7	8	9
12.	Felspar. Potassium-Sodium-Calcium (and/or Barium) Silicate	White, grey, pink, green or brown	Massive, granular or lamellar. Monoclinic or Triclinic	6.0-6.5	2.5-3.0	Colourless	(001). (010) perfect	Occurs as large crystals in pegmatite which is mined

(Contd.)

1	2	3	4	5	6	7	8	9
13.	Diaspore $Al_2O_3.H_2O$	White, brown or grey	Commonly massive, bladed, foliated and scaly forms, or compact, also in Orthorhombic prismatic crystals	6.5	3.5	Colourless	(010) perfect	Occurs associated with bauxite
14.	Sillimanite Al_2O_3,SiO_2	White, brown, grey, green	Needle shaped crystals, wisp-like aggregate, fibrous. Orthorhombic	6.0-7.0	3.23	Colourless	(010) perfect	Occurs in schists or gneisses, high grade regionally metamorphosed argillaceous rocks
15.	Quartz SiO_2	Colourless, grey, white, pink, violet etc.	Massive, granular, sometimes stalactitic, crystals usually Hexagonal prisms	7.0	2.65	Colourless	Absent	Vitreous lustre, conchoidal fracture. Occurs in veins or pegmatites, intergrown with felspar, original constituent of more acid igneous rocks
16.	Andalusite Al_2O_3,SiO_2	Purple red, flesh red, pearl grey, brown, white	Massive, granular, columnar, crystals prismatic. Orthorhombic	7.5	3.1-3.3	Colourless	(110) poor	Occurs in metamorphosed rocks of clayey composition

(Contd.)

1	2	3	4	5	6	7	8	9
17.	Topaz Al$_2$SiO$_4$ (OH, F$_2$)	Wine yellow, white, greyish and sometimes blue or pink	Massive, granular, columnar, prismatic crystals. Orthorhombic	8	3.5-3.6	Colourless	(001) perfect	Occurs in druses, pegmatites and veins
18.	Diamond C	Colourless or white, sometimes yellow, red, green or very rarely blue or black	Octahedral crystals, having curved faces, commonly twinned. Also small rounded grains. Cubic	10	3.47-3.56	Colourless (not available).	(111) perfect	Brilliantly admantine lustre; Transparent, dark coloured translucent, occurs in ultra-basic, basic rocks or in alluvial deposits, placers, residual deposits

2. Coloured minerals

(i) Hardness being scratched with nails (Hardness 1 to 2.5)

1	2	3	4	5	6	7	8	9
19.	Vermiculite Mg$_3$Si$_4$O$_{10}$ (OH$_2$) in parts	Yellow, brown or green	Platy, tabular, granular, also radiating. Monoclinic	1.5-2.5	2.6-2.9	White	(001) perfect	Exfoliates on heating, generally 8 to 12 times, coppery or silivery lustre develops on heating. Occurs associated with Mg-rich rocks like ultra-mafites

(Contd.)

1	2	3	4	5	6	7	8	9
20.	Orpiment As_2S_3	Fine lemon yellow	Foliated or granular or massive. Monoclinic	1.5-2	3.4-3.5	Yellow	(010) perfect	Emits sulphurous and garlic fumes on heating, pearly and brilliant on cleavage faces
21.	Realgar As	Fine red or orange	Massive or granular. Monoclinic	1.5-2	3.56	Red or orange	(010) distinct	Resinous lustre, occurs associated with orpiment.
22.	Sulphur S	Yellow, often tinged with reddish or greenish colour	Massive, granular encrustation, crystals bounded by acute pyramids. Orthorhombic	1.5-2.5	2.07	Sulphur yellow	(110) and (111) imperfect	Resinous lustre, burns with blue flames and saffocating fumes of SO_2; Insoluble in water and acids, but dissolves in carbon-disulphide
23.	Cinnabar HgS	Red, sometimes brownish or dark coloured	Usually massive, granular, rhombohedra or prism. Hexagonal	2-2.5	8.09	Scarlet	(1010) perfect	Sublimate metallic mercury on heating and release of sulphur dioxide

(ii) Minerals being scratched with knife, but not with nails (Hardness 2.5 to 5.5)

1	2	3	4	5	6	7	8	9
24.	Lepidolite $KLi_2Al\,(Si_4O_{10})$ $(OH, F)_2$	Rose red, lilac, violet, grey, sometimes white	Platy, massive or in disseminated scales, massive. Monoclinic	2.5-4.0	2.8-2.9	White	(001) perfect	Pearly, translucent lustre, occurs in pegmatites, associated with tourmaline, topaz. etc. of pneumatolytic origin

(Contd.)

1	2	3	4	5	6	7	8	9
25.	Rhodochrosite $MnCO_3$	Various shades of rose red, yellowish grey and brownish	Usually massive, globular, botryoidal or encrusting, Hexagonal rhombohedral	3.5-4.5	3.45-3.7	White	(1011) Perfect	Vitreous, inclining to pearly lustre, dissolves with effervescence in warm HCl
26.	Azurite $2CuCO_3.Cu(OH)_2$	Deep azure blue	Usually massive or earthy, modified prisms. Monoclinic	3.5-4.0	3.7-3.8	Light blue	(201) good	Associated with other oxidised copper minerals-malachite, etc. Dissolves in HCl with effervescence
27.	Malachite $CuCO_3.Cu(OH)_2$	Bright green	Massive, encrusting, stalactitic or stalagmitic, mammillated or botryoidal surface, also granular or earthy. Monoclinic	3.5-4.0	3.9-4.0	Light green	(201) good	Dissolves with effervescence in HCl
28.	Fluorite CaF_2	Colourless, white, green, purple, amethyst, yellow or blue	Cubic crystals, massive compact, coarsely or finely granular. Isometric	4.0	3.18	White	(111) perfect	Illuminated in ultra-violet light. Occurs in hydrothermal veins and replacement deposits. Also in veins of pneumatolytic origin

(Contd.)

1	2	3	4	5	6	7	8	9
29.	Siderite $FeCO_3$	Pale yellowish or buff, brownish and brownish-black or brownish red	Rhombohedral crystals. Massive, granular. Hexagonal	3.5-4.5	3.7-3.9	White	(1011) perfect	Acts slowly by cold HCl acid, but in hot HCl it effervesces briskly
30.	Sphalerite ZnS	Black, brown yellow or white	Crystals often twinned, massive and compact, occasionally botryoidal or fibrous. Cubic, tetrahedral	3.5-4.0	3.9-4.2	White to reddish brown	(110) perfect	Resinous to admantine lustre. Dissolves in HCl with evolution of H_2S
31.	Cuprite Cu_2O	Different shades of red	Crystals octahedron and rhombdodecahedron. Also massive, earthy or capillary	3.5-4.0	5.8-6.15	Brownish red and shining	(111) imperfect	Admantine or submetallic to earthy. Occurs in weathered zone of copper lode
32.	Kyanite $Al_2O_3SiO_2$	Blue, green, white	Generally bladed, sometimes radiating. Triclinic	4-7	3.6-3.7	White	(100) perfect	Hardness varying on different faces, lustre pearly on cleavage faces
							(010) good	Characteristics of metamorphosed argillaceous rocks under high stress and moderate temperature

(Contd.)

1	2	3	4	5	6	7	8	9
33.	Monazite (Ce, La, Yt) PO_4 with some ThO_2 and SiO_2	Pale yellow to dark reddish brown	Small flattened crystals, massive or as rolled grains. Monoclinic	5.5	5.27	White	(001) imperfect	Heavy residues in sediments, placer deposits
34.	Goethite FeO(OH)	Brownish black, some- times yellowish or reddish	Massive, stalactitic and fibrous, flattened prisms. Orthorhombic	5-5.5	4-4.4	Brow- nish yellow or ochre yellow	(010) good	Adamantine, opaque lustre, distinguished from limonite by being crystalline

(iii) More harder minerals (Hardness more than 5.5)

1	2	3	4	5	6	7	8	9
35.	Rhodonite $MnO.SiO_2$	Flesh red or light brownish red, greenish or yellowish when impure	Massive cleavable, tabular crystals. Triclinic	5.5-6.5	3.4-3.6	White	(110) perfect	Occurs as veinstone, sometimes used for ornamental work when cut and polished
36.	Zircon ZrO_2SiO_2	Reddish brown, pale yellow, grey, greenish, colourless	Prismatic crystals, rounded detrital grains. Tetragonal	7.5	4.7	Colour- less	(110) indistinct	Lustre adamantine, transparent to opaque, concentrated in placers
37.	Beryl $3BeO.Al_2O_3.6SiO_2$	Green, blue, yellow, white	Prismatic crystals, massive, crystalline. Hexagonal	7.5-8	2.7	White	(0001) indistinct	Mainly in mica-peg- matites. Emerald (eme- rald green) and Aqua- marine (pale blue) are gemstone varieties

(Contd.)

1	2	3	4	5	6	7	8	9
38.	Corundum Al_2O_3	Grey, greenish, reddish, dull white, sometimes colourless	Barrel shaped or pyramidal crystals, massive and granular. Hexagonal rhombohedral	9.0	3.9-4.1	Colour-less	None. Separation planes (0001)	Vitreous lustre. Conchoidal or uneven fracture. Occurs in silica poor contact metamorphic rocks

Class B : Metallic minerals - (i) Minerals scratched with nails (Hardness 1 to 2.5)

1	2	3	4	5	6	7	8	9
1.	Graphite C	Iron-grey to dark steel grey	Usually in scales, laminae or columnar masses, sometimes granular or earthy. Hexagonal	1-2	2-2.3	Black shining	(0001) perfect	Metallic lustre, feels cold like metal. Thin laminae flexible, sectile. Marks paper, occurs in metamorphic rocks of regional or contact metamorphic origin
2.	Molybdenite MoS_2	Lead grey or blue-grey	Usually in scales, also massive, foliaceous and sometimes granular. Hexagonal	1-1.5	4.7-4.8	Greenish lead-grey	(0001) perfect	Sectile and almost malleable, metallic lustre and heaviness are distinguishing characters, marks paper.
3.	Stibnite Sb_2S_3	Lead-grey	Radiating crystals or with columnar or bladed structures, elongated prisms, sometimes granular Orthorhombic	2.0	4.5-4.6	Lead-grey	(010) perfect	Occurs in quartz-veins, also associated with lead-zinc sulphides, dolomite, calcite and barytes veins

(Contd.)

1	2	3	4	5	6	7	8	9
4.	Covellite CuS	Indigo-blue	Massive, crystals platy, Hexagonal	1.5-2.0	4.6	Black	(0001) perfect	Occurs in the secondary enrichment zones of copper lodes
5.	Pyrolusite MnO_2	Iron grey or dark-steel grey	Usually massive or reniform, sometimes fibrous and radiating, pseudomorphs after manganite. Orthorhombic	2-2.5	4.8	Black or bluish black	Indistinct	Occurs in secondary manganese deposits
(ii) Minerals being scratched with knife, but not with nails (Hardness 2.5 to 5.5)								
6.	Galena PbS	Lead-grey	Crystal in Cubes & other forms. Also massive, granular.	2.5	7.4-7.6	Lead-grey	(100) perfect	Metallic lustre, occurs, associated with sphalerite
7.	Chalcocite Cu_2S	Blackish lead-grey, often with bluish or greenish tarnish	Usually massive, granular or compact, Orthorhombic	2.5-3.0	5.5-5.8	Bluish lead-grey	(110) poor	Formed by alteration of primary copper sulphides in zones of secondary enrichment
8.	Bornite Cu_5FeS_4	Copper red or pinchbeck brown, iridescent on exposure	Commonly massive, compact. Cubic	3.0	4.9-5.4	Pale greyish black, slightly shining	Absent or indistinct	Valuable ore of copper. Occurs associated with chalcopyite

(Contd.)

1	2	3	4	5	6	7	8	9
9.	Chalcopyrite CuFeS$_2$	Brass yellow, sometimes iridescent when exposed	Generally massive, compact, Tetragonal	3.5-4	4.1-4.3	Greenish black	(011) indistinct	Crumbling when cut with knife. Its brittle nature and non-malleability distinguishes it from gold
10.	Millerite NiS	Brass-yellow to bronze-yellow	Capillary crystals sometimes as columnar coatings. Hexagonal	3.0-3.5	5.3-5.5	Greenish black	(1011) (0112) perfect	Occurs in veins with other Nickel and Cobalt minerals. Also in nodules in clay-ironstone
11.	Pentalandite 2FeS.NiS	Bronze-yellow	Usually massive or granular, Cubic	3.5-4.0	5.0	Black	Absent	Occurs associated with pyrrhotite
12.	Manganite MnO(OH)	Iron-black or dark steel grey	Prismatic crystals in bundles, also columnar. Orthorhombic	4.0	4.2-4.4	Reddish brown or black	(010) perfect	Submetallic lustre, occurs with other manganese oxides. In veins
13.	Psilomelane Hydrated oxide of manganese	Iron-block to dark steel grey	Amorphous, massive botryoidal, reniform and stalactitic	5.0-6.0	3.7-4.7	Brownish black	Absent	Submetallic lustre, occurs as secondary deposits of manganese, fomed due to aleration of rocks containing manganese-bearing minerals

(Contd.)

1	2	3	4	5	6	7	8	9
14.	Wolframite (Fe,Mn)WO$_4$	Chocolate-brown, dark greyish-black, reddish-brown	Tabular prismatic crystals. Also massive and bladed. Monoclinic	5.0-5.5	7.1-7.9	Chocolate brown	(010) perfect	Occurs in pneumatolytic vein. Also placer deposit
15.	Niccolite NiAs	Pale copper-red	Usually massive. Hexagonal	5.0-5.5	7.3-7.6	Pale brownish black	Absent	Associated with cobalt, silver and copper veins
			(iii) More harder minerals (Hardness more than 5.5)					
16.	Pyrite FeS$_2$	Bronze-yellow to pale brass yellow	Cubic and pyritohedron crystals, often striated faces. Also massive, nodular, disseminated grains. Cubic system	6-6.5	4.8-5.1	Brownish black	Indistinct	Harder than chalcopyrite and difference in streak
17.	Anatase TiO$_2$	Brown, indigo-blue or black	Crystals slender, acute pyramidal or tabular. Tetragonal	5.5-6.0	3.82-3.95	Colourless	(001) and (011) perfect	Results as an alteration of other titanium-bearing minerals. Also in veins
18.	Rutile TiO$_2$	Reddish brown, red, yellowish or black	Crystals prismatic, frequently acicular & radiating, knee-shaped twins. Tetragonal	6.0-6.5	4.2	Pale brown	(110) and (100) poor	Metallic, admantine lustre

(Contd.)

1	2	3	4	5	6	7	8	9
19.	Ilmenite (FeO.TiO$_2$)	Iron-black	Platy or in scales. Also massive and sandy. Hexagonal	5.0-6.0	4.5-5.0	Black to brownish black.	Indistinct	Occurs in placer deposits
20.	Chromite FeO.Cr$_2$O$_3$	Iron-black and brownish black	Generally massive with granular or compact form. Crystals in octahedra. Cubic	5.5	4.5-4.8	Brown -		Occurs as primary mineral of ultrabasic igneous rock
21.	Braunite 3Mn$_2$O$_3$ MnO.SiO$_2$	Brownish black	Crystals of octahedral habit. Also massive, Tetragonal	6.0-6.5	4.75-4.82	Brownish black		In veins or in metamorphosed rocks
22.	Franklinite (Fe, Zn, Mn,) (Fe, Mn)$_2$O$_4$	Black	Octahera, rounded at the edges, also in grains and massive. Cubic	5.5-6.5	5.0-5.2	Black -		Feebly magnetic. Occurs associated with zinc minerals
23.	Magnetite Fe$_3$O$_4$	Iron-black	Octahedra crystals common. Also granular and massive. Cubic	5.5-6.5	5.18	Black	(111) poor	Magnetic, subconchoidal fracture
24.	Hematite Fe$_2$O$_3$	Steel-grey, earthy red	Flat crystals, generally massive, granular, dendritic. Hexagonal	5.5-6.5	4.9-5.3	Cherry red	Joint planes (0001) & (0112)	Occurs in sedimentary and metamorphic rocks

(*Contd.*)

1	2	3	4	5	6	7	8	9
25.	Columbite-Tentalite (Fe, Mn) (Nb, Ta)$_2$O$_6$	Grey, black or brown, sometimes iridescent	Prismatic or tabular crystals, often massive Orthorhombic	6.0	5.3-7.3	Dark red to black	(110) indistinct	Occurs in granitic pegmatites
26.	Arsenopyrite (FeAsS)	Tin-white inclined to steel-grey, Tarnishes to pale copper colour	Prismatic crystals, also massive. Orthorhombic	5.5-6.0	5.9-6.2	Dark greyish black	(101) imperfect	Arsenical fumes on heating
27.	Cobaltite CoAsS	Silver-white with a reddish tinge	Cubic or pyritohe-dral crystals, usually massive, granular and compact, Cubic	5.5	6.0-6.3	Greyish black	(100) perfect	Occurs in veins
28.	Smaltite CoAs$_2$	Tin-white to steel grey	Usually massive or reticulated, Cubic crystal	5.5-6.0	6.4	Greyish black	(100) and (111) distinct	Gives a sublimate of metallic arsenic on heating in closed tube
29.	Marcasite FeS$_2$	Pale bronze-yellow	Tabular crystals, often twinnend, radiating forms, nodular, massive, Orthorhombic	6.0-6.5	4.9	Greyish black	(101) imperfect	Forms at lower temperature than pyrite

(Contd.)

1	2	3	4	5	6	7	8	9
30.	Cassiterite SnO_2	Black or brown, rarely yellow or colourless	Prismatic crystals, knee shaped twins, also massive, fibrous or disseminated grains, Tetragonal	6.0-7.0	6.8-7.1	White to brownish	(110) imperfect	Occurs in quartz vein and granitic pegmatites and in placers
31.	Uraninite $2UO_3.UO_2$ (Pitchblende)	Velvet-black greyish or brownish	Commonly massive, botryoidal, or in grains. Cubic	5.5	6.4-9.7	Black with greenish or brownish tinge	Absent or indistinct	Occurs in granites, pegmatites or in high temperature veins with copper, lead and tin minerals
32.	Hausmannite Mn_3O_4	Brownish black	Pyramidal forms, frequently twinned, also massive and granular. Tetragonal	5-5.5	4.86	Chestnut brown	(001) indistinct	Occurs in veins or metamorphic rocks

Appendix II

Details of mineral production for 1995-96 to 1997-98 (excluding atomic and minor minerals)

Minerals	Units	1995-96 (April 95 to March 96)	1996-97 (April 96 to March 97)	1997-98 (April 97 to March 98)
METALLIC MINERALS				
1. Precious metals				
Gold	kg	2,036	2,710	2,603
Silver	kg	35,531	39,689	50,408
2. Ferrous and allied metals				
Iron-ore	000 tonne	67,423	68,173	70,237
Manganese-ore	000 tonne	1,836,705	1,832,811	1,597,000
Chromite	000 tonne	1,699,534	1,388,109	1,365,000
Tungsten (concentrate)	kg	6,451	3,826	-
3. Non-ferrous and allied metals				
Copper-ore	tonnne	4,736,870	3,895,153	4,473,000
Lead (concentrate)	tonne	61,583	60,329	57,000
Zinc (concentrate)	tonne	289,072	276,998	277,000
4. Light metal minerals				
Magnesite	tonne	345,077	371,109	383,000
Bauxite	tonne	5,564,425	5,930,640	5,844,000
NON-METALLIC MINERALS				
1. Mineral fuels				
Petroleum (crude)	000 tonne	34,517	31,480	33,772
Natural gas	million cu.m.	20,916	18,215	19,703
Coal	000 tonne	273,415	288,208	297,699
Lignite	000 tonne	22,144	22,540	23,047

(Contnd.)

Minerals	Units	1995-96 (April 95 to March 96)	1996-97 (April 96 to March 97)	1997-98 (April 97 to March 98)
2. Gemstones				
Diamond	carates	29,931	31,843	26,486
Corundum (Ruby)	kg	215	168	
Garnet (gem)	kg	602	654	679
Agate	tonne	542	400	226
Jasper	tonne	4,780	5,059	5,568
3. Abrasive minerals				
Corundum	kg	1,416	1,258	726
Garnet	tonne	62,306	42,296	54,857
Quartz	tonne	139,283	166,103	172,000
Chalk	tonne	147,066	124,382	114,000
4. Building materials and dimension stones				
Limestone	000 tonne	95,781	104,029	106,188
Lime kankar	000 tonne	307,050	330,154	372,922
Lime shell	000 tonne	105,973	80,015	74,002
Quartzite	000 tonne	116,085	109,961	75,000
Shale	000 tonne	302,599	453,172	543,859
Slate	000 tonne	9,696	7,826	9,220
Sand	000 tonne	1,673,439	1,671,446	1,756,000
5. Industrial minerals				
Mica crude	tonne	1,832	1,962	1,646
Mica scrap	tonne	1,240	1,439	1,013
Asbestos	tonne	23,844	26,101	21,653
Barytes	tonne	442,733	382,000	471,000
Talc (steatite)	tonne	540,570	462,152	-
6. Refractory minerals				
Fireclay	tonne	452,817	407,000	348,000
Graphite	tonne	136,263	117,527	100,800
Dolomite	tonne	3,717,541	3,400,000	2,830,000
Kyanite	tonne	8,944	7,055	5,976
Sillimanite	tonne	9,086	8,528	12,011
Diaspore	tonne	10,287	14,874	9,861

(Contd.)

Minerals	Units	1995-96 (April 95 to March 96)	1996-97 (April 96 to March 97)	1997-98 (April 97 to March 98)
7. Ceramine minerals				
Kaoline	tonne	831,098	775,000	545,000
Ball clay	tonne	507,681	492,000	325,000
Felspar	tonne	106,896	101,697	84,722
Wollastonite	tonne	96,017	93,066	108,817
8. Glass manufacturing minerals				
Silica sand	tonne	1,146,418	1,540,000	1,320,000
9. Fertiliser minerals				
Phosphorite	tonne	1,308,551	1,345,703	1 034,000
Apatite	tonne	10,777	9,186	-
Gypsum	tonne	2,195,111	2,210,000	2,006,000
10. Chemical minerals				
Fluorite (concentrate)	tonne	22,944	19,926	16,456
Sulphur	tonne	19,826	8,820	13,220
Pyrites	tonne	141,000	143,602	119,872
11. Mineral pigments				
Ochre	tonne	346,683	324,000	360,000

References

1. **Banerjee, A.K.**; 1979, *Origin of copper, lead and zinc deposits of India*. Geol. Surv. Ind., Miscellaneous Publication, No. 34, Pt II, pp. 1-8.

2. **Banerjee, Dilip Kumar**; 1976, *Mineral Resources of Purulia district, West Bengal*. Indian Minerals, Vol. 32, No. 1, pp. 27-49.

3. **Banerjee, Shyamadas** and **Satyanarayana, G.C.**; 1976, *Fluorite-lead-zinc mineralisation in the Chandidongri area, Durg district, Madhya Pradesh*. Geol. Surv. Ind., Bulletin, Series A, No. 41, 54 p.

4. **Basu, Aniruddha; Agarwal, N.K.** and **Gopalakrishnan**; 1989, *Geology of Umpirtha Base metal prospect, East Khasi Hills, Meghalaya*. Geol. Surv. Ind., Vol. 119, Pt. 2-8, pp. 34-42.

5. **Basu, A.N.; Kumar, Sunil** and **Sinha, K.K.**; 1982, *Uranium occurrences in the Pre-cambrian crystalline formations near Anak and Naukhara, Garo Hills, Meghalaya*. Geol. Surv. Ind., Records, Vol. 112, Pt. IV, pp. 12-16.

6. **Bassi, U.K.** and **Chopra, Suresh**; 1980, *Barite occurrence near Arsomang, Kinnaur district, Himachal Pradesh*. Indian Minerals, Vol. 34, No. 2, pp. 37-40.

7. **Basu, S.K.**; 1981, *Genesis and control of sulphide mineralisation in Sighana-Muradpur-Pacheri area with special reference to the Khetri Copper belt, Jhunjhunu district, Rajasthan*. Indian minerals, Vol. 35, No. 1, pp. 1-7.

8. **Basu, T.M.**; 1985, *Cassiterite and associated rare metal mineralisation in parts of Katekalayan area, Bastar district, Madhya Pradesh*. Geol. Surv. Ind., Records, Vol. 114, Pt. 6, pp. 99-106.

9. **Bateman, Alan M.**; 1959, *Economic Mineral Deposits*. Asia Publishing House, Bombay, 916 p.

10. **Bhatt, L.K.**, *et al.*; 1984, *Occurrences of magnesite, east of Shergol village, Kargil tehsil, Ladakh*. Geol. Surv. Ind., Special Publication, No. 15, pp. 141-146.

11. **Bhattacharya, C; Talpatra, A.K.**; and **Bose, S.S.**; 1984, *Integrated geochemical approach for tracing gold mineralisation in parts of*

Singhbhum and Ranchi districts, Bihar, India. Geol. Surv. Ind., Records, Vol. 114, Pt. 2, pp. 1-14.

12. **Chande, V.D.** and **Baswal, T.K.**; 1989, *Some salient features of tungsten mineralisation in Rajasthan.* Indian Minerals, Vol. 43, No. 1, pp. 31-38.

13. **Chattopadhyay, B.** and **Bhattacharya, S.**; 1986, *The nickeliferous magnetite associated with the Naga Hills ophiolites at Phokpur, Tuensang district, Nagaland, India.* Geol. Surv. Ind., Records, Vol. 114, Pt. IV, pp. 15-24.

14. **Chattopadhyay, P.B.**; 1989, *Petrology and petrochemistry of the kimberlites of Majhgawan, Panna district, M.P. and Wajrakarur and Lattavaran, Anantpur district, A.P.* Geol. Surv. Ind., Records, Vol. 119, Pt. 2-8, pp. 115-143.

15. **Dana, E.S.**; 1949, *A textbook of Mineralogy.* Fourth Edition, Revised and enlarged by William E. Ford; John Wiley & Sons, Inc., London, 851 p.

16. **Dar, K.K.**, and **Phakhe, A.V.**; 1964, *On occurrence of beryl and other minerals in pegmatites in India.* Report of the XXIInd Session, I.G.C., India, Pt. VI, pp. 213-224.

17. **Das, Sanjay**; 1977, *A note on prospecting of Amjhore pyrite, Rohtas district, Bihar with discussion on the origin of the deposit.* Indian Minerals, Vol. 31, No. 4, pp. 8-23.

18. **Datta, L.N.**; 1978, *An appraisal of ceramic and refractory raw material of Rajasthan.* Proc. Vol., All India Seminar on Raw Materials for Glass, Ceramic and Refractory Industries, Shillong, Oct., 1977.

19. **Dayal, B.**; 1966, *Limestone deposits of Manipur State and their suitability for cement manufacture.* Indian Minerals, Vol. 20, No. 1, pp. 68-77.

20. **Deshpande, M.L.**; 1978, *Gemstones and semi-precious stones.* Indian Minerals, Vol. 32, No. 1, pp. 1-17.

21. **Deshpande, M.L.**; 1979, *Tin-ores, Geology, Resources and Exploration.* Indian Minerals, Vol. 33, No. 3, pp. 10-22.

22. **Deshpande, M.L.**; 1975, *Geology of the diamondiferous gravels in a portion of Krishna river basin, Andhra Pradesh.* Indian Minerals, Vol. 29, No. 3, pp. 1-9.

23. **Deshpande, M.L.**; 1976, *The cassiterite - lepidolite bearing pegmatites, Bastar district, Madhya Pradesh.* Indian Minerals, Vol. 30, No. 1, pp. 67-74.

24. **Deshpande, M.L.**; 1978, *Lithium resources in India.* Indian Minerals, Vol. 32, No. 4, pp.. 41-47.

25. **Dey, A.K.**; 1983, *Geology and Mineral Resources of Jashpur, Raigarh district, Madhya Pradesh.* Geol. Surv. Ind., Memoirs, Vol. 114, 88 p.

26. **Dunn, J.A.**; 1942, *The Economic Geology and Mineral Resources of Bihar Province.* Geol. Surv. Ind., Memoirs, Vol. 78, 238 p.

27. **Dutt, N.V.B.S.**; 1975, *Note on Fuller's earth deposits of Hyderabad district, Andhra Pradesh.* Indian Minerals, Vol. 29, No. 4, pp. 39-44.

28. **Dutt, N.V.B.S.**; 1981, *Geology and Mineral Resources of Andhra Pradesh.* Ramesh Printers, Hyderabad, 205 p.

29. **Ghosh, Arbinda**; 1976, *A note on the polymetallic sulphide mineralisation in Askot area, Pithoragarh district, Uttar Pradesh.* Geol. Surv. Ind., Records, Vol. 107, Pt. 2, pp. 1-11.

30. **Ghosh, P.K.; Prasad, U.** and **Banerjee, A.K.**; 1977, *Geology and Mineral Resources of the part of the "Abhuj Marh" area, Bastar district, M.P.* Geol. Surv. Ind., Vol., 108, Pt. 2, pp. 182-188.

31. **Ghosh, P.K., Prasad, U.** and **Ghosh, R.B.**; 1963, *Hematite Iron-ores and associated banded rocks in Madhya Pradesh, Bihar and Orissa.* Geol. Surv. Ind., Records, Vol. 92, Pt. 2, pp. 239-252.

32. **Ghosh, Sibdas; Datta. A** and **Chandrasekaran, V.**; 1980, *A note on the chromite occurrences of Gammon Sirohi area, Manipur east district, Manipur.* Indian Minerals, Vol. 34, No. 4, pp. 7-14.

33. **Ghosh, Sibdas** and **Ghosh, B.K.**; 1983, *A note on Chhenclapathar tungsten prospect, Bankura district, West Bengal.* Indian Minerals, Vol. 37, No. 4, pp. 1-10.

34. **Gupta, Anupendu**; 1986, *Ore mineral assessment at Lawa gold prospect, Singhbhum district, Bihar, India.* Indian Minerals, Vol. 40, No. 3, pp. 17-30.

35. **Gupta, S.N.** and **Mathur, S.M.**; 1987, *Emerald deposits of Rajasthan and their future prospects.* Indian Minerals, Vol. 41, No. 4, pp. 31-38.

36. **Hatch, F.H.**; 1901, *The Kolar Gold Field bearing a descriptions of quartz mining and gold recovery as practised in India.* Geol. Surv. Ind., Memoirs, Vol. 33, Pt. 1.

37. **Jaggi, T.N.** *et al.*; 1986, *Mussoorie - Phosphorite and its utilisation.* Geol. Surv. Ind., Miscellaneous Publication, No. 41, Pt. IV, pp. 221-227.

38. **Jain, S.S.** and **Mohan, M.**; 1984, *Prospects of tungsten mineralisation in Kuhi-Chapegarhi area, Nagpur district, Maharashtra.* Geol. Surv. Ind., Records, Vol. 113, Pt. 6, pp. 141-149.

39. **Jog, R.G.; Agarwal, N.** and **Ghosh, Sibdas**; 1987, *A review note on the Rãnga valley polymetallic prospect, Lower Subansiri district, Arunachal Pradesh.* Geol. Surv. Ind., Records, Vol. 115, Pt. III and IV, pp. 45-64.

40. **Kanungo, S.C.**; 1976, *Graphite mineralisation in Titlagarh graphite belt, Bolangir and Kalahandi districts, Orissa.* Indian Minerals, Vol. 30, No. 4, pp. 1-5.

41. **Kar, P.** *et al.*; 1975, *Exploration of the tungsten mineralisation in Bankura district, West Bengal.* Indian Minerals, Vol. 29, No. 2, pp. 1-30.

42. **Kar Ray, M.K.**; 1973, *Geology with special reference to the modes of occurrence and control of mineralisation of Sargipali lead deposit, Sundargarh district, Orissa.* Indian Minerals, Vol. 27, No. 3, pp. 37-47.

43. **Karunakaran, C** and **Murthy, S.R.N.**; 1974, *Diamonds.* Indian Minerals, Vol. 28, No. 4, pp.. 23-37.

44. **Kashkari, R.L.**; 1984, *On the evaporites of Tsokar lake, Ladakh district, Jammu and Kashmir State.* Geol. Surv. Ind., Special Publication, No. 115, pp. 105-112.

45. **Krishna Rao, S.V.G.**; 1984, *Bauxite occurrences of Surguja district, Madhya Pradesh, adjoining Palamau and Ranchi districts of Bihar.* Geol. Surv. Ind., Special Publication, No. 14, pp. 243-246.

46. **Krishnaswamy, V.S.** and **Shanker, Ravi**; 1982, *Scope of development, exploration and preliminary assessment of Geothermal Resource Potentials of India.* Geol. Surv. Ind., Records, Vol. 111, Pt. 2, pp. 17-38.

47. **Kurien, T.K.** *et al.*; 1976, *Barytes deposit of Mangampeta, Cuddapah district, Andhra Pradesh.* Indian Minerals, Vol. 30, No. 3, pp. 7-16.

48. **Kurien T.K.**; 1980, *Geology and Mineral Resources of Andhra Pradesh.* Geol. Surv. Ind., Bulletin, Series A, No. 44, 242 p.

49. **Mahadevan, T.M.**; 1986, *Geological criteria for exploration and development of mica deposits.* Geol. Surv. Ind., Special Publication, No. 18, pp. 189-198.

50. **Malik, A.K.**; 1989, *Platinum group of elements in Sittampundi Complex, Tamil Nadu.* Indian Minerals, Vol. 43, No. 2, pp. 104-116.

51. **Meenakshisundaram, S., Subramanian, K.S.** and **Vemban, N.A.**; 1976, *Molybdenite mineralisation associated with meta-ultrabasic rocks and granites in Tamil Nadu.* Geol. Surv. Ind., Miscellaneous Publication, No. 23, Pt. II, pp. 580-591.

52. **Mukherjee, A.K.**; 1988, *Handbook of Iron-ore.* Geol. Surv. Ind., Bulletin, Series A, No. 51, 56 p.

53. **Mukherjee, N.K.**; 1976, *Exploration for copper at Singhbhum Copper belt, Bihar.* Indian Minerals, Vol. 30, No. 3, pp. 82-106.

54. **Mukherjee, N.K.** *et al.*; 1978, *Refractory minerals in India - status and future potentiality.* Geol. Surv. Ind., Proc. Vol., Seminar Raw Materials for glass, ceramic and refractory Industries, Shillong, Oct., 1977.

55. **Munshi, R.L.**; 1978, *Bentonite deposit near Bisu Kalan, Barmer district, Rajasthan.* Indian Minerals, Vol. 32, No. 2, pp. 46-49.

56. **Murthy, Y.G.K.**; 1980, *Certain aspects of the occurrence of fluorite in India and its exploration.* Geol. Surv. Ind., Records, Vol. 113, Pt. 5, pp. 1-31.

57. **Muthuraman, K.** and **Ramanamurthy, Venkata A.**; 1984, *Bauxite deposit in Waki plateau, Kolhapur district, Maharashtra.* Geol. Surv. Ind., Special Publication, No. 14, pp. 276-281.

58. **Narang, J.L.** *et al.*; 1979, *Position of the Malanjkhand sulphide mineralisation in the tectonic history of India.* Geol. Surv. Ind., Miscellaneous Publication, No. 34, Pt. II, pp. 113-116.

59. **Narayanaswami, S.; Ziauddin, M.** and **Ramachandra, A.V.**; 1960, *Structural control and localisation of gold-bearing lodes, Kolar Gold Field, India.* Econ. Geol, Vol. 55, pp. 1429-1459.

60. **Narainswamy, S.**; 1978, *Tectonic - metallogenic provinces in the Precambrian shield of India.* Geol. Surv. Ind., Miscellaneous Publication, No. 34, Pt. III, pp. 14-30.

61. **Neelakantam, S.** and **Roy, Sabyasachi**; 1979, *Barytes deposits of Cuddapah basin.* Geol. Surv. Ind., Records, Vol. 112, Pt. 5, pp. 55-64.

62. **Nene, S.G., Rizvi, M.J.** and **Chatterjee A.K.**; 1984, *Relationship of basalts with bentonites in Kuchch, Gujarat.* Geol. Surv. Ind., Special Publication, No. 14, pp. 252-258.

63. **Pareek, H.S.**; 1977, *Limestone deposits of north western Rajasthan.* Indian Minerals, Vol. 31, No. 4, pp. 24-32.

64. **Pareek, H.S.**; 1978, *Geological setting, petrography and origin of phosphorites of the Himachal Himalaya.* Indian Minerals, Vol. 32, No. 1, pp. 22-47.

65. **Pareek, H.S.**; 1984, *Pre-quarternary Geology and Mineral Resources of north-western Rajasthan.* Geol. Surv. Ind., Memoirs, Vol. 115, 99 p.

66. **Parthasaradhi, E.V.R.** and **Appavadhanulu, K.**; 1976, *Tungsten mineralisation near Burugubanda, East Godavari district, Andhra Pradesh.* Geol. Surv. Ind., Miscellaneous Publication, No. 23, Pt. II, pp. 570-578.

67. **Prakash, P.** and **Wadhwan, S.K.**; 1984, *Sillimanite corundum mineralisation in Sonapahar area, West Khasi hills district, Meghalaya.* Geol. Surv. Ind., Records, Vol. 13, Pt. IV, pp.. 81-88.

68. **Poulose, K.V.**; 1972, *Tile clay deposits of Kerala.* Indian Minerals, Vol. 26, No. 4, pp. 80-84.

69. **Prasad, R.N.** and **Prasannan, E.B.**; 1976, *Asbestos-Barytes-Steatite mineralisation in the lower Cuddapah of Andhra Pradesh.* Geol. Surv. Ind., Miscellaneous Publication, No. 23, Pt. II, pp. 560-569.

70. **Prasad, U.** and **Muzumdar, M.K.**; 1966, *A note on the hydrobiotites of the Bhagalpur and Santhal Parganas districts, Bihar.* Indian Minerals, Vol. 20, No. 1, pp. 45-48.

71. **Prasad, U**; 1974, *Vermiculite of Jaled area, Hazaribagh district, Bihar.* Indian Minerals, Vol. 28, No. 4, pp. 38-43.

72. **Prasad, U**; 1975, *Base metal mineralisation in the Gajadhar area, Giridih district, Bihar.* Indian Minerals, Vol. 29, No. 1, pp. 41-47.

73. **Prasad, U**; 1975, *Andalusite deposit near Nagar Untari, Palamau district, Bihar.* Indian Minerals, Vol. 29, No. 4, pp. 32-38.

74. **Prasad, U**; 1975, *Lead and copper occurrences in Bhagalpur district, Bihar.* Indian Minerals, Vol. 29, No. 2, pp. 65-72.

75. **Prasad, U**; 1975, *Base metal mineralisation in Phaga area, Bhagalpur district, Bihar.* Geol. Surv. Ind., Records, Vol. 106, Pt. 2, pp. 177-188.

76. **Prasad, U**; 1976, *Geology and Petrochemistry of a part of Hesatu-Belbathan polymetallic mineralised belt, Eastern Bihar.* Geol. Surv. Ind., Memoirs, Vol. 107, 130 p.

77. **Prasad, U**; 1976, *Bae metal mineralisation in Charkipahari-Toolsitanr area, Santhal, Parganas district, Bihar.* Indian Minerals, Vol. 30, No. 1, pp. 75-83.

78. **Prasad, U**; 1977, *Newly located K-felspar occurrence in Chatra subdivision, Hazaribagh district, Bihar.* Indian Minerals, Vol. 31, No. 3, pp. 9-14.

79. **Prasad, U.** and **Singh B.**; 1981, *Geochemical studies of dunite with reference to chromite mineralisation, Indus ophiolite belt, Ladakh, Jammu and Kashmir state.* Geol. Surv. Ind., Records, Vol. 112, Pt. 8, pp. 118-124.

80. **Prasad, U.**; 1982, *Role of beryllium and lithium in base metal mineralisation of Phaga area, Bhagalpur district, Bihar.* Geol. Surv. Ind., Special Publication, No. 8, pp. 79-84.

81. **Prasad, U**; 1984, *Geochemical prospecting in the glacial terrain of Kashmir Himalaya.* Geol. Surv. Ind., Special Publication, No. 15, pp. 101-104.

82. **Prasad, U**; 1985, *Ophiolites of India*, Geol. Surv. Ind., Records, Vol. 115, Pt. 2, pp. 13-54.

83. **Prasad, U**; 1986, *Pegmatites in parts of Chatra sub-division, Hazaribagh district, Bihar and their economic potentiality.* Geol. Surv. Ind., Special Publication, No. 18, pp. 131-134.

84. **Prashra, K.C.**; 1981, *New finds of huge limestone in the Tons region of Simla district, Himachal Pradesh, and Dehradun district, Uttar Pradesh.* Geol. Surv. Ind., Special Publication, No. 6, pp. 63-68.

85. **Pyror, T.**; 1923, *The underground geology of the Kolar Gold Field.* Trans. Inst. Min. Met., Vol. 33, pp. 95-135.

86. **Puri, S.N.** and **Sinha, P.K.**; 1980, *On the search for tin in Bihar.* Indian Minerals, Vol. 34, No. 3, pp. 1-11.

87. **Raghunandan, K.R.**; 1981, *Exploration for copper, lead and zinc ores in India.* Geol. Surv. Ind., Bulletin, Series A, No. 47, 222 p.

88. **Raha, P.K.**; 1975, *Crystalline magnesite deposit in Jammu limestone near Katra, Jammu - its nature and origin.* Indian Minerals, Vol. 29, No. 3, pp. 18-24.

89. **Raina, V.K.**; 1987, *A note on sulphur occurrence in volcanoes of Bay of Bengal.* Indian Minerals, Vol. 41, No. 3, pp. 79-86.

90. **Rajarajan, K.**; 1976, *Prospects of platinum and tin in India.* Indian Minerals, Vol. 30, No. 3, pp. 82-106.

91. **Rajaraman, S.** and **Deshpande, M.L.**; 1978, *Status of assessment of diamond resources in Andhra Pradesh (1967-1973).* Indian Minerals, Vol. 32, No. 2, pp. 39-45.

92. **Raman, P.K.; Prakash, P,** and **Khan, Md. Imam Ali**; 1975, *Galikonda bauxite deposit, Visakhapatnam district, Andhra Pradesh.* Indian Minerals, Vol. 29, No. 1, pp. 20-26.

93. **Rao, C.S. Raja**; 1982, *Coalfields of India.* Geol. Surv. Ind., Bulletin, Series A, No. 45, Vol. I to IV.

94. **Rao, M.G.** and **Raman, P.K.**; 1979, *The east coast bauxite deposits of India.* Geol. Surv. Ind., Bulletin, Series A, No. 46, 24 p.

95. **Roy Chowdhury, M.K.** *et al.*; 1968, *Report on the detailed investigation of some bauxite deposits of the Amarkantak area, Madhya Pradesh.* Geol. Surv. Ind., Bulletin, Series A, No. 28, 209 p.

96. **Roy Chowdhury, S.**; 1977, *On the lead-zinc mineralisation around Kushbani, Purulia district, West Bengal.* Geol. Surv. Ind., Records, Vol. 108, Pt. 2, pp. 36-41.

97. **Sahasrabudhe, Y.S.**; 1978, *Bauxite deposits of Gujarat, Maharashtra and parts of Karnataka.* Geol. Surv. Ind., Bulletin, Series A, No. 39, 163 p.

98. **Sahoo, R.K.**; 1976, *Mineralogy of Sukinda ultramafics with particular reference to the associated chromites and nickeliferous laterites.* Indian Minerals, Vol. 30, No. 4, pp. 11-12.

99. **Sankarsan Roy,** *et al.*; 1973, *A note on the nickel mineralisation in the Sukinda valley, Cuttack district, Orissa.* Indian Minerals, Vol. 27, No. 1, pp. 48-52.

100. **Santra, D.K.** and **Dange, M.N.**; 1979, *Assessment of limestone deposit near Tidding, Lohit district, Arunachal Himalaya.* Indian Minerals, Vol. 32, No. 3, pp. 1-11.

101. **Sarma, K.J.; Sundaram, S.M.** and **Sekharam, A.**; 1989, *Low grade manganese occurrences in Diguva Medangi and Chikkapara areas, Vizianagaram district, Andhra Pradesh.* Indian Minerals, Vol. 43, No. 1, pp. 24-30.

102. **Sehgal, M.N.** and **Prasad, U.**; 1986, *Basemetal prospects in Kashmir - An appraisal.* Geo. Surv. Ind., Miscellaneous Publication, No. 41, Part IV, pp. 50-66.

103. **Sen, S.N.**; 1974, *Integrated exploration for diamond in India.* Indian Minerals, Vol. 28, No. 1, pp. 20-23.

104. **Sen. S.N.** and **Prasad, U.**; 1975, *Copper in India.* Indian Minerals, Vol. 29, No. 1, pp. 1-12.

105. **Sen, T.K.** and **Hore, M.K.**; 1981, *A note on geology and sulphide mineralisation in Matasula area, Jaipur district, Rajasthan.* Indian Minerals, Vol. 35, No. 1, pp. 13-18.

106. **Shanker, Ravi**; 1987, *Status of Geothermal exploration in Maharashtra and Madhya Pradesh.* Geol. Surv. Ind., Records, Vol. 115, Pt. 6, pp. 7-29.

107. **Sharma, Subhas** and **Agnihotri, S.K.**; 1981, *Mineral Assets of Himachal Pradesh.* Geol. Surv. Ind., Special Publication, No. 6, pp. 42-52.

108. **Sharma, V.P.**; 1979, *Sulphide mineralisation in north-west Himalaya and its relation with magmatic activity.* Geol. Surv. Ind., Miscellaneous Publication, No. 34, Pt. II, pp. 143-150.

109. **Singh, Gurdev** and **Tewari, A.P.**; 1981, *New occurrences of barytes in district Sirmur, Himachal Pradesh.* Indian Minerals, Vol. 35, No. 2, pp. 30-33.

110. **Singh, K.N.**; 1989, *Tin-tungsten mineralisation in Dudhatoli Gneissic Complex, Chamoli and Pauri districts, Uttar Pradesh.* Indian Minerals, Vol. 43, No. 2, pp. 143-150.

111. **Singh, M.P.** and **Sinha P.K.**; 1974, *On the occurrence of gypsum in the Kurgiakh area, Ladakh, Jammu and Kashmir.* Indian Minerals, Vol. 28, No. 4, pp. 71-77.

112. **Sinha, B.N.**; 1963, *Geology of the manganese ore deposits between Kori and Ratanpur in Bilaspur district, Madhya Pradesh.* Geol. Surv. Ind., Records, Vol. 92, Pt. 2, pp. 253-264.

113. **Sinha Roy, S.** *et al.*; 1985, *Limestone resources of the Vindhyan sequence, Rajasthan.* Indian Minerals, Vol. 39, No. 1, pp. 28-41.

114. **Sivdas, K.M.** *et al.*; 1985, *Lead and copper deposits of the Agnigundala Mineralised belt, Guntur district, Andhra Pradesh.* Geol. Surv. Ind., Memoirs, Vol. 118, 102 p.

115. **Smith, P. Bosworth**; 1889, *Report on the Kolar gold field and its southern extension.* Govt. Press, Madras. Cited from Geol. Surv. Ind., Bulletin, Series A, No. 38, 48 p.

116. **Soni, M.K.**; 1981, *On base metal mineralisation in Damoh and Chattar districts, Madhya Pradesh.* Geol Surv. Ind., Miscellaneous Publication, No. 50, pp. 201-204.

117. **Srinivasachari, K.**; 1976, *Chromite deposits of India.* Indian Minerals, Vol. 30, No. 2, pp. 16-21.

118. **Srivastava, B.S.** and **Nanda, M.M.**; 1984, *Coal resources of Jammu area, Jammu and Kashmir state - An appraisal.* Geol. Surv. Ind., Special Publication, No. 15, pp. 163-176.

119. **Srivastava, R.N.** and **Kapoor, A.K.**; 1981, *Argentiferous lead mineralisation in Birgana village, Pauri Garhwal district, Uttar Pradesh.* Indian Minerals, Vol. 35, No. 1, pp. 8-12.

120. **Srivastava, S.N.P.**; 1986, *'Bihar ka bhuvigyan tatha khanij sadhan'.* Bihar Hindi Academy, Patna, 688 p.

121. **Subramanyan, M.R.** and **Ghosh, Prabir Kumar**; 1972, *Dolomite deposits of the Jainti area, Jalpaiguri district, West Bengal.* Indian Minerals, Vol. 26, No. 4, pp. 55-67.

122. **Sundaram, S.M.** *et al.*; 1989, *Uranium mineralisation in Vempalle dolomite and Pulivendla conglomerate/quartzite of Cuddapah basin, Andhra Pradesh.* Indian Minerals, Vol. 43, No. 2, pp. 98-103.

123. **Tewari, A.P.**; 1981, *Geology of Himachal Pradesh.* Geol. Surv. Ind., Special Publication, No. 6, pp. 1-17.

124. **Tripathi, C.**; 1984, *Bauxite in Madhya Pradesh.* Geol. Surv. Ind., Special Publication, No. 14, pp. 266-275.

125. **Valdiya, K.S.**; 1987, *Environmental Geology.* Tata McGraw Hill Publishing Comany Ltd., New Delhi, 583 p.

126. **Vemban, N.A.** and **Nagarajaiah, R.A.**; 1974, *The geology and manganese-ore deposits of the Manganese belt in Madhya Pradesh and adjoining parts of Maharashtra.* Geol. Surv. Ind., Bulletin, Series A, No. 22, 192 p.

127. **Viswanatha, M.N.**; 1982, *Economic potentiality of gem tracts of southern India and other aspects of gem exploration and marketing.* Geol. Surv. Ind., Records, Vol. 114, Pt. 5, pp. 71-89.

128. **Ziauddin, M.** and **Narayanaswami, S.**; 1974, *Gold resources of India.* Geol. Surv. Ind., Bulletin, Series A, No. 38, 186 p.

129. ————————; *Progress of the mineral industry of India.* 1906-1955. The Mining, Geological and Metallurgical Institute of India, Golden Jubilee Commemoration Volume, The Newman's Printing Works, Calcutta, 567 p.

130. ————————; *Minerals and Genesis of Pegmatites.* Report XXII Session of International Geological Congress, India, 1964, Pt. VI, 260 p.

131. ————————; 1966, *Base metals (Part II).* Geo. Surv. Ind. Miscellaneous Publications, No. 16, 822 p.

132. ————————; 1973, *Mineral Wealth of Uttar Pradesh.* Indian Minerals, Vol. 27, No. 4, pp. 1-21.

133. ————————; 1973, *Mineral Resources of Punjab State.* Indian Minerals, Vol. 27, No. 1, pp. 33-40.

134. ————————; 1974, *Geology and Mineral Resources of the States of India.* Geo. Surv. Ind., Miscellaneous Publication, No. 30, Pt. I to X.

135. ————————; 1977, *Exploration and Development of non-ferrous metals in India.* Geol. Surv. Ind., Miscellaneous Publication, No. 27, 455 p.

136. ————————; 1977, *Exploration in some major coal fields of India.* Geol. Surv. Ind., Miscellaneous Publication, No. 35, 583 p.

137. —————————; 1986, *Himalayan Geology, Mineral Resources,* Geol. Surv. Ind., Miscellaneous Publication, No. 41, Pt. VI, 503 p.

138. —————————; 1986, *Proceedings National Seminar on Mineral Exploration, Challenges and Constraints.* Directorate of Geology and Mines, Bihar, 293 p.

139. —————————; 1993, *Indian Minerals Year Book.* Vol. 2, Indian Bureau of Mines, Nagpur.

140. —————————; 1994, *Indian Minerals Year Book.* Vol. 1, Indian Bureau of Mines, Nagpur.

141. —————————; 1985-89, *World Mineral Statistics.*

142. —————————; *Commodity Summaries,* USBM.